The Resistance Vasculature

Vascular Biomedicine

The Resistance Vasculature, *edited by John A. Bevan, William Halpern, and Michael J. Mulvany,* **1991**
(A publication of the University of Vermont Center for Vascular Research)

The Resistance Vasculature

*A publication of the
University of Vermont Center for Vascular Research*

Edited by

John A. Bevan

University of Vermont, Burlington, Vermont

William Halpern

University of Vermont, Burlington, Vermont

Michael J. Mulvany

Aarhus University, Aarhus, Denmark

 Humana Press • Totowa, New Jersey

Library of Congress Cataloging-in-Publication Data

The Resistance Vasculature : a publication of the University of
    Vermont Center for Vascular Research / edited by John A. Bevan,
    William Halpern, Michael J. Mulvany.
        p. cm. -- (Vascular biomedicine)
    Includes index.
    ISBN 0-89603-212-4
    1. Vascular resistance. I. Bevan, John A., 1930- .
II. Halpern, William. III. Mulvany, M. J. (Michael J.)
IV. University of Vermont. Center for Vascular Research.
V. Series.
    [DNLM: 1. Blood Vessels--physiopathology. 2. Vascular Resistance-
physiology. WG 106 R433]
QP110.V38R47 1991
612.1'33--dc20
DNLM/DLC
for Library of Congress   91-20884
                                                                 CIP

Contents

Preface

The idea that a book might be brought out on resistance arteries occurred independently at about the same time to each of the editors. When this was realized, it confirmed for us that the time was right for such an endeavor. At about the same time the trustees of the University of Vermont approved the establishment of the Vermont Center for Vascular Research. Since one of the goals of the Center is to bring out publications in vascular science, and since all three editors have or have had close association with the University and much of the earliest in vitro work on small blood vessels occurred there, it seemed natural that the first publication of the Center should be this book, *The Resistance Vasculature*!

The direct study of small blood vessels is a natural outcome of the recognition that the function of resistance arteries cannot be learned from the study of the larger ones. The development of new techniques that made such studies feasible allowed this to become a practical goal. It is obvious that the location of these vessels, found as they are between the larger conduit arteries and the smaller distribution and exchange vessels of the microcirculation, determine their physiological role and importance. Thus, except under very unusual circumstances, and then for only specific reasons, any work carried out on vessels larger in diameter has been excluded.

The Resistance Vasculature aims to provide an up-to-date account of resistance artery research, and the authors have as far as possible presented a review of their respective interests. We hope that the volume will not only prove a valuable source of reference, but will also provide a framework for future research in this rapidly growing field.

We particularly want to thank Tammy Provencher, who was responsible for the organizational aspects of the preparation of the manuscripts, ensuring that style requirements and deadlines were met and so on. We suspect that her kindly, but matter-of-fact and persuasive Vermont accent is known to a number of the authors!

John A. Bevan
William Halpern
Michael J. Mulvany

Contributors

Christian Aalkjaer · *Department of Pharmacology and Danish Biomembrane Research Center, University of Aarhus, Aarhus, Denmark*

James B. Bassingthwaighte · *Center for Bioengineering, University of Washington, Seattle, WA*

John A. Bevan · *Department of Pharmacology and Vermont Center for Vascular Research, University of Vermont College of Medicine, Burlington, VT*

H. Glenn Bohlen · *Department of Physiology and Biophysics, Indiana University Medical School, Indianapolis, IN*

Joseph E. Brayden · *Department of Pharmacology and Vermont Center for Vascular Research, University of Vermont College of Medicine, Burlington, VT*

Shu Chien · *Department of Applied Mechanics and Engineering Sciences, Bioengineering, AMES-Bioengineering, University of California, San Diego, LaJolla, CA*

William M. Chilian · *Department of Medical Physiology, Microcirculation Research Institute, Texas A & M University College of Medicine, College Station, TX 7*

Brian R. Duling · *Department of Physiology, University of Virginia School of Medicine, Charlottesville, VA*

Jeff C. Falcone · *Department of Medical Physiology, Microcirculation Research Institute, Texas A & M University College of Medicine, College Station, TX*

Björn Folkow · *Department of Physiology, University of Goteborg, Goteborg, Sweden*

William Halpern · *Department of Physiology and Biophysics and Vermont Center for Vascular Research, University of Vermont College of Medicine, Burlington, VT*

Michael A. Hill · *Department of Medical Physiology, Microcirculation Research Institute, Texas A & M University College of Medicine, College Station, TX*

Phillip M. Hutchins · *Department of Physiology and Pharmacology, Wake Forest University, The Bowman Gray School of Medicine, Winston-Salem, NC*

Paul C. Johnson · *Department of Physiology, University of Arizona College of Medicine, Tucson, AZ*

Hermes A. Kontos · *Department of Medicine, Division of Cardiology, Medical College of Virginia, Virginia Commonwealth University, Richmond, VA*

Chiu-Yin Kwan · *Smooth Muscle Research Program and Department of Biomedical Sciences, McMaster University Health Sciences Center, Hamilton, Ontario, Canada*

Ismail Laher · *Department of Pharmacology and Vermont Center for Vascular Research, University of Vermont College of Medicine, Burlington, VT*

Julia M. Lash · *Department of Physiology and Biophysics, Indiana University Medical School, Indianapolis, IN*

Susan E. Luff · *Baker Medical Research Institute, Alfred Hospital, Prahran, Victoria, Australia*

Colleen D. Lynch · *Department of Physiology and Pharmacology, Wake Forest University, The Bowman Gray School of Medicine, Winston-Salem, NC*

J. Jeffrey Marshall · *Department of Medicine, Division of Cardiology, Medical College of Virginia, Virginia Commonwealth University, Richmond, VA*

Takamichi Matsuki · *Department of Physiology, University of Virginia School of Medicine, Charlottesville, VA*

Janie Maultsby · *Department of Physiology and Pharmacology, Wake Forest University, The Bowman Gray School of Medicine, Winston-Salem, NC*

John G. McCarron · *Department of Physiology, University of Massachusetts, Worcester, MA*

Gerald A. Meininger · *Department of Medical Physiology and Microcirculation Research Institute, Texas A & M University College of Medicine, College Station, TX*

David E. Mohrman · *Department of Physiology, University of Minnesota School of Medicine, Duluth, MN*

Michael J. Mulvany · *Department of Pharmacology and Danish Biomembrane Research Center, Aarhus University, Aarhus, Denmark*

Tetsuya Nakamura · *Department of Physiology, Eastern Virginia Medical School, Norfolk, VA*

Tim O. Neild · *Department of Physiology, Monash University, Clayton, Victoria Australia*

Mark T. Nelson · *Department of Pharmacology and Vermont Center for Vascular Research, University of Vermont College of Medicine, Burlington, VT*

George Osol · *Department of Obstetrics and Gynecology and Vermont Center for Vascular Research, University of Vermont College of Medicine, Burlington, VT*

Russell L. Prewitt · *Department of Physiology, Eastern Virginia Medical School, Norfolk, VA*

John M. Quayle · *Department of Pharmacology and Vermont Center for Vascular Research, University of Vermont College of Medicine, Burlington, VT*

Geert W. Schmid-Schönbein · *Department of Applied Mechanics and Engineering Sciences, Bioengineering, AMES-Bioengineering, University of California, San Diego, LaJolla, CA*

Steven S. Segal · *Noll Laboratory for Human Performance Research, Pennsylvania State University, University Park, PA*

Richard Skalak · *Department of Applied Mechanics and Engineering Sciences, Bioengineering, AMES-Bioengineering, University of California, San Diego, LaJolla, CA*

Edward G. Smith · *Department of Physiology, Eastern Virginia Medical School, Norfolk, VA*

Cornelis van Breemen · *Department of Pharmacology, University of Miami School of Medicine, Miami, FL*

Donna H. Wang · *Department of Physiology, Eastern Virginia Medical School, Norfolk, VA*

F. Eugene Yates · *Department of Medicine, Physiological Monitoring Unit, University of California, Los Angeles, Los Angeles, CA*

Benjamin W. Zweifach · *Department of Applied Mechanics and Engineering Sciences, Bioengineering, AMES-Bioengineering University of California, San Diego, LaJolla, CA*

Chapter 1

Vascular Resistance

Structural vs Functional Basis

Benjamin W. Zweifach

General Considerations

The term "peripheral resistance" represents a measure of the physical hindrance encountered by the blood as it is transported through the successive conduits (arteries, capillaries, veins) that make up the vascular system. The physical driving force in the vascular system, the blood pressure, and the resulting volumetric flow rate, are the two dynamic modalities that are used to provide a quantitative estimate of vascular resistance. Another term frequently used in the physiologic literature with respect to the flow in appendages or organs is conductance, which represents the reciprocal of the resistance for the whole array of vessels from arteries, through the capillaries and the effluent veins. Various approaches have been taken to separate the resistance for the vascular tree as a whole into at least three major components—arterial, capillary, and venous. A distinction between the relative contribution of the larger arterial vessels as opposed to that of their hierarchical ramifications is useful because of substantial differences and complexity of the controls for the proximal and distal portions of the system.

From *The Resistance Vasculature*, J. A. Bevan et al., eds. ©1991 Humana Press

Definitions

Calculations of the physical hindrance encountered in the vascular system have been dealt with in an analagous way to the impedance encountered in an electrical circuit.[1] An important difference, however, stems from the fact that in an electrical circuit, the impedance or resistance is a fixed entity, whereas in the vascular system, the level of hydraulic resistance not only varies at different segmental levels, but is under continuous modulation because of the need to bring pressure and flow in accord with changing needs of specific tissues or with different types of bodily activities. Such adjustments are in essence achieved by active modulation of lumen dimensions of muscular blood vessels, and by the redistribution of the circulation among the in-series and in-parallel circuits that make up the peripheral vascular bed. Under physiological conditions, the activation of selected functional controls serves to normalize the effects of acute perturbations, whereas in the face of a persistent perturbation, structural remodeling of the vessel wall and the segmental alignment of the vessels is brought into play, presumably as compensatory measures.[2]

It is obvious that the vascular system cannot be viewed as a static entity since it is capable of considerable adaptation at the individual vessel level, and as an organic unit under conditions of physiological stress or disease. Implicit in dealing with the design characteristics of the vascular system is the assumption that this organic unit, as well as its constituent parts, are organized to carry out a specific bodily function. The newer techniques of molecular biology and immunology have brought to light challenging information concerning the cellular constituents of the vascular system. The significance of the data is clouded by the fact that diverse endogenous substances can produce the vascular responses observed under *in situ* conditions. A useful frame of reference for bringing the cellular and physiologic data together is the concept of homeostasis that was formulated some 75 years ago by Walter B. Cannon[3] to deal with the basic function of the so-called autonomic nervous system. Cannon concluded that the single overriding need for the organism was to maintain a stable internal environment (the milieu intérieur) in order to make it possible for the parenchymal cells to carry out their selective metabolic activities at a prescribed level. It is this basic homeostatic mechanism that is the domain of the circulatory system as a whole, and in particular, the microcirculation proper.

A balanced tissue environment is achieved in individual organs under normal circumstances by distinctive structural features and a hierarchy of functional controls that enable the terminal vascular tree to operate as an independent organic unit. The basic element of vascular homeostasis is the maintenance of exchange between the blood and the interstitial compartments at a prescribed level by continuously adjusting the surface area available for such exchange in line with the changing volume of blood that is being delivered. No single set of changes in vascular dimensions or resistance can account for this complex homeostatic function. Indeed a whole family of mediator pathways have been found to be implicated in these basic vascular adjustments.

Although for the sake of expediency, the various segmental branchings can be considered as finite structural entities, generalizations concerning the relevance of changes in particular structural features to the overall resistance for the entire consortium of peripheral vessels cannot readily be drawn because of its highly reticulated nature involving in-series as well as in-parallel pathways in which the hierarchical constituents have different lumenal dimensions, different pathlengths, and different wall characteristics.

Sites of Resistance

The term "resistance vessels" is commonly used in the literature to designate the family of vessels—operationally the hierarchy of arterial branchings—that represents the major site of resistance. Inasmuch as the arterial conduits range in size from several hundred microns down to 10 μm, there are limits to treating data for this entire hierarchy of vessels as a lumped parameter in which the smaller arteries are simply a scaled-down version of the larger ones.[4] Conclusions drawn from averaged values for total arterial resistance are valid in a phenomenological framework, but must be buttressed by detailed anatomical information at the network level for further analysis.

A comparatively fixed volume of blood is made available for systemic circulation by the pumping action of the left ventricle of the heart and is carried by the aorta and a number of large central conduits for distribution to the various organs of the body by way of a succession of arterial branches of diminishing size.[5] Inasmuch as the actual capacity of the total vascular envelope is greater than the volume of blood in circulation, the flow of blood must be apportioned regionally so that an increased perfusion in one area is counterbalanced by an equivalent decrease in other areas. The segments of the vascular tree that are involved in fundamental adjustments of this kind have only been sketchily identified. Calculations based on cardiac output measurements show that the greater part of the total vascular resistance is encountered along the regional arteries and their branchings, which range in diameter from hundreds of microns down to 10–12 µm. The principal physical or structural feature that determines the level of resistance encountered in a given segment is the relative cross-sectional area of the vessels forming the pathway along which the blood is being transported (lumen diameter and vessel number).

Basic Mathematical Relationships

Numerical estimates of resistance (R) for individual vessels or groups of vessels have been calculated on the basis of the relationship established by Poiseuille for the flow of fluids in tubes involving the driving force, the net effective blood pressure or pressure drop (ΔP), and the resulting volumetric flow rate (Q), where $R = \Delta P / Q$. The hydraulic resistance for a given tissue has also been calculated by taking into account the average length (L), the number of vessels in parallel (N), the average diameter (D), and blood viscosity [η] using the following expression: $R = K \times L \times \eta / (N \times D^4)$. In situations where vascular hemodynamics are being monitored on the basis of changes in blood flow (Q), $Q = \Delta P \times N \times D^4 / [K \times L \times \eta]$, in which K is a constant, P = pressure, D = diameter, L = vessel length, η = viscosity of blood. In general, the pressure component in the formulation for resistance in general has a monotonic relationship to changes in vessel diameter, whereas the flow component is related exponentially to changes in lumenal diameter. The fact that D^4 relationships tend to be well preserved overshadows the effects of other variables. As a consequence of this power law relationship, even small changes in diameter scale up the problem so as to make it difficult to set up an experimental procedure that would permit a sufficiently sensitive analysis of perturbations in the vascular system on the basis of changes in volumetric flow.

Resistance values have been calculated for the entire vascular system,[6] for the vascular network in selected organs,[7] and for individual vessels.[8] When resistance values are calculated for individual vessels whose dimensions are well characterized, the application is straightforward and in line with the Poiseuille formulation for flowthrough tubes. However, when the resistance is calculated for a mass of tissue or even for the network between a given artery and its corresponding vein, the implications of the change in resistance with respect to cause and effect are not readily discernible.[4] Total resistance for the vascular tree may remain constant despite the fact that the distribution of the segmental resistances within the peripheral circulation has been shifted. In vivo resistance data has been supplemented with information obtained by the perfusion of whole organs or appendages under conditions of controlled pressure and flow,[9] and by investigations on excised vessels (200–300 µm arteries).[10]

Lautt, in a recent communication,[11] has raised the question regarding whether it is more appropriate in dealing with pressure-flow relations for the broad array of vessels that make up whole organs to calculate changes in conductance rather than resistance since at low flow rates, resistance is too high to be measured accurately. The argument is largely a semantic one. Lautt defines the pressure gradient as $P_a - P_v$ for an entire network with multiple inflows, and outflows instead of the conventional dP/dx gradient used in a local context with respect to the Poiseuille formulation. Although the formal difference in such usage seems minor, some interpretational confusion can arise.

Dynamic Considerations

The active readjustment of "vascular resistance" in the face of perturbations in pressure and flow has been approached by macroscopic as well as microscopic techniques. A common practice has been to use the relative reduction in systemic pressure along the ramifications of the peripheral vascular tree as a yardstick for estimating the rank order of the hierarchical segmental resistance on the assumption that the resistance and pressure bear a linear relationship to one another.[12,13] Alterations in the distribution of resistance have been described through the rate of dissipation of pressure along the arterial ramifications.

The pressure profiles recorded for different tissues show significant differences. An example of the pressure distribution in segmental ramifications of the arterial system is shown in Fig. 1 for a skeletal muscle, the spinotrapezius of the rat. Systemic pressure for such normotensive controls ranged between 100 to 105 mmHg under general anesthesia. The scatter plot indicates the pressures in individual vessels ranging from 100 to 150 μm feed arteries through the microvascular bed. Note that the major reduction in pressure in this tissue occurs within the arteriolar branchings (as indicated by the rectangle demarcated by the broken lines).

The systemic pressure shows only a modest 10–15% reduction during the transport of the blood along the large conduits into the regional feed arteries (100–150 μm) of other muscles, such as the tenuissimus, gracilis, and sartorius. Once these arterial vessels enter the tissue that they supply, their branchings interanastomose with one another to form a meshwork of arcade arterioles. It is within this meshwork of arterioles that a substantially higher vascular resistance becomes apparent, as reflected by a net pressure drop across the hierarchy of aracade arterioles of between 30 to 40 mmHg (a 25–30% reduction below the feed artery level). The subsequent selective apportionment of the arteriolar bloodstream into the myriads of capillary vessels is carried out by an array of side-arm branches (4–6/arcade segment) that form long tapered pathways before they blend in with the true capillaries. Inasmuch as these side-arm branches terminal arterioles are only one-half to one-quarter the size of the parent trunk, the restricted entry conditions across this branching configuration allow for a further abrupt reduction in pressure.

In Fig. 2, a comparison is made between pressures in normotensive controls in the Dahl salt-sensitive strain and the WKY strain of rats. Note that the slope for the segmental pressure drop relative

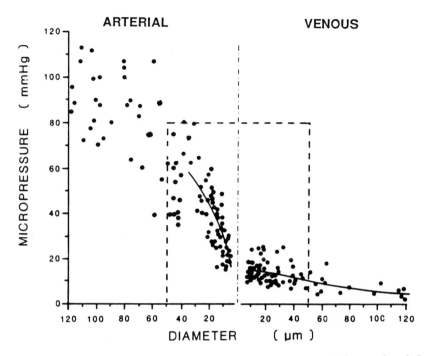

Fig. 1. Scatter plot of intravascular pressures in arterial vessels of the spinotrapezius muscle preparation of normotensive controls (WKY). The rats are under Alfathesin anesthesia . Mean arterial blood pressure is 95–105 mm Hg. Note the broad range of pressures in the larger arteries with no obvious trend. Cubic spline curve fit is shown for the pressure reduction in the microvessels.

to the feed artery pressure level begins to diverge in the feed artery region of the Dahl strain (a 25% greater pressure drop than in the WKY normotensives). The difference between the two profiles is progressively narrowed along the arteriolar branchings.

Ancillary approaches dealing with pressure relationships have also been utilized for this purpose, including the effect of induced changes in pressure on the flow through exteriorized organs and appendages[14,15] and the interplay of transcapillary fluid filtration vs absorption in isolated structures, as well as under in vivo conditions.[16] The functional contribution of the sequence of resistances across organs or tissues is much more complicated in view of the substantial structural heterogeneity.

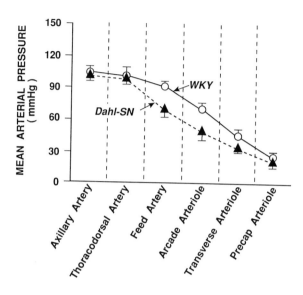

Fig. 2. Comparison of mean pressures in arterial vessels leading to a skeletal muscle for two strains of rats (Dahl and WKY) that have been used as controls in studies on hypertension. Points that are shown represent averaged values of 20–40 measurements (+ SD) in the spinotrapezius muscle preparation. A clear-cut difference exists beginning with the level of the branchings of the feed artery. From ref. 67, reprinted with permission.

In contrast to conventional physiologic approaches that provide data on an extensive array of vessels in masses of tissue, direct intravital microscopy deals with individual vessels that are usually selected at random and identified on the basis of an arbitrary alphanumerical branching order. Such studies for the most part are carried out on animals under general anesthesia. Both the macro and micro approaches provide meaningful information, but are beset with potential weaknesses. Vital microscopy measurements on randomly selected vessels cannot be used by themselves to characterize changes in the overall resistance of the network. A more realistic approach would be to have serial measurements along a number of well-defined pathways beginning with a discrete arteriolar location, and in turn, across successive branches leading into the capillary exchange side of the network.[17] Similar reservations exist with respect to studies on excised arteries,[18] in view of the fact that neither the structural properties nor the responses of any test vessel need necessarily be representative of comparable blood vessels in other regions.

Although some degree of anatomical innervation can be demonstrated throughout the vascular tree,[19,20] centrally mediated, baroreceptor-type reflex adjustments appear to have their maximal effect on the larger arteries.[21] Still unresolved is the extent to which the more peripheral ramifications of the arterial system in different regions are affected by such neurogenically mediated controls. Hutchins and coworkers,[22] in studies on the effects of bilateral occlusion of the common carotid artery, found that the response involved some of the microvessels but not others. Conversely, there is no well-documented evidence concerning the extent to which adjustments of local tissue origin are spread in a retrograde direction to involve the more proximal arteries. Meininger and coworkers have selectively modified pressure and flow levels in perfused appendages in an attempt to determine the relative importance of adjustments of systemic origin (large vessel modulation) vs local tissue modulation (arterioles) with respect to the overall resistance for the hind limb. Although they concluded that myogenic-type of autoregulatory adjustments take precedence over peripheral adjustments induced by vasoconstrictor agents, the data in such experiments deal with the dynamic behavior of an ensemble of vessels in which shifts in segmental contributions were not identified.

Shibata and Kamiya,[22] who examined the mutual interaction between central and local type control mechanisms, concluded that local controls (e.g., reduced PO_2) can modulate the efficacy of centrally mediated adjustments (carotid artery occlusion), particularly in the terminal arterioles (<25 μm). In view of the closed nature of the vascular system, some type of integration of central and local controls must exist, presumably in the intermediate vessels represented by the feed arteries just proximal to the tissue proper. More definitive information is needed, especially regarding sustained adjustments.[23,24]

As a consequence of these many uncertainties, investigators have proposed a number of alternative approaches for the analysis of the hydraulic resistance in the peripheral vascular system. Borders and Granger,[25] for example, have suggested that the physical hindrance in the vascular system should be calculated on the basis of energy costs—so-called "power dissipation" (Pd), where Pd = ΔP (the net driving pressure) × Q (volumetric flow rate). Emphasis is thus placed on the actual energy utilized in moving the blood along the vascular tree relative to the volumetric flow involved without entering into details of vessel alignment or structure. From this point of view, the dissipation of energy or power is greatest in the larger arterial vessels because of the mass of the blood being transported and the simple dichotomous nature of the branchings in the larger segments. Power and resistance, however, represent different components with different physical connotations. Although in a physiological context power dissipation (Pd) or energy conservation is an important biological hallmark, it is not clear whether Pd provides a more meaningful parameter for weighing the significance of changes in segmental hindrance in response to physiological or abnormal adjustments.

Nellis and Zweifach[1] adapted the methods used to calculate the impedance in electrical circuits to determine resistance for successive segments of the terminal vascular bed. Operationally, flow through individual branchings of the arterial tree in the mesentery was interrupted by occlusion with a microneedle, and upstream and downstream pressures were recorded until zero flow was present. This maneuver can be treated as a short circuit in a simple electrical analog of the vascular network. Serial measurements made it possible to break the total resistance for the microvascular network into at least three major subclassifications: arterial, capillary, and venous, on the basis of the slopes of the pressure-flow plot during the occlusion procedure. When the vessels of the mesentery or omentum were studied, the highest resistance was encountered on the arterial side for the 30–35- μm array of vessels.

In a comparable context, pressure and flow values for a given set of vessels (i and j) can be used to calculate the resistance (R), on the basis of the pressure drop ($\Delta P = P_i - P_j$) between them and the blood flow at either of the two points (Q_i or Q_j) . Two parametric resistance lines can be identified for each vessel—one related to flow at the upstream reference point and the other for the downstream flow rate. The resistance calculated in this way will be proportionately higher in the smaller reference vessel.

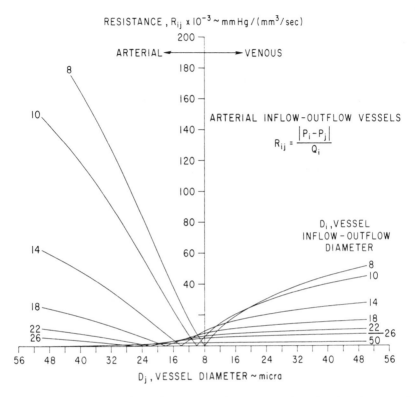

Fig. 3. Parametric plots of resistance between pairs of points in vascular network based on the distribution of pressure and flow in the mesentery. Numbers to the right and left of each of the lines represent the size of the arteriole for which resistance was calculated relative to the other vessels. (Zweifach and Lipowsky, 1978).

As shown in Fig. 3 (adapted from Zweifach and Lipowsky, 1978),[26] the resistance relative to each selected point is projected to the right or to the left. The resistance line to the left is the value between an arterial vessel and a given smaller microvessel. Conversely, the line on the right indicates successive resistances as one moves into the confluent vessels on the venous side of the network. Inasmuch as flow proceeds from left to right, the resistance from a 14-μm precapillary to a larger feed vessel, such as a 40-μm arteriole, represents the inflow resistance at the 14-μm precapillary, whereas the resistance between the same 14-μm precapillary and a 40-μm venule represents the outflow resistance at that point. As a result of the sharp drop-off in flow rate on the arterial side, the R distribution lines rise much more steeply than the corresponding lines for venous vessels. When comparable distribution plots are devised for other tissues, such as skeletal muscle, it becomes obvious that the sites of major resistance are distributed differently in each.

Determinants of Vascular Resistance

Structural Aspects

Vessel Wall Characterisitics

The various structural elements that make up the hierarchical branchings of the arterial vessels determine not only their viscoelastic properties but their response to active stimuli.[27] The coat of smooth muscle in the wall is progressively attenuated from three circular layers in the 100–150-μm feed arteries down to two layers along their immediate branches within the tissue, and then to a single layer at the level of the 25–30-μm terminal arterioles.[28,29] The prominent elastic lamella characteristic of the intima of the larger arteries becomes increasingly fenestrated and fragmented at the level of the 30–50-μm sized arterioles until it is no longer seen in cross-sections of the 20–25-μm arterioles.[30] This succession of structural changes is associated with a modification of the compliance of the arterial ramifications to the extent that the elastic properties of the wall in the terminal arteriolar segments become increasingly dependent on the state of the smooth muscle cells *per se.*

The "wall thickness to lumen ratio"[31] reflecting the relative thickness of the wall of various sized arteries becomes progressively higher in the smaller arteriolar segments. Since the inner lumenal circumference of these arterioles is significantly less than the outer perimeter of the vessel, the inner aspect of the wall bears a significantly higher proportion of the strain imposed by the intravascular pressure. This uneven distribution of the strain in the wall of the arterioles is in contrast to the more uniform distribution of wall stress in the larger feed arteries. Distinctive differences in the elements that bear the strain may contribute to the different types of structural adaptation that appear under the stress imposed by higher pressure in the large and small arterial vessels.[32,33]

When vessel dimensions in hypertensives are compared with control values on the basis of their dilated or completely relaxed state, it can be seen that in contrast to the trend for wall thickening in the larger arteries, the relaxed small arterioles, on average, are as much as 30–40% wider than in controls. These arterioles display an augmented tone[34-37] whereas in the arteries, which undergo wall hypertrophy[38] and hyperplasia,[39] there is little or no evidence of a change in vascular tone.

Terminal Vascular Bed

In view of the substantial evidence that the ramifications of the arterial system within the tissue proper can act collectively as an organic unit independent of the circulation at large,[40] attempts have been made to formulate a structural module of the microvasculature that in principle represents a prototype of the network wherever the terminal vascular bed serves primarily a nutritive function.

In principle, such a microcirculatory module has been conceived to consist of the array of vessels interposed between a supply artery and the corresponding venule.[41] There is some question whether the feed artery can be considered to be an intrinsic component of this organic unit. Detailed two-dimensional reconstruction of the meshwork of terminal arterioles and their ramifications has been possible in thin flat skeletal muscle preparations and in flat tissues, such as the retina[42] or the mesentery of visceral appendages.[43] However, the manner in which the activities of contiguous modules reenforce one another at a higher three-dimensional level of organization has not, as yet, been clearly defined.[43]

When the topological features of the terminal vascular bed as an organic entity are taken into account, it can be seen that the arrangement of the microvascular network is dominated in tissues, such as skeletal muscle, by arteriole to arteriole interconnections[43] that in effect set up a scaffold from which large numbers of thin muscular side-branches are distributed. The latter are long tapered pathways and in turn deliver as many as 4–6 capillary-sized offshoots, which are also deployed as in-parallel branchings. As a consequence of this structural alignment, almost 75–80% of the arteriolar offshoots within the tissue represent in-parallel circuits. Most analyses have been concerned with the effect of such junctional configurations on the red blood cell flux distal to this point. The selective distribution of blood through varying numbers of side-arm offshoots by itself, however, is an important physical determinant of the peripheral resistance.

The fact that the parent trunk as well as the individual branches have the capacity to undergo active adjustments provides a mechanism for the selective modulation of precapillary pressure and volumetric flow, which is the primary homeostaic responsibility at the exchange level of the microcirculation.[46] Emphasis in this regard has been placed on the modulation of pre- and post-capillary resistance levels by calculating the effects that shifts in the so-called pre- to post-capillary resistance ratios have on transcapillary exchange under different conditions.[47]

A number of investigators[48–50] have analyzed the makeup of the terminal vascular bed on the basis of optimal design theory, according to which the efficient operation of the network as a whole depends on the maintenance of a D^3 diameter relationship with respect to changes in flow. In the past, analyses of structural remodeling have emphasized adaptive changes involving the vessel wall proper relative to the wall stress imposed by the blood pressure. However, under conditions in which persistent perturbations in pressure-flow relationships develop, it is equally important to note that there is a structural remodeling of the network constituents, presumably to bring the geometric features of the network into line with optimal design principles. In view of the increasing evidence that the endothelial cells are an important factor in controlling arterial vessel dimensions and resistance through the release and modification of vasoactive mediators, a number of investigators, including Griffith et al.,[51] suggest a key role for endothelial reactions in maintaining microvascular dynamics at the structural as well as the functional level. Such a flow-dependent mechanism could serve to coordinate physical and biochemical relationships in line with the power law dependance governing changes in vessel diameter.

Functional Aspects

Overlapping Controls

The dimensions of the feed arteries and their arcade intercon-nections within the tissue are under the influence of a family of overlapping controls concerned with separate aspects of vascular homeostasis. These mechanisms include the sympathetic nervous system,[52] baroreceptor reflexes,[53] endothelial cell transducer activ-ity with respect to vasoactive materials,[54,55] as well as fluctuations in the shear rate of the blood itself.[56] Just as the large arteries adjust the regional apportionment of the circulating blood volume, the meshwork of interconnecting arterioles within the tissue can be considered operationally as a reservoir for the apportionment of blood to discrete capillary arrays within the tissue. This distribu-tion of blood within the terminal vascular bed is in the main under the influence of local regulatory mechanisms that are activated by fluctuations in tissue metabolism.

Intrinsic and Extrinsic Vascular Tone

The contractile elements that modulate vascular dimensions are kept under a variable state of active shortening (operationally referred to as "tone"), depending on the size and hierarchical loca-tion of the vessel. It is possible to obtain a reasonable estimate of vascular tone in anesthetized animals by measuring the difference between the width of optimally relaxed blood vessels and their diameter under steady-state conditions.[57] The levels of arterial tone[58] in general bear an inverse relationship to vessel size (Fig. 4). The 15–30-μm terminal arterioles are kept under comparatively high levels of tone, whereas the 75–100-μm arterioles have the lowest levels. The resistance of the more peripheral arterioles is thus to a considerable extent dependent on the level of smooth muscle tone that undergoes substantial time-dependent variations, fluctuating gradually between an upper and lower level over periods ranging from 15 to 40 min.[59] Tone in a set of arterioles from a given parent vessel is not necessarily the same, so that under steady-state condi-tions some of the arterioles that make up a designated branching order are in a partially or completely narrowed state, whereas oth-ers are essentially at maximal relaxation. The overall phenomenon

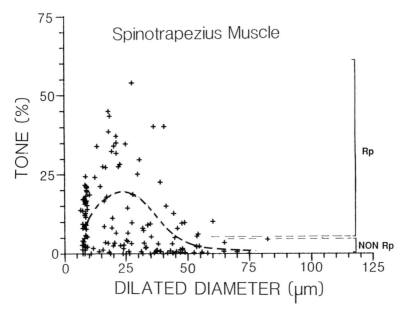

Fig. 4. Arteriolar tone expressed as a function of vessel diameter (steady-state vs maximal diameter after dilation). The trend in the data points was approximated by cubic spline curve-fitting (broken line). Responding (Rp) and nonresponding (Non Rp) arterioles following topical application of the vasodilator papaverine in the spinotrapezius of rat (Schmid-Schönbein et al., 1987). Reprinted with permission.

is apparently not random, since the average tone values for the population of vessels forming a particular subgroup of vessels consistently fall within a predictable range under normotensive conditions. When animals are subjected to acute perturbations, or to various experimental manipulations, the mix of vascular tone values within the population shows a characteristic shift in either a positive or negative direction.[60,61] For example, frequency distribution plots of tone values for the arcade type of arterioles and their transverse arteriole branches in the spinotrapezius muscle are shifted in hypertensives (SHR), so that the averaged values for the group of vessels are significantly above normal.

The tone exhibited by the smooth muscle of the larger arterial vessels can be adjusted by intrinsic as well as extrinsic factors.[62] Recent work suggests that the intrinsic or "basal" tone in particular segments of the arterial tree can be modified in a selective way through localized changes in the status of the endothelial lining cells in the chemical transduction of vasoactive mediators.[63] A number of reports in the literature[60,64] have demonstrated that an increase in blood shear rate, as occurs in hypertension, will by itself modify the release by the endothelial cells of vasoactive mediators that affect smooth muscle tone. It is interesting to note that Kamiya and coworkers[50] in setting up a model of the terminal vascular circulation, hypothesized that the primary dynamic variable around which peripheral circulatory controls were organized was the maintenance of the blood shear rate within a prescribed range.

Integration of Factors Influencing Segmental Resistance

In view of the range of vascular sequelae described in the literature, it is not surprising that diametrically opposite interpretations have been advanced concerning the mechanisms responsible for the development of a sustained elevation in vascular resistance in conditions such as hypertension; particular events are viewed either as a consequence of an already elevated systemic blood pressure, or they are considered to be one of the singular perturbations leading to the elevated blood pressure. Folkow and associates[10] developed as a working hypothesis a scenario in which the sustained increase in peripheral resistance becomes, over time, increasingly determined by a structural remodeling of the large arterial vessels. This general concept has received considerable support from various approaches to the problem.[65,66] The difficulty with the implications of the hypothesis advanced by Folkow et al.[4] is not whether or not such remodeling can or cannot account for the prevailing elevation in peripheral resistance in hypertension, but whether or not the changes found to be associated with one set of vessels are necessarily representative of those in comparable vascular segments in other parts of the body. The assumption has been made that the higher wall stress imposed by the elevated blood pressure leads to an equivalent increase in wall thickness in all comparable arteries. However, neither dynamic (pressure profiles) nor structural evidence are available to support such a contention. We have been unable,[30] in regional arteries that supply five different skeletal muscles in SHR and in Dahl-S rats and are presumably subjected to the same displacement of the blood pressure, to show evidence of structural adaptation (hypertrophy or hyperplasia).

Extensive intravital studies of various exteriorized tissues[67-69] have led other investigators to adopt as an alternative to this point of view the primary contribution of the arteriolar portion of the vascular tree to peripheral vascular resistance. An observation of singular importance favoring the key role played by arteriolar readjustments concerns the fact that despite the substantial structural and functional heterogeneity of vascular organization, the arteriolar ramifications of all of the system exhibit an intrinsic capacity to make adjustments of pressure and flow needed to maintain the physical and chemical integrity of the tissue milieu. Even minor perturbations in the tissue milieu lead to autoregulatory readjustments of these arterioles.[62] This intrinsic capacity in the face of acute perturbations is reenforced by the ability of the microvascular network to undergo extensive structural remodeling in a compensatory mode when confronted by chronic perturbations.

It should be pointed out that the comparatively low physical hindrance encountered in the larger arterial segments of the vascular system under a particular set of circumstances can, however, be significantly modified under other circumstances. For example, under steady-state conditions in both normotensive WKY and SHR hypertensive rats, the arterial driving force is reduced by only a modest 15–20% during the transport of the blood along the successive branchings leading to the feed artery just proximal to the spinotrapezius muscle.[2] On the other hand, in a salt-sensitive model of hypertension (Dahl-S strain of rats) with an equivalent systemic blood pressure, there is as much as a 40–50% reduction in pressure across the same set of arteries, an almost twofold greater change.[70]

The rapidity with which autoregulatory responses appear at different levels of vascular organization makes it difficult to determine the extent to which changes identified at one vascular level can be viewed as primary or secondary factors in readjustments of peripheral resistance.[71] This issue cannot readily be resolved on the basis of conventional intravital microscopy protocols because of the acute nature of such experiments. Inasmuch as the vascular system is a closed circuit, changes at every level are in the long run interactive so that many different scenarios can develop involving a mix of large vessel and microvascular changes that, over time, eventually modify the distribution of blood until peripheral homeostasis is undermined.

References

[1] Nellis, S. H. and Zweifach, B. W. (1977) *Circ. Res.* **40,** 546–556.

[2] Zweifach, B. W. and Lipowsky, H. H. (1985) *Handbook of Physiology, Section 2: The Cardiovascular System, vol. 4: The Microcirculation* (Renkin, E. M. and Michel, C. C., eds.), American Physiological Society, Bethesda, MD, pp. 251–307.

[3] Cannon, W. B. (1929) *Physiol. Rev.* **9,** 399–431.

[4] Folkow, B. and Neal, E. (1971) *Circulation,* Oxford University Press, London, pp. 73–97.

[5] Berne, R. M. and Levy, M. N. (1981) *Cardiovacular Physiology,* 4th Ed., C.V. Mosby Co., St. Louis, MO, pp. 58–62.

[6] Lais, L. T., Schaffer, R. A., and Brody, M. J. (1974) *Circ. Res.* **35,** 764–774.

[7] Proctor, K. G. and Busija, D. W. (1985) *Amer. J. Physiol.* **249,** H34–H41.

[8] Bevan, J. A. (1985) *Prog. Appl. Microcirc.* **8,** 7, 8.

[9] Folkow, B., Grimby, G., and Thulesius, O. (1958) *Acta Physiol. Scand.* **44,** 255–272.

[10] Warshaw, D. M., Mulvany, M. J., and Halpern, W. (1979) *Circ. Res.* **45,** 250–259.

[11] Lautt, W. W. (1989) *Microvasc. Res.* **37,** 230–236.

[12] Fronek, K. and Zweifach, B. W. (1975) *Am. J. Physiol.* **228,** 791–796.

[13] Bohlen, H. G., Gore, R. W., and Hutchins, P. M. (1977) *Microvasc. Res.* **13,** 125–130.

[14] Meininger, G. A., Lubrano, V. M., and Granger, H. J. (1984) *Circ. Res.* **55,** 609–622.

[15] Mueller, S. M. (1983) *Hypertension* **5,** 489–497.

[16] Pappenheimer, J. M. and Maes, J. P. (1942) *Am. J. Physiol.* **137,** 187–199.

[17] Mulvany, M. J. and Korsgaard, N. (1983) *J. Hypert.* **1,** 235–244.

[18] Mellander, S. and Johansson, B. (1968) *Pharmacol. Rev.* **20,** 117–196.

[19] Marshall, J. M. (1983) *J. Physiol.* **332,** 169–186.

[20] Lombard, J. H., Hess, M. E., and Stekiel, W. J. (1984) *Hypertension* **6,** 520–535.

[21] Hutchins, P. M., Bond, R. F., and Green, H. D. (1974) *Microvasc. Res.* **7,** 321–325.

[22] Shibata, M. and Kamiya, A. (1985) *Microvasc. Res.* **30,** 333-345.

[23] Delashaw, J. B. and Duling, B. R. (1986) *Microvasc. Res.* **36**, 162–171.

[24] Segal, S. S. and Duling, B. R. (1986) *Circ. Res.* **59**, 283–290.

[25] Borders, J. and Granger, H. (1968) *Hypertension* **8**, 184–191.

[26] Zweifach, B. W. and Lipowsky, H. H. (1978) *Cardiovascular Systems Dynamics*, (Baan, J., Nordergraff, A., and Raines, J., eds.), MIT Press, Cambridge MA, 2, pp. 187–193.

[27] Schmid-Schönbein, G. W., Skalak, T. C., and Sutton, D. W. (1988) *Microvascular Mechanics* (Lee, J. S. and Skalak, T. C., eds.), Springer-Verlag, New York, pp. 65–99.

[28] Engelson, E. T., Schmid-Schönbein, G. W., and Zweifach, B. W. (1985) *Microvasc. Res.* **30**, 29–44.

[29] Engelson, E. T., Schmid-Schönbein, G. W., and Zweifach, B. W. (1986) *Microvasc. Res.* **31**, 356–374.

[30] Schmid-Schönbein, G. W., Delano, F. A., Chu, S., and Zweifach, B. W. (1990) *J. Microcirc. Clin. Exp.* **9**, 47–66.

[31] Nordborg, C., Ivarsson, H., Johansson, B. B., and Stage, L. (1983) *J. Hypertens.* **1**, 333–338.

[32] Lee, R. M. K. W., Forrest, J. B., Garfield, R. E., and Daniel, E. E. (1983) *Blood Vessels* **20**, 72–91.

[33] Aalkjaer, C., Heagerty, A. M., Peterson, K. K., Swales, J. R., and Mulvany, M. J. (1987) *Circ. Res.* **61**, 181–186.

[34] Grande, P. O., Borgstrom, P., and Mellander, S. (1979) *Acta Physiol. Scand.* **107**, 365-376.

[35] Bohlen, H. G. (1986) *Hypertension* **8**, 181–183.

[36] Lynch, C., Rodick, V., and Hutchins, P. M. (1989) *Microvasc. Res.* **38**, 164–174.

[37] Lombard, J. H., Hess, M. E., and Stekiel, W. J. (1986) *Am. J. Physiol.* **18** (*Heart Circ. Physiol.*), H-761–H-764.

[38] Mulvany, M. J., Hansen, P. K., and Aalkjaer, C. (1978) *Circ. Res.* **43**, 854–864.

[39] Owens, G. K. and Schwartz, S. M. (1982) *Circ. Res.* **51**, 280–289.

[40] Wiedemann, M. P. (1967) *Physical Bases of Circulatory Transport: Regulation and Exchange*, (Reeve, E. B. and Guyton, A. C., eds.), Saunders, Philadelphia, PA, p. 307.

[41] Schmid-Schönbein, G. W. and Skalak, T. C. (1987) *Microvasc. Res.* **34**, 385–393.

[42] Kohoner, E. M. and Dollery, C. T. (1970) *Eur. J. Clin. Invest.* **1**, 167–171.

[43] Lipowsky, H. H. and Zweifach, B. W. (1974) *Microvasc. Res.* **7**, 73–83.

[44] Hsiung, H. H., Merickel, M. B., and Skalak, T. C. (1990) *FASEB. J.* (Abstracts II) **4**, A-1256.

[45] Schmid-Schönbein, G. W., Skalak, T. C., and Sutton, D. W. (1988) *Micro-*

vascular Mechanics (Lee, J. S. and Skalak, T. C., eds.), Springer-Verlag, New York, pp. 65–102.

[46] Vawter, D. L., Fung, Y. C., and Zweifach, B. W. (1974) *Microvasc. Res.* **8**, 44–52.

[47] Gore, R. W. (1974) *Circ. Res.* **34**, 581–591.

[48] Murray, C. D. (1926) *Proc. Natl. Acad. Sci. USA* **12**, 207–213.

[49] Mayrovitz, H. N. and Roy, J. (1983) *Am. J. Physiol.* **245** *(Heart Circ. Physiol.)*, H-1031–H-1038.

[50] Kamiya, A. and Tatsuo, T. (1980) *Am. J. Physiol.* **239** *(Heart Circ. Physiol.)*, H-14–H-21.

[51] Griffith, T. M., Edwards, D. H., Davies, R. R., Harrison, T. J., and Evans, K. T. (1987) *Nature* **329**, 442–445.

[52] Touw, K. B., Haywood, J. R., Shaffer, R. A., and Brody, M. J. (1980) *Hypertension* **2**, 408–418.

[53] Gorden, F. J., Matsuguchi, H., and Mark, A. L. (1981) *Hypertension* **3** **(suppl. 1)**, 135–141.

[54] Furchgott, R. F. (1983) *Circ. Res.* **53**, 557–573.

[55] Daniel, T. O. and Ives, H. E. (1989) *News in Physiol. Sci.* **4**, 139–142.

[56] Olesen, S. P., Clapham, D. E., and Davies, P. F. (1988) *Nature* **331**, 168–170.

[57] Hutchins, P. M., Dusseau, J. W., Marr, M. C., and Greene, P. W. (1982) *Microvascular Aspects of Spontaneous Hypertension* (Henrich, H. and Huber, H., eds.), Bern, FRG, pp. 41–53.

[58] Schmid-Schönbein, G. W., Zweifach, B. W., Delano, F. A., and Chen, P. C. Y. (1987) *Hypertension* **9**, 164–171.

[59] Murphy, R. A. and Miros, S. (1983) *Arch Int. Med.* **143**, 1001–1010.

[60] Koller, A. and Kaley, G. (1990) *Am. J. Physiol.* **258** *(Heart Circ. Physiol. 27)*, H-916–H-920.

[61] Meininger, G. A. and Trzeciakowski, J. P. (1990) *Am. J. Physiol.* **258** *(Heart Circ. Physiol. 27)*, H-1032–H-1041.

[62] Grande, P. O., Lundvall, J., and Mellander, S. (1977) *Acta Physiol. Scand.* **99**, 432–477.

[63] Hinjosa-Laborde, C., Greene, A. S., and Cowley, A. W., Jr. (1988) *Hypertension* **11**, 685–691.

[64] Rubanyi, G. M., Romero, J. C., and Vanhoutte, P. M. (1986) *Am. J. Physiol.* **250**, H-1145–H1149.

[65] Mulvany, M. J. (1983) *Blood Vessels* **20**, 1–22.

[66] Owens, G. K. (1988) *Hypertension* **11**, 198–207.

[67] Zweifach, B. W., Kovalcheck, S., Delano, F., and Chen, P. (1981) *Hypertension* **3**, 601–614.

[68] Bohlen, H. G. (1987) *Hypertension* **9**, 325–331.

[69] Joshua, I. G., Weigman, E. L., and Harris, P. D. (1984) *Hypertension* **6,** 61–67.

[70] Delano, F. A., Schmid-Schönbein, G. W., and Zweifach, B. W. (1990) *FASEB Proc.* (abstracts, 11), A-1249.

[71] Duling, B. R., Hogan, R. D., Langille, B. L., Segal, S. S., Vatner, S. F. Weigelt, H., and Young, M. A. (1987) *Fed. Proc.* **46,** 251–263.

Chapter 2

The Resistance Vasculature

Functional Importance in the Circulation

Björn Folkow

Introduction

In the same way that the relationships among voltage, current, and resistance in parallel- and series-coupled electric circuits used to be outlined in schematic diagrams according to Ohm's law, similar principles are useful in visualizing the principal hemodynamic events in the cardiovascular system, though here the related Poiseuille's law prevails. When the events in the various cardiovascular compartments are so approached, the capillaries stand out as the obvious key feature, simply because the diffusion and filtration–absorption exchange across their walls fulfill the essential purpose of the circulatory system, i.e., the maintenance of an optimal *milieu intérieur* for the tissue cells. All other parts of the system, as well as the superimposed neurohormonal control mechanisms, finally serve to keep capillary flow, pressure, and perfused surface area well adjusted to the prevailing exchange needs in the various tissues. It is mainly when some of these must transiently be given priority that centrally governed redistributions are induced between the parallel-coupled systemic circuits.

From *The Resistance Vasculature*, J. A. Bevan et al., eds. ©1991 Humana Press

This safeguarding of the capillary exchange processes naturally demands a variety of regulatory vascular functions, which all are executed by specialized smooth-muscle cells. Therefore the parallel-coupled circuits, which all are dimensioned to suit the particular demands of the tissues they subserve, can in turn be subdivided into a set of functionally defined series-coupled sections,[1] as schematically outlined in Fig. 1. These are only in part congruent with the conventional, morphologically defined parts of the vascular beds,[2] but they may better express their various hemodynamic functions:

I. Arterial windkessel and conduit vessels (largely the aorta and main arterial tree), whose specialized luminal and wall designs with minimal drop in mean arterial pressure (MAP) transform the intermittent expulsion from the heart into a more even blood delivery to the flow-regulating sections (II and III) of the various circuits.

II. Precapillary resistance vessels (largely all microvascular sections proximal to the capillaries), which, because of their low inner radius (r_i) and relatively thick muscular walls (w), are specialized to provide by far the major part of the systemic resistance (R_A)—and the most adjustable one, thanks to the geometric amplifier inherent in their high w/r_i ratio [1-4]. Thereby, the resistance to flow can be profoundly changed by even modest smooth-muscle adjustments; hence, the changes in flow are further greatly amplified, since they vary proportionally to the fourth power of the r_i changes (Poiseuille's law). In addition, on intense constriction, the high w/r_i of the precapillary resistance vessels tends to unload inner-wall layers and thereby leads to bulging and crenation of the endothelial cells.[5] This further enhances the resistance increase, because the luminal transverse section area is then reduced much more than the inner circumference. As a consequence, the total frictional losses between the fluid layers in the parabolic flow profile become much higher than if the same luminal reduction had occurred at a maintained circular shape. Thus, this particular consequence of the high w/r_i in the precapillary microvessels further potentiates the flow reductions that can be accomplished by a given smooth-muscle shortening.

III. Precapillary sphincter vessels (largely corresponding to the "A1–A4" microvascular sections), which, besides contributing roughly the distal third of the precapillary resistance, also control how much of the capillary network is kept open to flow, and hence

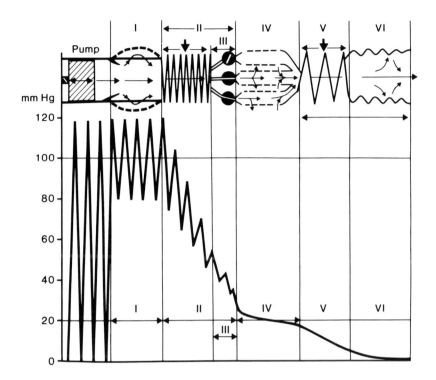

Fig. 1. A schematic subdivision of the various "parallel-coupled" systemic circuits into their "series-coupled," functionally specialized sections: I, arterial windkessel and conduit vessels; II, postcapillary resistance vessels; III, Precapillary sphincter vessels; IV, capillary exchange vessels; V, postcapillary resistance vessels; VI, venous capacitance and conduit vessels. Sections II and III are in this context the focus of interest, with respect to their main functions and how they, in turn, are to a great extent a matter of vessel design.

    exchange, by way of their smooth-muscle sphincter functions.[2-4] This affects not only the size of the exchange surface area and the time available for exchange, but also the diffusion distances between the blood stream and tissue cells.

IV.   Capillary exchange vessels (largely the true capillaries and adjacent venular sections), which, as mentioned, form the key cardiovascular section for both the diffusion exchange and the filtration–absorption events between the blood stream and tissues that occur across their walls; for lipid-soluble agents across the entire wall; for water-soluble agents via "small" and "large" pores; and for macromolecules, mainly via "large" pores.[4]

V. Postcapillary resistance vessels (mainly the muscle-endowed
    venular sections), which, though offering much less resistance to
    flow than Sections II and III, are nevertheless considerably impor-
    tant, since they serve as the denominator in the pre/postcapillary
    resistance ratio (R_A/R_V). Because of the much greater numerator (II
    and III) in this ratio, mean capillary pressure (P_c) is normally only 15–
    20% of MAP, but regulatory changes of P_c *(see below)* are obviously as
    readily induced by R_V changes as by R_A changes of equal percentage.
VI. Venous capacitance and conduit vessels (largely the entire venous
    side), which, thanks to a much wider bore than that of arterial coun-
    terparts, contain some 70% of the circulating blood volume. Thereby,
    even minor dimensional changes profoundly affect the filling, and
    hence the output of the heart, since this contains only about 10% of
    the blood volume.

From this functional division of the vasculature it is clear that
sections II and III, which are the main topic of this chapter and
book, jointly subserve some of the most important regulatory functions in the cardiovascular system: They are responsible for the
major part (R_A) of systemic resistance (R), which, together with
cardiac output (CO), determines the systemic perfusion pressure
(MAP), and also for control of the regional nutritional supply, as
well as its distribution over the capillary exchange network. In
addition, as discussed above, they form the "numerator" in the
R_A/R_V ratio, and are thus a major instrument in the neurohormonal
modulations of P_c, and hence of great importance for controlling
the fluid partition between the intra- and extravascular compartments. These functions of sections II and III are discussed below,
mainly from a general point of view, since subsequent chapters
deal with these vascular sections in the various circuits in great
detail and from individual points of view. For the same reasons,
reference is here given mainly to recent reviews.

Functional–Structural Interactions

General Aspects

The moment-to-moment control of vascular events is, of
course, executed by local and/or neurohormonal adjustments of
smooth-muscle activity in the mentioned regulatory sections. How-

ever, besides the obvious influence of this active factor (a) the hemodynamic outcome depends also on three passive factors, since they, too, can greatly influence the luminal dimensions of the respective sections, i.e., (b) their structural geometry $(r_i, w, \text{and } w/r_i)$, (c) their transmural pressure, and (d) their wall distensibility. In most studies of vascular events, only factor a is considered, although actually all four tend to become involved in most situations. Moreover, a change of any of them usually has consequences for all the others, so that none of them can be neglected in analyses of hemodynamic events, whether in physiology or in pathophysiology.[6,7]

For example, it is obvious that changes of smooth-muscle activity in Sections II and III inevitably also change their w/r_i relationships and thereby wall distensibility, while MAP and/or regional pressures are usually altered as well. Further, if exactly the same level of smooth-muscle activity is, e.g., maintained in two limbs when one is lowered well below, and the other raised above, heart level, the consequent differences in hydrostatic pressures will, by passive wall distension and recoil, lead to considerable differences in regional flow and blood contents. However, within seconds these local differences in transmural pressure usually induce precapillary myogenic counterregulatory adjustments *(see below)* so that blood flow, but hardly venous blood volume, returns toward the initial levels. Thus, in this case, an initial change of the passive factors c and d induce on the precapillary side active changes of a, whereby b and d are altered secondarily, though for functional reasons. However, such a situation may easily be mistaken for "nothing having changed in essentially stiff tubes.

Physical Considerations

The physically determined interactions between $a, b, c,$ and d were recently analyzed quantitatively by a hemodynamic approach in pair-perfused rat-hindquarter vascular beds, where normotensive (WKY) resistance vessels were compared with hypertensive (SHR) ones.[6,7] Although factor a surely can change flow and resistance dramatically within seconds, a structural resetting of b occurs much more slowly, for in rats it takes about a week to be completed. However, b is nevertheless, in the long run, quite a dynamic factor, and its change has very powerful hemodynamic consequences, as revealed in SHR. Actually, factors *a* and *b* are like "two sides of the same coin, which steadily interact and adapt to each other.[3,7] On the whole, vascular design tends to adapt to changes in transmural pressure in agreement with LaPlace's law, just as this law governs the construction principles used for high-pressure tubing in engineering. Thus, since

$$T = P \times r_i / w$$

where *T* represents tension per unit of wall layer and *P* represents transmural pressure, any change in *P* calls for corresponding changes in *w*, r_i to keep *T* constant.

Although the principal local response of precapillary resistance vessels to an acute pressure rise is smooth-muscle constriction, which functionally reduces r_i and increases *w* (and vice versa at a pressure reduction [*see below*, "functional autoregulation"]), the long-term local response of these vessels to a sustained pressure rise (lowering) is a structural r_i decrease (increase) and a wall (mainly media) thickening (decrease). This locally induced adaptation of precapillary vascular design can be called "structural autoregulation" as a long-term analog to functional autoregulation.[3,7] Thus, both serve to adjust r_i, *w*, and w/r_i to the altered pressure load in order to keep *T* constant; both tend to keep regional blood flow constant independent of MAP changes; and, even more important, both tend to keep P_c constant, thereby protecting the sensitive exchange vessels from undue pressure changes, and capillary-rich tissues from rapid edema formation.

It should further be realized that even modest changes of factor *b*, in terms of r_i, *w*, and w/r_i in the precapillary resistance vessels, can affect very profoundly the hemodynamic influences of factor *a* and, on the whole, the complex interactions among factors *a*, *b*, *c*, and *d*. For example, the mentioned comparison of WKY and SHR resistance functions[6] reveals how the structural r_i narrowing and w/r_i increase in established SHR hypertension are marked enough to fully match the MAP elevation. The reduced r_i implies that the baseline for smooth-muscle activation is reset upward, whereas the increases of *w* and w/r_i for geometric reasons result in vascular hyperreactivity, besides making the walls stiffer and stronger. This unspecific hyperreactivity is reflected as a steeper "resistance curve," which also starts from a raised baseline in SHR, whereas a selective increase (decrease) of smooth-muscle sensitivity shows up as a parallel left-hand (right-hand) shift of an otherwise unchanged resistance curve. Other experimental analyses show how these structural resistance adaptations are essentially confined to the precapillary side.[3,7]

Interactions

Figure 2 shows the complex interactions among factors *a*, *b*, *c*, and *d* in the SHR and WKY precapillary resistance vessels, where the in vivo MAP levels had differed by 45%. It is clear from this experimental comparison that the SHR vessels are stiffer, stronger, and, for geometric reasons, hyperreactive compared with the WKY ones. In fact, over the full range of smooth-muscle adjustments (i.e., from complete relaxation to maximal contraction) at their respective in vivo pressure, the range of resistance change in SHR is about 45% greater than in WKY, as shown in the right part of Fig. 2. When expressed as a range of flow changes, these prove to be almost precisely the same in SHR and WKY, even though the driving arterial pressure (MAP) and the average distending pressure on the precapillary side are 45% higher in SHR than in WKY. In other words, the raised systemic resistance in SHR can be maintained by a largely normal level of average smooth-muscle activity, though, of course, in vivo differences among various systemic circuits are likely, as is the case in normotension. Most findings in established human primary hypertension indicate a similar upward structural resetting of the precapillary resistance vessels.[3,7]

It should also be realized that, upon given reductions of smooth-muscle tone, as induced, e.g., by vasodilator drugs, the fall in resistance and in MAP will in absolute terms be greater in hypertensives than in normotensives, simply as a result of the "geometric amplifier" effect of the altered design in hypertensive resistance vessels. Too often, however, such consequences of the interactions between factors *a* and *b* have been mistaken as indicating primarily the presence of enhanced vascular smooth-muscle activity in hypertension. Figure 2 also shows in another way how

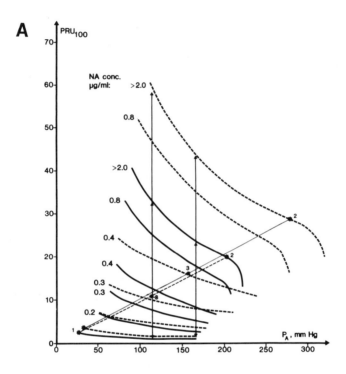

Fig. 2A. The relationships ("resistance lines") between arterial pressure (P_A) and flow resistance (PRU_{100}, i.e., mmHg/flow in mL/min/100 g), for paired, constant-flow perfused hindquarter vascular beds of adult normotensive (WKY) and spontaneously hypertensive rats (SHR). The curved "resistance lines" reflect the passive-elastic changes in resistance upon sudden alterations of the distending pressure, induced at different levels of stable vasoconstriction along the norepinephrine (NA) dose–response curves. In this type of diagram, the otherwise S-shaped constant-flow resistance curves show up as the slanting, straight lines (dashed, WKY; solid, SHR), with maximal dilatation at "1," maximal constriction at "2," and ED_{50} at "3." The diagram illustrates how the SHR precapillary resistance vessels, thanks to their narrower lumina in combination with their thicker walls compared with the WKY ones, are throughout stronger, stiffer, and for geometric reasons clearly "hyperreactive."

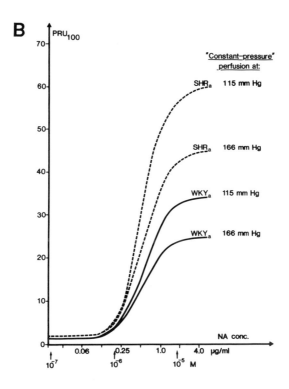

Fig. 2B. Four resistance curves for WKY and SHR hindquarter vessels, as they would appear if being "constant-pressure" perfused at their respective in vivo pressures of 115 and 166 mmHg. These curves are derived from the left panel by plotting the intercepts between the respective resistance lines and the vertical pressure lines at 115 and 166 mmHg. It is seen how the SHR resistance vasculature shows a geometrically based hyperreactivity in close proportion to their 45% higher perfusion and distending pressures in vivo. Thus, if instead expressed as flow changes (P/R) the SHR resistance vessels will, despite these higher pressures, cover the same range of flow changes as WKY for equal changes of smooth-muscle activity ("optimal structural upward resetting"). It is also shown how the SHR vessels would exhibit a greatly exaggerated hyperreactivity if operating at normotensive pressures, whereas the WKY vessels would exhibit an equally marked "hyporeactivity" if forced to operate at the SHR hypertensive pressure without compensatory structural adaptation. (From ref. 6, with kind permission of the Editor of *Acta Physiol. Scand.* and the authors.)

important factor b really is for the proper expression of a. Thus, the four deduced resistance curves in the right part show how the hemodynamic effects of changes in smooth-muscle activity would be greatly attenuated if WKY resistance vessels, without prior structural adaptation, were forced to operate at SHR pressure levels. Conversely, SHR resistance vessels would exhibit even more marked hyperreactivity if they were forced, without structural regression, to operate at normotensive pressure levels.

Another drastic example of the interactions between vascular function and design can be taken from the profound "physiological hypertension present in giraffes, where MAP at heart level in a 6-m-tall bull is about 275–300 mmHg to allow a proper perfusion of the brain 2.5–3 m higher.[3,7] Immense differences in arterial transmural pressures must be present in the giraffe between, e.g., cranial arteries (80–90 mmHg) and lowerleg arteries (400–500 mmHg). The vessels are, however, constructed accordingly, with w/r_i values four to five times higher in the arteries of the lower leg than in the cranial ones—a physical necessity for efficient smooth-muscle control of both regional and systemic resistances. On a less drastic scale, similar differences in regional w/r_i designs must be present in the corresponding arteries of humans. Such transmural pressure-dependent structural adaptations are, however, particularly obvious in human systemic veins, since in these veins the relative differences in hydrostatic pressures are much more pronounced than on the arterial high-pressure side. For example, the leg veins show w/r_i values much higher than veins at or above heart level, and in close proportion to the much higher transmural pressures present in the former veins in the erect position.[3,7,8] This also illustrates particularly well how the w/r_i design is primarily governed by the regional transmural pressure.

However, structural adaptations of the resistance vasculature in response to long-term functional alterations is related not only to changes in pressure, but also to sustained changes in tissue mass and/or metabolism. These processes are sophisticated and precise, and allow for independent changes of both r_i and w, even though the two adjust secondarily to each other to keep wall stress fairly constant, according to LaPlace's law. Thus, with tissue growth and/or with a sustained increase of tissue activity and metabolism, the resistance vessels show an average increase of r_i, associated with the formation of new capillaries, which implies that the distal "sphincter" sections also adapt their design. If MAP remains constant, w is secondarily adapted to the increased r_i in order to keep w/r_i also largely constant.[7,9]

If, however, both pressure and tissue demands increase, as is the case for the coronary circuit in hypertension with left ventricular hypertrophy, the coronary resistance vessels adapt their design to both these changes: coronary resistance at maximal vasodilatation (R_{min}) per unit of tissue mass then increases largely proportionally to the MAP elevation, as does w/r_i of the resistance vessels, while R_{min} adapts also to the increased bulk of myocardial tissue.[7] All these structural adaptation processes can, in rats, be completed in a week or so, and in human perhaps in a few months, humans having a metabolic turnover rate five to six times slower than that of rats.[3,7,9]

The hemodynamic function of the resistance vasculature is thus certainly not only a matter of their smooth-muscle activity a, though this no doubt accomplishes the most rapid and extensive changes in flow and resistance and is the obvious instrument for direct control. As outlined above, the hemodynamic influences of the rapidly adapted geometric design b of these vessels is also important, as is the impact of their wall distensibility c and transmural pressure d. Actually, a, b, c, and d continuously interact, and any change in one of them tends to change all the others, as exemplified above.

Principles of Smooth-Muscle Control in the Resistance Vasculature

General Aspects

As mentioned in the Introduction, the precapillary resistance and sphincter vessels subserve a number of general and local functions that are all of great importance. This demands not only a sophisticated system for remote control via neurohormonal links, but also one for strictly local control. Further, the interaction of these two control systems must be organized to largely satisfy both gen-

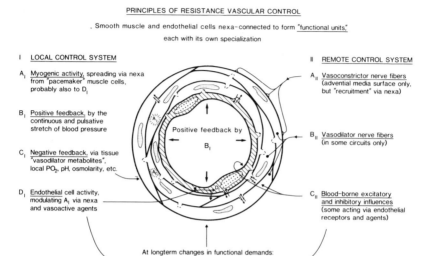

PRINCIPLES OF RESISTANCE VASCULAR CONTROL

. Smooth muscle and endothelial cells nexa-connected to form "functional units,"
each with its own specialization

I   LOCAL CONTROL SYSTEM

A_I Myogenic activity, spreading via nexa
from "pacemaker" muscle cells,
probably also to D_I

B_I Positive feedback, by the
continuous and pulsative
stretch of blood pressure

C_I Negative feedback, via tissue
"vasodilator metabolites",
local PO_2, pH, osmolarity, etc.

D_I Endothelial cell activity,
modulating A_I via nexa
and vasoactive agents

Positive feedback by
B_I

II   REMOTE CONTROL SYSTEM

A_{II} Vasoconstrictor nerve fibers
(advential media surface only,
but "recruitment" via nexa)

B_{II} Vasodilator nerve fibers
(in some circuits only)

C_{II} Blood-borne excitatory
and inhibitory influences
(some acting via endothelial
receptors and agents)

At longterm changes in functional demands:
"STRUCTURAL AUTOREGULATION"
as modulated by "trophic" influences

Fig. 3. Schematic illustration of the local and remote control systems of
the resistance vasculature. For details, *see text.*

eral and local demands in most situations, but it must also allow
for special priorities to prevail, at least transiently, e.g., in states of
emergency.

The local control system, obviously subserving primarily regional tissue needs, corresponds in a way to "community rule" in
human societies; the superimposed remote system, conveyed via
nerves and hormones from high centers that also respond to information from a variety of intero- and exteroreceptors, has much in
common with "governmental rule."[10] In both cases, the local rule
ordinarily prevails, with only mild overall supervision, but in emergencies even drastic central interferences can give priority to more
pressing demands of general importance.

Since the remote control system in this balance operates by
modulating links in the local system, the local system will be outlined first. Some general principles of these two control systems
and their actions on the resistance vasculature are schematically
illustrated in Fig. 3, as further outlined below.

The Local Control System

Since several tissues can drastically change their metabolism, e.g., skeletal muscles and the myocardium, the resistance and sphincter vessels of their circuits must be able to change both the blood supply and its capillary distribution over wide ranges. Here the state of maximal dilatation sets the upper limit for supply at a given perfusion pressure and blood viscosity, and even in heavy exercise MAP is seldom raised more than 20–30%. Since the nutritional demands in the resting steady-state are, for most tissues, only a fraction of the maximal ones, regional blood flows must be kept reduced accordingly so that the work load for the heart is kept as low as possible in the maintenance of a fairly stable perfusion pressure.

This naturally calls for an often quite marked smooth-muscle tone in the resistance and sphincter vessels, which thereby establishes a correspondingly large blood flow reserve for the respective tissues, which can be mobilized to the extent needed by appropriate inhibition of this basal resting tone. Evidently, the greater the difference between resting and maximal metabolism in the various tissues, the more pronounced the basal smooth-muscle tone on the precapillary side tends to be when at rest.[11,12]

The question arises about how this basal tone in systemic resistance and sphincter vessels is established, how it can be suitably inhibited during increased tissue activity, and how it is modulated by remote control mechanisms, both for the continuous regulation of the relationships among CO, R, and MAP and for allowing more extensive redistributions of flow, e.g., in emergency situations.

In essence, the basal smooth-muscle tone in these precapillary microvascular sections is myogenic in origin, just as the rhythmic activity of the heart is, in the end myogenic. This concept was proposed as early as 1902 by W. M. Bayliss, *(see* refs. 11,12), who suggested that the distension offered by the blood pressure acted as a triggering stimulus, whereas the vasodilator metabolites produced by the tissues had an inhibitory action. However, mainly because of technical difficulties, the presence of such an organization of the local blood flow control could not be shown experimentally until some 50 years later, and since then progress has been rapid. Recently, a symposium was devoted to the myogenic mechanisms and their importance for resistance-vessel control in health and disease, which should be consulted for details.[13] To simplify matters it may be concluded that at least some of the smooth-muscle effectors in these microvascular sections function like largely tonically active stretch, or perhaps better, tension receptors. Transmural pressure here acts as a facilitating positive-feedback influence, whereas local tissue-produced vasodilator factors serve as a potentially powerful negative feedback (Fig. 3).

Even if only some of the microvascular muscle cells in this system serve as myogenic pacemakers, the many intermuscular nexa connections will readily spread this activity so the entire media functions more or less as a unit. It appears as if pressure serves mainly to increase the rate of rhythmic myogenic activity at the cellular level, presumably explaining why the finest precapillary arborizations with discontinuous muscle arrangements usually exhibit asynchronous rhythmic contraction–relaxation cycles ("vasomotion"). In more proximal sections, it may well be that several pacemaker cells together contribute to create a relatively more stable tone by myogenic spread and fusion of initially asynchronous excitation along the microvascular tree.[11,12]

As mentioned, many of the tissue-produced metabolic products exert vasodilator influences, as does lowered local oxygen tension and pH, and thus serve as a potentially powerful negative feedback to the myogenic activity. Therefore, increased tissue activity, or reduced blood supply, leads to further accumulation of such local chemical factors and hence to a corresponding vasodilatation (the "metabolic factor"). Thus, if arterial pressure (and hence regional blood flow) is increased, the myogenic and metabolic factors act in concert to increase precapillary resistance, and vice versa at pressure reduction. These two factors thus together establish functional autoregulation of blood flow. However, an even more important consequence of this locally induced precapillary adjustment is that it tends to keep pressure in the pressure-sensitive capillary exchange section largely constant. The capillaries are thus protected during, e.g., transient MAP elevations from rapid edema formation.

On the whole, the pressure-dependent myogenic vascular tone and the steadily adjusting vasodilator metabolites together serve to keep tissue oxygen supply well balanced to local metabolic needs, whether these demands and/or the arterial pressure are altered.[11,12] For decades , however, it has been debated whether the myogenic or the metabolic factor is the important element, when actually both are equally important in their own right for this local control system: one generates the basal vascular tone and hence the blood flow reserve, while the other serves the respective tissues as the physiological mobilizer of this blood flow reserve.

Thus, in such situations, the myogenic and metabolic factors work in harmonious cooperation to form the basic framework of the local control system. When instead both arterial and venous pressures are equally increased (or lowered), as is the case, e.g., when limbs are passively shifted in relation to the heart level, and thus a change is induced primarily of the transmural vascular pressures, the myogenic factor dominates the hemodynamic situation, since there is then no initial change in either perfusion pressure or tissue metabolism. However, the myogenic and metabolic factors can also be put directly against each other—namely, if venous pressure is selectively increased, which reduces the perfusion pressure, and hence the nutritional flow, but increases transmural pressure. The net hemodynamic outcome will depend to some extent on the circuit and tissue involved, and on such factors as the current levels of tissue metabolism, myogenic reactivity, vascular sensitivity to local dilator factors, and so on.

Even though the interactions of the key elements in the local control system, as principally outlined above, may appear simple and straightforward, the situation is in reality far more complex and sophisticated. For example, in the precapillary microvessels, Rhodin[2] showed in the 1960s that there are numerous nexa between the endothelial and the smooth-muscle cells, which strongly suggests that these two cell types ordinarily act as a functional unit.[2,11] Furthermore, as pioneered by Furchgott,[14] the endothelial cells also produce important vasodilator (e.g., nitrous oxide, prostacyclin) and vasoconstrictor (e.g., enthothelin) agents for vascular control. They are, moreover, the main site for a variety of receptors to biogenic agents, to which substances like acetylcholine and bradykinin are coupled to exert their vasodilator actions by causing an endothelial release of, e.g., nitrous oxide.[14,15] Again, this shows how these two cell types in reality form a functional unit, each with specializations of its own but with the motor responses executed by the muscle counterpart, as indicated in Fig. 3. For example, at least in some regions, the myogenic response to stretch might be reinforced by the presence of an intact endothelium.[16] Further, the endothelium may also respond to the flow-dependent shear stress and thereby release nitrous oxide.[15] When, e.g., distal resistance vessels are dilated by local release of tissue metabolites, such an endothelial mechanism may contribute to the secondary relaxation also occurring in more proximal conduit and resistance vessels ("ascending dilatation").

On the whole, the interest for these evidently quite sophisticated local control mechanisms, with the myogenic activity as a basic phenomenon,[11,12,17] has greatly increased in recent decades, as reflected by the many contributions to the mentioned symposium.[13] It has, for example, been possible to differentiate between "distal-phasic" and "proximal-tonic" elements in the regulation of basal myogenic tone.[18,19] There is also evidence that local axon reflexes, conveyed via vasoconstrictor fiber arborizations, may become activated at least by more extensive transmural pressure elevations and thus superimpose a reinforcing local-neurogenic element[20] on the basic myogenic resistance vessel response,[21] as observed, e.g., in human limbs. Knowledge about such aspects of the local control system is likely to advance rapidly, as also discussed elsewhere in this book.

During "resting" equilibrium, tonic smooth-muscle activity in the systemic resistance vasculature, as mentioned, is on average quite high, though with considerable differences between the circuits, depending on the specialized functions of the respective tissues. This is clearly revealed by the great differences in regional flow increases that occur upon local induction of maximal dilatation in the various circuits. For example, although flow can increase 10–15-fold in skeletal muscles or in salivary glands, five- to sevenfold in the myocardium or gastrointestinal tract, or three- to fourfold in the brain, it increases only some 50% or so in the kidneys.[13,22] Thus, the preglomerular vascular tone in the renal circuit in the resting steady-state is quite modest, mainly because the blood stream here also serves as the abundant raw material for the steadily ongoing excretory functions. However, on MAP increases, e.g., the renal preglomerular resistance is promptly enhanced, where the smooth-muscle cells function largely like contractile tension receptors.[23]

The question arises about how much of this high, "resting" basal tone in some regions of the resistance vasculature is in the intact organism of myogenic origin and how much can be ascribed to remote excitatory influences. Here, particularly, the vasoconstrictor fibers are in focus, since hormonal vasoconstrictor influences in this situation seem to be of fairly negligible importance. Regional blockade of these fibers then reveals that as much as 80–90%,[22] or even more, e.g., in the coronary and cerebral circuits[22,24,25] of this regionally quite pronounced basal vascular tone is, in fact, of local myogenic origin. Actually, in this situation the vasoconstrictor fiber activity seems to be of much greater hemodynamic importance for maintaining vascular tone on the venous capacitance side (where myogenic activity is negligible), hence ensuring a proper cardiac filling and output.[22] However, even slight tissue activation promptly reduces the high myogenic tone proportionally, e.g., in skeletal muscles,[11,13,22] thanks to the altered chemical environment, reflecting the fine balance between the two key elements in the local control system, as is also evident in the coronary and cerebral circuits.[24,25]

The Remote Control System

The question then arises when the resistance vasculature is dominated by the potentially powerful remote control system and how it then exerts its modulating influence, which in some situations and circuits, can entirely overrule the local resistance control.

As outlined in Fig. 3, the vasoconstrictor fibers to this part of the cardiovascular system make contact only with the adventitial surface of the outermost smooth-muscle layer.[1-3,22,26] However, since the muscle cells are functionally interconnected via nexa, the vasoconstrictor nerves can nevertheless recruit and fully command the entire media. This is particularly so at high rates of fiber discharge, since this extrinsic excitation rate then seems to far exceed that derived from myogenic pacemaker activities: a shift from local to centralized control then occurs. Although the vasoconstrictor fiber supply to the resistance vessels, e.g., in skeletal muscles, skin, kidneys, and the gastrointestinal tract, is extensive, it is far more modest in, e.g., coronary or cerebral resistance vessels. Therefore, even when uniform increases of fiber discharge occur, the blood supply to the latter circuits is favored at the expense of the others. Further, the net hemodynamic outcome is then also influenced by the counterregulatory effects of locally accumulated vasodilator metabolites during the neurogenic flow restrictions. Also, the potency of such local influences seems to vary between circuits, again favoring the myocardial and cerebral blood supply.

In addition, the relative balance between the remote-neurogenic control system and the local-myogenic/metabolic one seems to vary even along the resistance vasculature in any given circuit. In general, although the local control system more or less seems to dominate in the most distal resistance sections, the nervous influence seems to be strongest in proximal parts of the precapillary resistance vessels, at least concerning sustained effects. This relative differentiation is particularly evident in diving species, where even distal conduit arteries become so intensely constricted during diving reflexes that in several tissues virtually no blood reaches their microvascular sections.[27]

By way of the potentially most powerful sympathetic nervous system higher centers continuously exert mild adjustments of systemic resistance and capacitance vessels, as well as of cardiac output in the reflex maintenance of a fairly constant MAP. However, in emergencies the vasoconstrictor fibers can convey powerful and fairly generalized signals to restrict flow, such as during blood loss,[22] though this, for reasons discussed, still favors myocardial and cerebral blood supplies. They are, indeed, also capable of markedly redistributing flow between the parallel-coupled circuits, which is the case when differentiated discharge patterns are induced, such as during mental stress.[1,10,26]

In addition, the reflex modulations of tonic vasoconstrictor fiber activity to the systemic resistance vasculature also serves another important purpose. Particularly in skeletal muscles—the largest single tissue mass, with a consequently large interstitial fluid depot—even modest elevations of vasoconstrictor fiber activity considerably increase the pre/postcapillary resistance ratio, because of the w/r_i-dependent geometric-amplifier effects on the precapillary resistance. This correspondingly lowers P_c and thereby shifts the Starling equilibrium toward transcapillary fluid absorption, whereas reflex reduction of vasoconstrictor fiber activity favors outward filtration. In this way, the resistance vasculature in some tissues with large bulk also serves as an important tool for the reflex control of the fluid partitition between the intra- and extravascular fluid compartments. It should here be noted that this remote neurogenic control of P_c in the mentioned tissues then overrules the local autoregulatory mechanisms, since these, as mentioned, tend to keep P_c constant, exemplifying how the two control levels sometimes work in opposite directions.

Some vascular beds also have a supply of various types of vasodilator fibers to the resistance vessels, which then serve to induce a remote suppression of the basal myogenic tone. This can be exemplified by the sympathetic cholinergic vasodilator fibers to the precapillary resistance vessels in skeletal muscles of some species, which are then activated mainly as a link of the limbic-hypothalamic defense reaction.[3,10] Thereby the muscle blood flow can be increased even in an anticipatory way, so that the nutritional supply of the locomotor system is already enhanced before an attack or flight behavior ensues.

Hormonal Control

A few principal words, finally, about some of the various hormones[22] influencing the resistance vasculature (Fig. 3), such as vasopressin, adrenaline, angiotensin, medullipin, kinins, and so on, which are dealt with in more detail elsewhere. During "rest," most of them seem to be present only in quite low (for some, even in subthreshold) concentrations. However, when they reach higher concentrations, they certainly contribute to modulating the tone of the resistance vasculature, though usually still in a more low-key fashion than the more rapidly operating and potentially more powerful vasoconstrictor and vasodilator nerves. Some of these hormonal effects, such as the vasodilator actions of kinins, seem to be conveyed via the endothelial cells by means of nitrous oxide release, whereas, e.g., adrenaline can either enhance or suppress the myogenic tone by direct muscle actions that depend on the balance between α- and β-receptors in the various sets of resistance vessels. For example, in skeletal muscles, microvascular myogenic activity is suppressed by low adrenaline concentrations via β-mediated membrane hyperpolarization; at high concentrations, an α-mediated constriction takes over.

However, besides their direct functional influences on resistance vascular tone, both nerve transmitters and, particularly, some of the vasoactive hormones, such as angiotensin, seem to exert also trophic long-range influences on vascular design,[28] potentially of great importance for, e.g., the early establishment of the structural "upward resetting" in hypertension,[3,7,28] as also indicated in Fig. 3. In addition, both local and blood borne trophic factors are probably involved in many other situations in which structural adaptation of the resistance vessels occurs, such as at increases in tissue mass, during sustained increases in metabolism, during wound healing, and so on.

Summary

Since the aim of this introductory chapter has been to outline merely the principal functions of the resistance vasculature and to provide some examples, the reader is referred for further information to the subsequent chapters, which describe in more detail the specialization and great sophistication that characterizes the control of this important part of the cardiovascular system.

References

[1] Folkow, B. and Neil, E. (1971) *Circulation* (Oxford University Press, London).

[2] Rhodin, J. A. G. (1980) *Handbook of Physiology, Section 2: The Cardiovascular System, vol. 2: Vascular Smooth Muscle* (Bohr, D. F., Somlyo, A. P., and Sparks, H. V. Jr., eds.), American Physiological Society, Bethesda, MD, pp. 1–31.

[3] Folkow, B. (1982) *Physiol. Rev.* **62,** 347–504.
[4] Renkin, E. M. (1985) *Handbook of Physiology, Section 2: The Cardiovascular System,* vol. *4: Microcirculation* (Bohr, D. F., Somlyo, A. P., and Sparks, H. V. Jr., eds.), American Physiological Society, Bethesda, MD, pp. 627–687.
[5] Greensmith, J. E. and Duling, B. R. (1984) *Am. J. Physiol.* **247,** H687–H698.
[6] Folkow, B. and Karlstrom, G. (1984) *Acta Physiol. Scand.* **122,** 17–33.
[7] Folkow, B. (1990) *Hypertension* **16,** 89–101.
[8] Kügelgen, A von. (1955) *Z. Zellforsch. Mikrosk. Anat.* **43,** 168–183.
[9] Folkow, B. (1987) *J. Clin. Hypertens.* **3,** 328–336.
[10] Folkow, B. (1987) *Circulation* **76 (Suppl. I)** , I-10–I-19.
[11] Folkow, B. (1989) *J. Hypertens.* **7 (Suppl. 4),** S1–S4.
[12] Folkow, B. (1989) *Microvasc. Research* **37,** 243–255.
[13] Folkow, B., Hansson, L., and Johansson B. (eds.) (1989) *J. Hypertens.* **7 (Suppl. 4),** S1–S173.
[14] Furchgott, R. F. (1983) *Circ. Res.* **53,** 557–573.
[15] Vanhoutte, P. M. (1988) *Cerebral Vasospasm* (Wilkins, R. H., ed.), Raven, New York, pp. 119–128.
[16] Harder, D. R., Kauser, K., Roman, R. J., and Lombard, J. H. (1989) *J. Hypertens.* **7, (Suppl. 4),** S11–S15.
[17] Johnson, P. C. (1989) *J. Hypertens.* **7 (Suppl. 4),** S33–S39.
[18] Mellander, S. (1989) *J. Hypertens.* **7 (Suppl. 4),** S21–S30.
[19] Grände, P.-O. (1989) *J. Hypertens.* **7 (Suppl. 4),** S47–S53.
[20] Henriksen, O. (1977) *Acta Physiol. Scand.* **(Suppl. 450),** 1–48.
[21] Lundvall J. (1989) *J. Hypertens.* **7 (Suppl. 4),** S85–S91.
[22] Mellander, S. and Johansson, B. (1968) *Pharmacol. Rev.* **20 (3),** 117–196.
[23] Aukland, K. (1989) *J. Hypertens.* **7 (Suppl. 4),** S71–S76.
[24] Feigl, E. O. (1989) *J. Hypertens.* **7 (Suppl. 4),** S55–S58.
[25] Faraci, F. M., Baumbach, G. L., and Heistad, D. D. (1989) *J. Hypertens.* **7 (Suppl. 4),** S61–S64.
[26] Bevan, J. A., Bevan, R.D., and Duckles, S. P. (1980) *Handbook of Physiology, Section 2: The Cardiovascular System,* vol. *2: Vascular Smooth Muscle* (Bohr, D. F., Somlyo, A. P., and Sparks, H. V. Jr., eds.), American Physiological Society, Bethesda, MD, pp. 515–566.
[27] Schytte Blix, A. and Folkow, B. (1983) *Handbook of Physiology, Section 2: The Cardiovascular System,* vol. *3: Peripheral Circulation and Organ Blood Flow* (Shepherd, J. T. and Abboud, F. M., eds.), pp. 917–945.
[28] Lever, A. F. (1986) *J. Hypertens.* **4,** S515–524.

Chapter 3

Common In Vitro Investigative Methods

William Halpern

Introduction

The physiological and pharmacological behavior of resistance arteries are perhaps best observed in the intact vasculature. However, the design of such experiments is difficult when seeking comprehension of accountable mechanisms. For example, consider a pial artery observed through a window that decreases in diameter following a systemic pressure increase. Whether this decrease was caused by metabolic influences, flow changes, neural effects, or pressure changes arising from distal and/or proximal vascular adjustments is not easily determined. On the other hand, in vitro preparations afford control and measurement of the numerous variables and offer a more direct means for investigating these mechanisms. Given the importance of resistance arteries in the control of blood flow and their involvement in consequential pathologies, the majority of in vitro investigations are still made using large arteries. This is true despite little evidence to support a similarity of physiological regulatory mechanisms in conduit arteries and in the small artery or arterioles. This emphasis may arise from the difficulties of dissection, the availability of appropriate in

From *The Resistance Vasculature*, J. A. Bevan et al., eds. ©1991 Humana Press

vitro instrumentation, or the time-intensive nature of the data collection. These reasons seem hardly justifiable, since two experimental systems are in use today for studying excised arterial segments. One embodies a vessel supported on two wires, and the other comprises a cannulated and pressurized vessel. In this review, both methods are discussed, with greater emphasis on newer, cannulated techniques. It is suggested that the latter is a more physiological approach, although each can provide valuable information.

Wire-Mounted Vessels

The nearly circumferential arrangement of smooth-muscle cells within the resistance arterial wall is an anatomical constraint that precludes the use of viable, helically cut strip preparations. However, ring segments may be held on two wires passed through the lumen, allowing wall force exerted on the wires to be measured. The first wire-myograph design[1] suitable for applying this method was later technically refined.[2] Although used for isometric investigations of the mechanical properties of smooth-muscle cells,[3] the majority of applications encompass drug–receptor interaction, neurotransmitter release, and endothelial function. Ring-vessel segments can also be examined isotonically to better approximate their in vivo behavior by adjusting the circumference during changes in activation, in order to maintain a constant force.[4]

Stretch

An important question facing the investigator is how to set the initial stretch of the vessel so experimental responses may parallel its in vivo behavior. Stretch is achieved by moving the wires apart to increase vessel circumference; several approaches are used. Attaining the stretch at which force is first detected introduces large variations in circumference because it depends on the sensitivity of the transducer instrumentation and because of the exponential passive-force–circumference characteristic. Alternatively, a circumference matching the *in situ* dimension can be used. One can also find experimentally the stretch that maximizes the active response to some agonist; it may be necessary to ascertain whether this circumference is invariant for different agonists. Since an equivalent transmural pressure may be calculated from Laplace's relationship using the measured passive force and circumference, the degree of stretch may also be chosen to correspond to the known or estimated in vivo transmural pressure of the artery.[3]

Having established the circumference, the experimental procedure is generally carried out isometrically, involving some form of activation or deactivation of the smooth-muscle cells. Conventionally, the active response is obtained by subtracting the passive force from the total force response. One must be careful in this procedure when the total and passive forces are of the same order of magnitude, since this subtraction can subject the small difference to error. The validity of another technique, of using a circumference 10 or 20% less than that at the maximum agonist response, is based on data from arteries in which the active-force–circumference characteristic is such that the maximal active force is only slightly reduced and the passive force is substantially diminished.[5] However, these conditions do not necessarily pertain to all vessels. In sum, each method has its drawbacks, and, depending on the research aims, it may be desirable to use multiple methods.

Vessel Measurements

Dimensions are most easily determined by direct microscopic observation using a micrometer eyepiece calibrated against a stage micrometer. The circumference is readily calculated from the distance between the wires and the wire diameter.[5] Wall tension is expressed as mN/mm (force per wall per unit of vessel length). This calculation normalizes vessels of different lengths, and wall stress (mN/mm^2 is used to further account for vessels of different wall thicknesses. The wall thickness may be obtained by focusing on the wall as it wraps around the wires, using a higher microscope magnification than is used for the other dimensions. However, this part of the wall has been shown to be somewhat thinner than the upper and lower parts of the wall, which produce the force, and therefore, this method will lead to an overestimation of wall stress.

Nonperfused, Cannulated Vessels

Cannulated and pressurized vessels assume a nearly circular cross-section as they do in vivo, unlike the two flat sheets of tissue formed when vessels are mounted in a wire myograph. Other desirable features that make this technique attractive and place the vessels closer to a physiological state are

1. Allowance for diameter to change and the vessel to manifest any functional changes arising from alterations in the geometry of the wall;
2. Maintenance of endothelial cell integrity in the central portion of the vessel segment, where the wall is untouched;
3. Exposure of the vessel wall to a transmural pressure; and
4. Axial-length compensation for the retraction that occurs upon dissection.

The simpler application of this technique, using either two cannulae (one occluded) or one cannula and a clamp, will be discussed first. It should be noted that at any transmural pressure, diameter changes occur isobarically because the compliances of the associated tubing and pressure transducer accommodate the minute volume alterations of the vessel. The necessity for careful handling of the tissue cannot be overemphasized; although the subject will not be treated here, ample references are available.[6-8] In addition, vessels must be maintained in a chamber providing an adequate bathing-solution milieu; again, references are available.[7,9] In particular, control of pH is vital, since small arteries are exquisitely sensitive to changes in this parameter, constricting to elevations and dilating to reductions from physiological levels. The common types of cannulae in use and the aspects of diameter and pressure measurement discussed below are also applicable when flow is considered.

Cannulae

A practical design of a double-barrel pipet specifically for cannulating vessels of arteriolar dimensions (5–100 µm), and details of their construction and the vacuum system needed, has been described elsewhere.[9] A more recent modification eliminates the internal cannula and thereby lowers the flow resistance in perfusion experiments.[8] However, double pipets require care in handling and are more difficult to make than single cannulae, which can be drawn either from glass, using a micropipet puller,[10] or from polyethylene tubing.[7] One limitation in using these pipets is the skill required for cannulating vessels smaller than about 80 μm. Some larger arteries cannulated at room temperature develop a spontaneous tone at 37°C, reducing their lumen to 40 or 50 μm.[11,12]

Dimensional Measurements

In principle, mechanical diameter gages devised for large vessels could be modified for resistance vessels.[13,14] However, they impose a force restraint on the vessel and, for this reason, optical techniques are preferable. Optimum visualization of the vessel is essential for precise and accurate measurements, and it is well worth the effort to achieve the right combination of microscope and tissue illumination. Lumen size is perhaps the most important measurement, since flow is drastically influenced by changes of this dimension in small arteries. The spectrum of recording methods ranges from simply writing down the measurements obtained from microscopic observation to automatic optical recording evices. The most straightforward on-line approach is to use a micrometer eyepiece, as described for the wire-mounted vessels, or an image-splitting eyepiece.[15,16] Analog voltage outputs using manual tracking of the diameter can be derived by attaching a DC-excited potentiometer to the shafts of these devices.[16] Similarly, a potentiometer attached to a pair of calipers has been used to record diameters from off-line, video-taped vessel images.[17] However, the more rapidly the dimensions change (as from myogenic activation or spontaneous vasomotion), the more advantageous are the continuous, automatic instruments. Photoelectric transducers with appropriate optics have been used to image the outside diameter of in vitro vessels as small as 600 μm.[14,18] Their resolving power of about 0.5 μm and their high-frequency response are desirable attributes, but their inability to determine lumen size may explain why they have not been modified for smaller vessels. However, a new fluorescent photoelectric method has been devised to continuously and rapidly measure lumenal cross-sectional area of 100 to 200-μm vessels.[8] The principle of this technique takes advantage of the fact that the amount of fluorescent light emitted by a dye in the vessel lumen is proportional to the volume, hence to the mean cross-sectional area. A different technique employs a photodiode array to sense vessel boundaries, and has been used to record simultaneously diameters of two *in situ* vessels of 100–1000 μm.[19] Conceivably, this idea would be applicable for in vitro vessels.

Diameter measurements made from video images are popular; these systems use both a television camera attached to the microscope viewing port and accessory specialized electronics. The images can be displayed on a video monitor, recorded, and analyzed either on or off line. Some require operator tracking and others are self-following. Certain limitations apply to their application: For example, absorption of light by vessels greater than about 400 μm leads to difficulty in obtaining lumen measurements and may restrict detection to the outside diameter only. Also, thickening of the wall and infolding of the endothelium[20] occurs when vessels constrict excessively, so the inside wall edge, hence the diameter, cannot be detected with certainty.

A video micrometer (Colorado Video, Boulder, CO) superimposes two lines, or markers, the separation of which can be varied and located anywhere on the video image of the vessel. The observer by moving two controls, places the markers to match the borders of the vessel. Turning these controls, either while watching the monitor during the experiment or on playback of the stored video image, generates a DC voltage signal output proportional to the line spacing that has been calibrated in microns for recording. This instrument has been widely used for isolated arterioles in the range of 20–80 μm and is suitable for larger diameters.[6,9,12,21] Another instrument, the image-shearing monitor (Instruments for Physiology and Medicine, San Diego, CA) horizontally delays a portion of the vessel image to an extent controlled by the operator, so the left edge of a vessel overlays the right edge, in a manner similar to that of the optical image-shearing device previously mentioned. A different device using operator tracking has recently been developed for use in microvascular-bed studies, but which would be usable for isolated preparations.[22] It features a single control for adjusting marker spacing, and another to accommodate lateral vessel movement. The device resembles those above in that operator skill is required, especially for accurate real-time measurements during vasomotor activity. An automatic tracking video dimension analyzer designed specifically for isolated resistance vessels permits measurement of both lumen diameter and wall thickness simultaneously (Living Systems Instrumentation, Burlington, VT). This device provides digital readout and analog voltage signals of these variables every 17 ms. The instrument highlights the horizontal scan line chosen for dimensional measurement representing preset trigger levels at the borders of the vessel wall.[23] In this way, the operator can make a visual judgment of the validity of the dimensions being measured. Although a different continuous tracking video dimension analyzer (Instruments for Physiology and Medicine) uses the highlighting feature mentioned, its design permits only one dimension to be measured, such as the inside or the outside diameter. Furthermore, a novel digital technique uses a frame grabber installed in a personal computer; it was devised for optical diameter measurement of vessels *in situ*, and is limited to 80 ms time resolution.[24] Considering that newer digital circuitry has increased speed and memory, the time resolution of this method could be expected to improve and the method possibly applied to in vitro experiments.

Pressure

Prior to establishing a pressure in the cannulated vessel, the distal end of the vessel should first be left open so that all air and remnant blood is removed and replaced with saline. To prevent over-pressurizing and excessive flows, only 1–3 mmHg of pressure should be applied to the vessel segment, using a micrometer syringe, motorized servo syringe, or a reservoir raised above the level of the vessel. A pressure transducer (preferably a semiconductor having a small displacement volume) and a recorder or readout device connected to the proximal side of the inflow cannula will be useful in this regard. The outflow cannula is then closed and the pressure raised to perhaps 100 mmHg. A steady transducer reading will confirm the absence of leakage caused by side branches that are undetected or poorly tied. At this point, the vessel is ready for experimentation.

Perfused, Cannulated Vessels

Once flow is initiated in the cannulated, pressurized preparations, attention must be given to new measurements. At the same time, relatively unexplored avenues for research are opened regarding the interaction of perfusate and control mechanisms of the resistance-vessel wall. The most common method for perfusing resistance vessels is through two cannulae, as already discussed. However, single-cannula and three-cannulae arrangements have also been used. For example, a single proximal branch of a dissected tree was cannulated and the perfusate collected from the terminal segment (50–250 μm).[25] Changes in perfusion pressure were interpreted as changes in the flow resistance of the distal vessel. A similar approach was used for electrophysiological studies 174-μm-diameter guinea pig ear arteries perfused at constant pressure from a reservoir in which the distal ends of the segment were partially blocked by small pins to restrict the flow.[26] In another study, all three ends of a vessel tree containing a single branch were cannulated.[27] The branch emanated just distal to the proximal cannula in the main segment and was used for the measurement of vessel pressure during flow conditions.

In order to perfuse a cannulated vessel, a pressure differential (P_1-P_2) must be established between the pressure (P_1) at the proximal end of the input cannula and the pressure (P_2) at the distal end of the outflow cannula.[28] Because there are frictional energy losses associated with flow, pressure gradients develop along this series connection of cannulae and vessel. Hence, some suggestions of how to calculate or measure flow resistance (R) are given below, along with some techniques for measurement of flow ($\dot{Q}$) and determination of vessel pressure (P_v). Approximate physiological levels of these variables for the particular test preparation can be gleaned from various microvascular studies. For example, arteries with lumens of 80–100 μm experience physiological blood flows of about 2–5 μL/min, so equivalent saline flows should lie between 6 and 15 μL/min, and transmural mean pressures from normotensive animals fall in the range of 45–80 mmHg.[12,17,26] The analytical expression that will be most helpful in determining the parameters and variables in this system is the Poiseuille-Hagen equation for flow:

$$\dot{Q} = (\Delta P \, \pi \, r^4)/8 \, \eta \, L \qquad (1)$$

Here, ΔP is the difference in pressure between two ends of a tube or vessel of uniform size; r is the radius; η is the viscosity of the perfusion fluid; and L is the length of the segment. Certain limitations apply to the use of this relationship, but these can generally be met for the isolated, cannulated preparation.[29] Selecting a vessel with, between the tips of the cannulae, an axial length (L) of at least 3–4 times the lumen size will probably satisfy the criterion for laminar flow.[9] Since volume flow is

$$\dot{Q} = \Delta P/R \qquad (2)$$

the flow resistance is

$$R = (8 \, \eta \, L)/\pi r^4 \qquad (3)$$

At 37°C , water has a viscosity of 6.92×10^{-3} poise, which is approximately that of physiological saline and one-third that of blood at this temperature. Hence, perfusion with saline will produce the same endothelial shear stress at three times the volume flow of blood. In more practical R units,

$$R = \Delta P/\dot{Q} = 3.5 \times 10^3 \, L/\phi^4 \, \text{mmHg}/\mu L/\text{min} \qquad (4)$$

Here, the diameter ϕ ($2r$) and L are expressed in μm. Thus, for example, a 1000-μm-long vessel having a lumen of 100 μm has a flow resistance of 0.035 mmHg/μL/min. That is, a flow of 1 μL/min will result in a 0.035-mmHg pressure difference between the proximal and distal ends of this vessel. Scaling of these equations is readily done to afford values for larger or smaller vessels. There is no single way in which the pressures and flows must be established, and the investigator has many options for conducting experiments under constant, or variable, flow and pressure. The techniques described below are intended to be a guide, not a prescription.

Flow

The simplest way to determine flow is to weigh the effluent for a timed interval or to collect it in a tube attached to a strain gage, where it may be recorded,[25,27] However, care should be taken to avoid evaporation of the low volumes collected. A drop counter (40 μL/drop) has been used for measuring flow through 800-μm arteries,[30] but it may be difficult to obtain uniform drop size at low flows.

Flow meters offer another measurement possibility. One design consists of a precision glass tube (Gilmont, Great Neck, NY) in which flow in the range of 0.2–120 μL/min is indicated by the height to which a ball rises within a tapered tube. Alternatively, measurements of the timed change in volume of a graduated capillary tube held horizontally at the distal end of the flow system can provide the information. Calibration of these tools using the perfusing solution is recommended.

Flow velocity and diameter measurements used in microvascular bed studies allow the calculation of flow (*see* Chilian's chapter, this volume). This approach has been used for isolated arteries by perfusing red blood cells through a given set of cannulae at known pressure differentials. Subsequent experiments made without this particulate matter, but with the same cannulae and pressure differentials, provide the flow information.[12] As discussed below, application of Eq. 2 will give reasonable estimates of flow from the measured pressure difference (e.g., 1–15 mmHg) between P_1 and P_2 when input and output cannulae have flow resistances much larger than that of the vessel.

Flow Resistance

Glass cannulae will maintain fixed flow resistances during experiments, but the vessel resistance, R_v, will vary as a function of its lumen size. If the resistances of the input and output cannulae (R_1 and R_2, respectively) are much larger than R_v, variations of R_v will not affect either the established flow or any of the pressures. However if $R_v \gg (R_1 + R_2)$, $\dot{Q}$ and P_v will be influenced by diameter changes when experiments are made with (P_1-P_2) fixed; likewise, P_v will change under constant-flow conditions. The more probable situation is that the three flow resistances are of comparable magnitude. Hence, it is important to measure $\dot{Q}$ and P_v. As discussed below, obtaining a pair of cannulae that are nearly matched in resistance simplifies the estimation of vessel pressure.

One method for measuring cannula hydraulic resistance is to insert the cannula carefully and tightly into a section of PE 160 tubing.[28] This tubing is connected to a suspended bottle, proximal and distal pressure transducers, and a constant flow syringe pump. Before initiating flow, all air is removed from the system and the bottle raised to produce a pressure that approximates the transmural pressure of the vessel in subsequent experiments. A steady flow is then initiated and the proximal pressure recorded. This is done for various flows to cover the range of interest. The resistance is determined from a graph of the difference in pressure vs flow. A typical flow resistance for a cannula of 100 μm od and 85 μm id is 0.08 mmHg/μL/min. Measurement of the electrical resistance of drawn glass cannulae by techniques common in electrophysiology has also been used to match cannulae.[12]

Pressure

Knowledge of P_v is essential in flow experiments, because its constancy is necessary when flow-induced effects, *per se*, are to be separated from myogenic mechanisms. Servo-micropressure null measurement by puncture of the vessel wall is the direct means to do this, although wall damage may release vasoactive agents that could influence the outcome of the experiment. An alternative, but indirect, approach is to use cannulae matched in flow resistance for obtaining an estimate of vessel pressure. In this situation, under a steady flow, P_v will be the mean of P_1 and P_2. Usually, a small pressure difference (<15 mmHg) exists between P_1 and P_2, thereby requiring careful calibration of the pressure transducers. Connection of a differential pressure transducer to measure (P_1-P_2) will improve this determination.[30]

As in most experimental procedures, assiduous attention to details is also required for resistance vessel flow studies. A number of these have been mentioned, and some are discussed in the literature cited.[9,31] Other problems include leaks in hose fittings and valves, air trapped in the system, improper illumination of the vessels, and poor water quality used for solutions.

Summary

A single isolated vessel chosen for study from the myriad attainable inevitably leads to a functional variability arising from the heterogeneity imposed by nature. In spite of this, studies using both ring- and cannulated-vessel preparations have already provided significant insights into vascular control mechanisms at the level of the small vessel, which importantly affect blood flow and tissue distribution. Differences in behavior are being found between these techniques, suggesting that the cannulated vessel offers more flexibility for study and is more representative of in vivo physiological performance.[28,32] It is anticipated that the available methods will increasingly be applied to afford a better understanding of hemodynamic control mechanisms, the role of endothelial factors, and insights into mechanisms gone awry in human vascular pathologies.

Acknowledgment

This work was supported in part by the National Institutes of Health grant HL 17335.

References

[1] Bevan, J. A. and Osher, J. V. (1972) *Agents Actions* **2**, 257–260.

[2] Mulvany, M. J. and Halpern, W. (1976) *Nature* **260**, 617–619.

[3] Halpern, W., Mulvany, M. J., and Warshaw, D. M. (1978) *J. Physiol.* **275**, 85–101.

[4] Nilsson, H. and Sjöblom, N. (1985) *Acta Physiol. Scand.* **125**, 429–435.

[5] Mulvany, M. J. and Halpern, W. (1977) *Circ. Res.* **41**, 19–26.

[6] Duling, B. R. and Rivers, R. J. (1986) *Microcirculatory Technology* (Baker, C. H. and Nastuk, W. L., eds.), Academic, New York, pp. 265–280.

[7] Osol, G. and Halpern, W. (1985) *Am. J. Physiol.* **249** (*Heart Circ. Physiol.* **18**), H914–H921.

[8] van Bavel, E. T. (1989) Thesis, University of Amsterdam, Offsetdrukkerij Haveka B. V., Alblasserdam.

[9] Duling, B. R., Gore, R. W., Dacey, R. G. Jr., and Damon, D. N. (1981) *Am. J. Physiol.* **241** (*Heart Circ. Physiol.* **10**), H108–H116.

[10] Halpern, W., Mongeon, S. A., and Root, D. T. (1984) *Smooth Muscle Contraction* (Stephens, N. L., ed.), Dekker, New York, pp. 427–456.

[11] Mackey, K. and Halpern, W. (1989) *Neurotransmission and Cerebrovascular Function I* (Seylaz, J. and MacKenzie, E. T., eds), Elsevier, Amsterdam, pp. 115–118.

[12] Kuo, L., Davis, M. J., and Chilian, W. M. (1990) *FASEB J.* **4**, A1258.

[13] Dobrin, P. B. (1973) *Am. J. Physiol.* **225**, 659–663.

[14] Papageorgiou, G. L. and Jones, N. B. (1985) *J. Biomed. Eng.* **7**, 295–300.

[15] Halpern, W. and Osol, G. (1985) *Prog. Appl. Microcirc.* **8**, 32–39.

[16] Baez, S. (1966) *J. Appl. Physiol.* **21**, 299–301.

[17] Johnson, P. C. and Intaglietta, M. (1976) *Am. J. Physiol.* **231**, 1686–1698.

[18] Schabert, A., Bauer, R. D., and Busse, R. (1980) *Pflugers Arch.* **385**, 239–242.

[19] Sakaguchi, M., Ohhashii, T., and Azuma, T. (1980) *Pflugers Arch.* **388**, 261–265.

[20] Greensmith , J. E. and Duling, B. R. (1984) *Am. J. Physiol.* **247** (*Heart Circ. Physiol.* **16**), H687–H698.

[21] Davis, M. J. and Gore, R. W. (1989) *Am. J. Physiol.* **256** (*Heart Circ. Physiol.* **25**), H630–H640.

[22] Goodman, A. H. (1988) *Innov. Tech. Biol. Med.* **9**, 350–356.

[23] Halpern, W., Osol, G., and Coy, G. S. (1984) *Ann. Biomed. Eng.* **12**, 463–479.

[24] Neild, T. O. (1989) *Blood Vessels* **26**, 48–52.

[25] Uchida, E., Bohr, D. F., and Hoobler, S. W. (1967) *Circ. Res.* **21**, 525–536.

[26] Keef, K. and Neild, T. O. (1982) *J. Physiol.* **331**, 355–365.

[27] Halpern, W., Osol, G., Brown, B., and Tesfamariam, B. (1985) *Vascular Neuroeffector Mechanisms* (Bevan, J. A., Godfraind, T., Maxwell, R. A., Storlet, J. C., and Worrel, M., eds.), Elsevier, Amsterdam, pp. 277–283.

[28] Halpern, W. *Blood Vessels*, in press.

[29] Blick, E. F. and Stein, P. D. (1972) *Med. Res. Eng.* **11**, 27–31.

[30] Nakayama, K. (1987) *Tohoku J. Exp. Med.* **152**, 259–267.

[31] Moulds, R. F. W. (1983) *Gen. Pharmacol.* **14**, 47–53.

[32] Speden, R. N. (1985) *Experientia* **41**, 1026–1028.

Chapter 4

In Vivo Measurements
of Resistance Artery Dynamics

William M. Chilian

Introduction

It has been slightly over 300 years since Marcello Malpighi and Antoni van Leeuwenhoek used microscopic techniques to observe microvascular pathways and follow the path of red cells. in microvessels.[1] Although the microscope utilized by these pioneers was simple by today's standards, consisting of only a single lens, the innovative approach and unique measurements have provided the basis for contemporary studies utilizing sophisticated techniques of intravital video microscopy. In the more recent past, measurements of microvascular pressure were completed about 60 years ago by Landis.[2] Compared to contemporary methods and techniques, the pressure measuring device used by Landis was simple, but it provided the first direct measurements of intravascular capillary hydrostatic pressures. These measurements are difficult even using today's technology. It is my intent in this chapter to summarize microvascular methodological procedures used for acquisition of measurements of diameters, red cell flow velocities, and microvascular pressures in resistance arteries. Specialized

From *The Resistance Vasculature*, J. A. Bevan et al., eds. ©1991 Humana Press

applications and preparations involving these parameters will also be discussed. It is beyond the scope of this review to discuss all possible variables that can be measured to reflect function of small arteries *in situ,* and the reader is directed to *Microcirculatory Technology* by Baker and Nastuk[3] as a comprehensive overview.

An ideal experiment involved in analysis of resistance artery tone would be one in which the experimental variables are measured without perturbation to the vessel. Unfortunately, this is a difficult proposition in acute preparations because most microvascular procedures and measurements are accomplished following surgery, anesthesia, and suffusion with crystalloid buffer solutions that are designed to mimic interstitial fluid. Any of these interventions could influence basal microcirculatory hemodynamics. Another difficulty with in vivo microvascular techniques pertains to the many experimental assumptions, of which several may be untested. For instance, studies utilizing pharmacological interventions assume equilibrium of drug concentrations at the effector site and the tissue fluid compartments with a suffusion. Thus, the concentration of a drug in a suffusion is assumed equal to the concentration that is influencing behavior of certain receptors. This may not be true because of diffusion, metabolism, or removal of the drug by blood. In many studies measurements of one microvessel, e.g., small artery, per vessel class is completed; this presents a problem, because of the inherent heterogeneity existing within microvascular networks. Such heterogeneous behavior of microvessels invalidates many assumptions involving homogeneous pressure and flow profiles of microvessels at a certain vascular level. Many of these problems regarding vasoactive influences of the microcirculatory preparation, or assumptions about drug concentrations can be obviated by utilizing other approaches to study regulation of resistance arteries, such as in vitro methodology (please refer to chapter by Halpern) or by studying the microcirculation of conscious animals (please refer to chapter by Hutchins). Collectively, studies of the resistance artery function in acute, in vivo preparations are not without problems and limitations, but it is worth emphasizing that a plethora of useful information, expanding knowledge of basic microvascular regulatory mechanisms, has been elucidated by these approaches.

Measurement of Diameter

Principles

The caliber of resistance arteries, or any microvessels, is an important variable to assess many characteristics and properties of the microvasculature. For instance, diameters of vessels are used as distinguishing characteristics for vessels within a network of arterioles, capillaries, resistance arteries, and so on. Also, physiological and pharmacological interventions that influence microvascular diameter engender understanding of regulatory mechanisms. Intravital miroscopic techniques are used most frequently for in vivo or *in situ* measurements of microvascular diameters. The integral component of intravital microscopy is, of course, the microscope and, more specifically, the compound microscope, which is used in concert with transillumination or epi-illumination systems (Figs. 1 and 2). The compound microscope consists of two lenses: an objective lens that collects the light and a second lens, the eyepiece from which an image is visualized either by eye or video system (Fig. 1). The objective lens forms a primary image, which is then observed with the second lens, the eyepiece. In video microscopy, the image is transmitted to the video camera tube or charged-coupled array (Fig. 2).

The microscopic image is formed by light coming from the observed object. Light can be modulated by the object that is under observation, as in bright-field microscopy in which an object modulates the emitted light. An image can also be formed by light directly coming from an object, which characterizes fluorescence microscopy. In this situation, a barrier-excitation filter is used to activate a fluorochrome at one wavelength of light, but a different wavelength is emitted. Using various excitation-barrier filters enables use of different fluorochromes for microvascular research, such as fluorescein, rhodamine, and so on.

The objective lens is possibly the most critical component of a compound microscope that determines magnification and resolution of the entire intravital microscopic system. Total magnification is the product of magnification of the objective lens and the viewing system. Using total magnification, theoretical limits of reso-

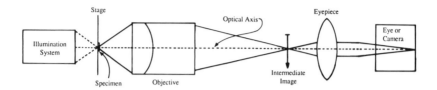

Fig. 1. A schematic of the compound microscope. An observer visualizes an intermediate image of the specimen, projected from the objective lens, either directly (eye) or from a monitor via a video camera.

lution can be determined, such as pixel sizes on the video screen, but this is not completely comparable to experimental resolution. Resolution of the microscope is determined by the combination of properties of the objective lens, illumination, and optical properties of the tissue.[4] For instance, thin tissues such as the mesentery, in which transillumination is possible, will generally allow better resolution than thick tissues in which epi-illumination is only possible. Many different types of objectives lenses are currently used for intravital microscopic procedures, but objectives with higher numerical apertures will enhance resolution and the quality of the images. Resolving power of an objective lens, which is the resolution of the system solely determined by properties of the objective lens and the illuminating system, can be related to numerical aperture and illumination by the following expression[5]

$$y = 0.61\lambda/n \sin \alpha \qquad (1)$$

where y is the resolving power, λ is the wavelength of light, n is the numerical aperture, and α is one-half the angle between the observed object and the objective lens. The relationship between a and resolving power is related to working distance. Long working distance lenses have smaller values for α than lenses with short working distances, i.e., less resolving power, whereas the converse *generally* holds for objectives with shorter working distances. From Eq. (1), it is apparent that the wavelength of light involved in the diameter measurement, i.e., light that is reflected by the red cell column in epi-illumination or refracted by the vessel wall in transillumination, in part, determines the resolving power of the objective lens. Another aspect of the illumination system that influ-

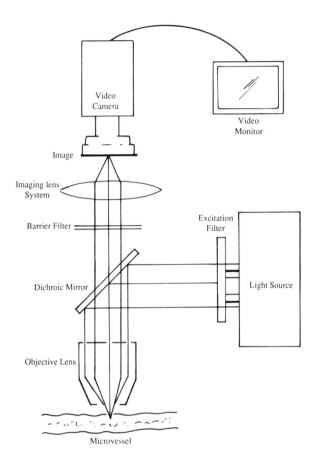

Fig. 2. A schematic of an epi-illumination microscope, in which the image is transmitted to a video camera, and can be displayed on a video monitor. Epi-illumination systems can be used in conjunction with fluorescence microscopy using proper excitation and barrier filters for different fluorochromes.

ences resolution is the condenser, which focuses the light coming from the source in the plane of the object. The numerical aperture of the condenser influences resolution of the optical system, according to the following expression[5]

$$RA = NA_{OBJ} + NA_{COND/\alpha} \qquad (2)$$

where RP is resolving power, α is the wavelength of light, and NA_{OBJ} and NA_{COND} are numerical apertures of the objective lens and condenser, respectively. Thus, properties of the objective lens, such as numerical aperture and working distance, and of the illumination system determine resolving power and, inevitably, resolution of the intravital microscopic system.

The ideal light source should have sufficient radiation energy in the required spectral region. For instance, xenon emits a relatively uniform energy distribution between 400 and about 600 nm, whereas mercury offers more intense levels of radiation, but at select wavelengths of about 400, 420, 550, and 580 nm.[4] A halogen lamp emits much lower energy levels, but has virtually no emissions lower than 500 nm.[4] Xenon lamps offer an additional advantage in that they can be used as stroboscopic light sources. This is extremely important for visualizing moving objects, in which a stroboscopic light source can be synchronized to the moving object and create the illusion that the object is motionless or moving in slow motion. Lasers are gaining popularity for certain unique applications in intravital microscopy, such as activation of certain fluorochromes in relatively thick preparations, photobleaching, or even photo-induced damage to cells. There is no single best light source, rather it is the one that is particularly well adapted for the preparation and experimental conditions.

Techniques

After the microscopic image of the blood vessel has been formed, the diameter can be measured by a variety of techniques. There are techniques such as image shearing and image splitting, that are almost 30 years old.[6,7] These systems create a second image of the blood vessel, which can be moved laterally relative to the original image. The dimension of the blood vessel is measured by displacing the second image until the opposing edges juxtapose. A disadvantage of this technique is that microvessels tend to be surrounded by rather complex structures, which impair the ability to accurately detect edges. This is an important factor in any automated system using image shearing, in which erroneous signals can be introduced into the automated procedures. The accuracy and repeatability of measurements using manual techniques is primarily a function of the ability of an operator to maintain the proper alignment of the vessel walls, and ignore the spurious signals. Unfortunately, distinguishing edges of the vessels is not a simple matter of black and white, rather differences between the wall and tissue are gradients of gray, and usually do not appear as step changes in optical density. Another related problem concerns electrical noise within the measuring system. Intensified cameras (SIT, ISIT) have lower signal-to-noise ratios than those that are nonintensified (Newvicon, Vidicon). Even the monitor, video recorder, and frame digitizers all contribute a component of noise; thus, care should be exercised to purchase components with the best possible signal-to-noise ratios.

Other manual methods for measuring diameter of microvessels also depend on the ability of the operator to distinguish the edge of the vessels. For instance, some methods depend on an operator positioning raster lines or positioning cursors along the edges of blood vessels.[8,9] With practice, a trained observer appears to be able to make reproducible measurements of microvascular diameters. For instance, in our experience, positioning of cursors along the edges of blood vessels that are illuminated by a fluorochrome usually vary about 3%, indicating that this is the error in our measurements.[9]

Measurements of microvascular diameters have also been accomplished using automated techniques. These approaches are primarily based on procedures that detect defraction patterns of the edge of the microvessel into an electrical signal. For example, on a TV screen, a dark color is associated with a lower voltage than a light color. This principle can be used to automatically detect the edge of a vessel, which, using incident light microscopy, appears dark in relation to the lumen; thus, the vessel wall will be represented by a lower voltage output than the vessel lumen. This technique was developed over 20 years ago by Wiederhielm[10] and Johnson[11] to provide an automated method for measuring internal diameter of blood vessels. Since tracking is virtually instantaneous, these automated methods are very useful for the detection of transient phenomena such as vasomotion.

Specialized Applications

A specialized application for diameter measurements involves the measurement of the caliber of small arteries in a moving tissue. Measurements of vascular diameters in a moving tissue are complicated, because the extent of motion is magnified to the same extent as the microvessel from which measurements are being made; consequently, only a blur or an out-of-focus image can be visualized using a conventional approach. Because of the motion owing to cardiac contraction, microvascular studies of the heart lagged behind studies of the microcirculation in other organ systems. Some initial approaches for the study of the coronary microcirculation in the beating heart were devised by Martini and Honig,[12] in which high speed cinematography was used to continuously film the heart, then retrospective frame-by-frame analysis was used to select the cine frames that were in focus. The major deficiency of this method was that the data reduction was a very tedious process. More recently, Nellis and colleagues devised a technique for the measurement of coronary microvascular diameters in the freely beating right ventricle using transillumination.[13] To compensate for cardiac motion, these investigators flashed a strobe light once per heart cycle, and the light was transmitted through a fiberoptic that was inserted into the right ventricle and the images were captured with a synchronized cine camera (one camera frame per flash). The result of these procedures was that when viewed through microscope during stroboscopic illumination, the epicardial microcirculation of the right ventricle appeared motionless and, therefore, could be studied.

We have modified the Nellis approach for the study of the left coronary vasculature using epi-illumination and high frequency ventilation.[14] Illumination of the epicardial surface of the myocardium is accomplished with a stroboscopic xenon light source, synchronized to the cardiac cycle, and a Leitz Ploemopak (any epi-illumination system would most likely be satisfactory). As the strobe flashes at the same point in successive cardiac cycles, the heart and microvasculature appear to be motionless, because the epicardium is in view for a short instant (15–30 µs) at the same point in each cycle. The illusion of motionless is accomplished by a computer receiving the hemodynamic input from left ventricular pressure, and the appropriate software used to trigger strobe light. The Ploemopak epi-illumination system can also be used for fluorescence microscopy, which enables resolution of fine details of microvessels. Also, the fluorochrome is a plasma marker; consequently, the entire lumen of a vessel is illuminated, enabling measurements of internal diameters.

Figure 3 illustrates fluorescent images of small coronary arteries obtained from a beating heart preparation. Note, the microvessels are illuminated by the fluorochrome (FITC-dextran), and the background myocardium appears dark. Also, note the irregular appearance of the lumen, which is characteristic of coronary microvessels with tone. If bolus injections of the fluorochrome are utilized, differentiation between arteries and veins is also possible. Specifically, a bolus injection of the fluorochrome into the left atrium or large coronary artery enables sequential illumination of arterioles and venules, because of the transit time associated with plasma flow. Using the above epi-illumination stroboscopic fluorescence techniques, we have been able to measure diameters of small arteries in the beating heart under a variety of conditions, such as norepinephrine infusion,[14] a_1- and a_2-adrenergic activation,[15] and autoregulation,[16] to examine regulatory mechanisms at different levels of the coronary microcirculation in the beating heart.

Conclusions: Diameter Measurements

The accuracy of measurements of microvascular diameter depend on many factors: anatomical features of the blood vessels, types of optical components, electronic noise in the system, and the data processing method. Resolution is critically dependent on the total magnification of the system, but it is worth noting that with any increase in magnification, contrast decreases. Systems that are based on video microscopy also are limited by intrinsic resolution of the camera tube, which varies depending on the tube type. Generally, most camera tubes have about 525 lines per video frame, but the number of lines decreases with intensified camera tubes.

The measurements of microvascular diameter obtained by the various measuring techniques must be interpreted mindful of the procedures utilized to acquire the information. For instance, measurements of in vivo blood vessels utilizing bright field micros-

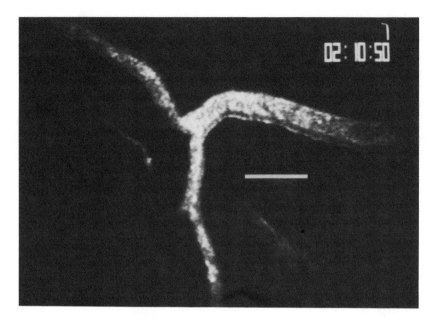

Fig. 3. Images of small coronary arteries from a beating heart using fluorescence microscopy. Note, the small arteries appear white and the background myocardium appears dark. The inset bar represents 100 μm.

copy do not represent measurements of intraluminal microvascular diameter, rather one actually measures the diameter of the red cell column. Fluorescence microscopy can be utilized to obtain measurements of internal diameter of a blood vessel since the fluorochrome illuminates the plasma, but this technique also suffers from some limitations. Specifically, streaming of the fluorochrome will often result in polarized images with one edge of the vessel illuminating to a much greater extent than the other side. Also, because of the low light level emissions of a fluorochrome, intensified tube cameras are a necessity, but these types of cameras suffer from relatively poor resolution. Thus, there is a trade-off between bright field microscopy, not providing an empirical measurement of diameter, and fluorescence microscopy, providing diameter measurements but requiring special equipment for low levels of light. There is no single answer to which intravital microscope should be utilized for diameter measurements, rather unique aspects of the preparation, illumination system, and microscope must be considered before an individual decides on using bright field or fluorescence microscopy.

Microvascular diameter measurements are made routinely in a variety of organ systems, ranging from the beating heart,[14-16] and virtually every other organ system.[17] By estimating resistance or hindrance from such measurements, it is assumed that the cross-section of the blood vessel is circular. Small blood vessels depart significantly from this shape, often being oval or containing many intraluminal folds;[18] therefore, extrapolations to hydraulic resistance or vascular hindrance of a microvessel based on a diameter change should be exercised with caution. A more conservative and better experimental approach is to relate microvascular diameter to another measured variable, such as microvascular pressure or flow velocities; only then will the physiological significance of a diameter measurement be elucidated.

Measurements of Flow Velocities

Principles

There are several different methods utilized to measure red cell flow velocities in a variety of experimental preparations. Methods have ranged from image tracking of single cells during successive frames using high-speed cinematography or video microscopy. These approaches, however, are limited to situations in which the course of a single cell can be followed; thus, its usefulness is primarily limited to measurements of red cell velocities in capillaries. Using fluorescent particles as the marker for flow velocities, these techniques can be applied to much larger vessels, such as resistance arteries, because the path of a single particle can be followed among the large numbers of flowing red cells. The principal limitation of these approaches is that they are tedious to use, because velocities must be calculated off-line. There have been developments that enable on-line measurements of red cell velocities in a variety of relatively large microvessels, and some of these approaches will be discussed in the following subsection.

A problem with most measurements of red cell or particle flow velocity pertains to the sample volume of measurements.[19] Because of the asymmetric distribution of flow velocities and of flowing particles in microvessels, it becomes critical to determine the sample volume of the method, i.e., axial location of the flowing particles or cells. Because of the parabolic flow profile,[19] flow velocities in the center stream are much higher than on the edge of the vessel. Moreover, the center line flow velocity is critical for experimental interpretations, because this velocity parameter is intimately related to the volume flow.[20]

Techniques

The dual slit method was among the first to be developed for on-line measurement of red cell velocities.[19,21] In this procedure, a microvascular preparation is transilluminated and the image is projected onto a screen. The screen contains two slits through which light is passed to two photomultiplier tubes. During the flow of blood, red cells generate a characteristic pattern that can be detected by the photomultipliers, with the upstream detector first detecting the pattern, then the downstream one. The time delay between the patterns to move from one photomultiplier to another is correlated to the blood flow velocity, which can be calculated as the quotient of the time delay and the distance between the two detectors. For on-line measurements, a correlation technique that involves multiplication of sequential points of the outputs from the two photomultipliers is accomplished.[21] Limitations of the dual slit technique pertain to the orientation of the two detectors, i.e., the detectors must be close together for the pattern of red cells to be maintained. Also, the sample volume for the photomultiplier tubes is high, so that an average blood flow velocity for a large vessel is obtained and measurements of peak (or center line) red cell velocity cannot be made. A correction factor of 1.6 has been used to multiply the velocity from the dual slit method to represent the true center line blood flow velocity in vessels greater than 15 μm in diameter.[19,20,22] The dual slit technique has been modified by Intaglietta to use video signals of two areas of a microvessel as it appears on the television monitor.[23] The limitation of the dual slit video technique is the framing rate of a camera, which limits the maximum detectable velocity to approx 2 mm/s. Small arteries may frequently have flow velocities in excess of 2 mm/s, but by use of newer technologies in which the video framing rate of a camera can be increased, or by using stroboscopic light sources, the maximal detectable velocity may be increased.

Another video technique that has been utilized to measure microvascular red cell velocities involves the use of prisms or photodiode arrays.[24,25] Basically, an image of a microvessel is projected on a photoelectric sensor that is positioned behind a grating. As the image of the red cells passes over the grating, a signal is generated in which the frequency is proportional to the velocity of the red cells. Frequency of the signal is critically dependent on spacing between the gratings and the total magnification of the system. Most of these systems are not directional, in that antegrade or retrograde flow cannot be resolved, but Slaaf et al. recently developed a three-stage grating technique that is sensitive to direction of flow.[26] The more recent variation of the grating method involves the use of linear arrays of photodiodes, with the currents from the alternating photodiodes summed separately, and frequency analysis is performed on the differential current between the odd and even diodes.[25] Recently, the prism grating technique has been modified by Borders and Granger.[27] This improvisation, using an optical Doppler principle, shows excellent accuracy at velocities up to 100 mm/s and a flat frequency response to 10 Hz in vessels as large as 100 μm in diameter. Goodman has further modified this technique for the use of low-light levels in conjunction with fluorescent red cells and intensified camera systems.[23] This latter technique should prove to be very beneficial for use of measurements of velocity in thick tissues where only epi-illumination is possible.

Specialized Applications

The previously mentioned techniques are suitable for velocity measurements in stationary tissue, but in the beating heart specialized adaptations of these techniques are necessary, although it is worth mentioning that the original techniques devised for measuring red cell velocities in the coronary microcirculation have traditionally involved immobilization of the tissue.[29,30] A more recent technique devised by Ashikawa and colleagues utilized a floating objective to keep the myocardium in continual focus during the course of a cardiac cycle,[31] but once again, movement of the myocardium was greatly restrained by insertion of a light pipe and restraining needles. Nellis and colleagues have devised a technique for measurement of flow velocities in the microcirculation of the freely beating heart.[32] To accomplish this difficult task, stroboscopic light source is flashed three times during a single video scan; thus, three images are captured on one video frame. Velocities are measured using fluorescence microscopic techniques to measure movement of microspheres (labeled with a fluorochrome) on a single video frame. These investigators also injected large fluorescent microspheres (25 μm in diameter), which lodge in precapillary vessels, to provide an index of background motion. The background movement is subtracted from the motion (velocity) of the smaller fluorescent spheres to yield measurements of flow velocities. We have utilized a similar approach to Nellis and colleagues to measure flow velocities in the epicardial microcirculation of the dog heart. Shown in Fig. 4 (top panel) is an illustration of small coronary arteries and veins, in which fluorescent microspheres were injected (bottom panel) and the stroboscopic light source was flashed twice during a video frame at a 5-ms interval. The split image on the bottom panel is the result of the dual flash. The upper portion shows the image was acquired during the initial flash, but the lower portion shows the results of image acquisition following two flashes. The fluorescent particles, therefore, appear twice in the bottom portion of the image (indicated by lettered brackets, A, B, and C). Velocity is then calculated as the quotient of the distance moved by the microspheres and the time interval between strobe flashes. Flow velocities in the arteries were in the range of 2000–4000 μm/s. A limitation of this technique is that fluorescent particles are used to estimate flow velocities, rather than red cells. Thus, an assumption is made that the rheological characteristics of the microspheres are similar to those of red cells, which is probably not entirely correct. This technique, however, could be readily adapted to measurement of red cell velocities, if the red cells were labeled with fluorescent markers. The principal limitation of the technique is that it is tedious and cumbersome to employ and there is not yet an automated method for the background subtraction

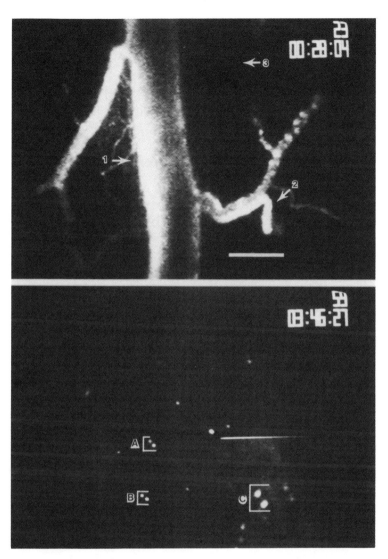

Fig. 4. Measurement of flow velocities in small arteries and veins. The top panel shows a small artery illuminated by FITC (1). The vein (2) is observed as a shadow, because it has not received the fluorochrome. A small arterial branch (3) is also shown. The bar represents 200 μm. The bottom panel shows fluorescent microspheres (2.5 μm in diameter) in the arterial vessels during double flashing of the strobe light per video frame. The upper portion of the bottom panel shows the results of a single flash, but the lower portion shows the results of the image with two flashes and, thus, the microspheres appear in pairs, indicated by the brackets and labels (A, B, and C). Pairs A and B are located in vessel 1, and pair C is located in artery #3. Velocity of flow can be calculated as the quotient of the distance traveled and the duration between the two flashes. Refer to text for details.

procedures. Hopefully this will be developed in the near future, thus enabling on-line measurements of flow velocities in the coronary microcirculation of the beating heart.

Conclusions: Flow Velocimetry

Several different techniques have been utilized for measurement of red cell velocities in microcirculatory studies. All of the discussed methods have their strengths and weaknesses, but it is worth emphasizing that no single technique is versatile enough to be utilized in the plethora of different microcirculatory preparations. For instance, it would not be possible to use the dual slit approach, which requires continuous transillumination, in the beating heart. Also, measurements of red cell velocity in the hepatic microcirculation would require a far different approach than that in the mesenteric microcirculation. One limitation of the red cell velocity measurements is that red cell velocity, and not actual volume flow, is measured. Measurements of volume flow would, of course, necessitate measurement of mean flow velocity and the cross-sectional area of the blood vessel, but unfortunately, techniques for integrating these two parameters are not yet currently available. Most likely measurements of vascular cross-sectional area will involve fluorescence techniques, in which the illumination of the fluorochrome inside the lumen of the blood vessel will be independent of inhomogeneity of vascular geometry.

Measurement of Pressures

Principles

Measurements of microvascular pressure impose a substantial problem for a number of reasons. First, the volume flow rate of microvessels is too small to satisfy compliance characteristics of most pressure measuring systems. Second, because of vascular resistance and the consequential pressure dissipation, pressures in the microcirculation only represent a low amount of mechanical energy. Because of these problems, traditional pressure measuring devices such as strain gages, which require a relatively large volume of blood to displace a compliant diaphragm or relatively high amount of mechanical energy were not applicable. Among the first measurements of microvascular pressure were those performed by Landis in 1926.[2] Landis filled micropipets with a dye and used micropuncture techniques to measure the capillary hydrostatic pressure of the frog. After impalement of the capillary, pressure of the capillary tube was increased by raising a reservoir until dye started to infuse into the capillary, then the pressure was subsequently reduced to a level in which infusion of the dye ceased. The height of the reservoir above the blood vessel was estimated as intraluminal pressure. The contemporary servo-null system for measuring microvascular pressures is based on principles similar to that devised by Landis; however, instead of visual determination of movement of a dye into a blood vessel, normalization of electrical resistance across the pipet is accomplished by means of electronics. Using the appropriate electronic amplification and detection methods, the frequency response of an active servo-null pressure measuring system is greater than 20 Hz.

Techniques

The specific manner in which the servo-null system works relates to maintenance of electrical resistance of a micropipet.[33,34] Micropipets are filled with a hyperosmolar saline solution (1–2 Osm). Compared to blood plasma, which has electrolyte concentrations of approx 290 mOsm, the saline in the micropipet has lower electrical resistance. As the micropipet is introduced into a blood vessel, the hydrostatic pressure of the microvessel forces the blood plasma into the tip of the micropipet; therefore, the electrical resistance of the micropipet increases because the relatively low conductive blood plasma displaces the high conductive saline. A sensitive device detects this change in electrical resistance and introduces an error signal, which is used to activate a pump and generate a counterpressure. By generating a counterpressure, the hyperosmotic saline solution displaces plasma from the pipet and normalizes the electrical resistance of the pipet. The counterpressure generated by the pump is equivalent to hydrostatic pressure of the micropunctured blood vessel and can be measured with a conventional strain gage manometer.

The major difficulty with servo-null pressure measurements relates to the micropipets. Each micropipette appears to possess its own unique attributes of capacitance and resistance. These unique qualities tend to complicate measurements of microvascular pressure, but such problems can be somewhat offset by proper training and experience in terms of making pipets as consistent as possible. The micropipets also pose a problem in terms of insertion into the tissue. Microvessels are often traumatized following insertion of the pipet and microthrombi often form at the tip of the pipet. This tends to both dampen the measured pressure and produce vasoactive responses in vessels from the substances released by the developing thrombus. Although all of these potential electrical and physiological problems complicate the use of the servo-null technique for measurements of microvascular pressures, this technique has proved invaluable for documentation of pressure gradients in the microcirculation and estimation of the distribution of microvascular resistance in a variety of organ systems.[13,17,35] Each organ system preparation is associated with its own set of individual problems and the reader is, therefore, encouraged to read the detailed descriptions of the servo-null technique.[33,34]

Specialized Applications

A specialized application of the servo-null technique involves measurements of microvascular pressure in the beating heart. This technique was originally developed by Nellis et al. for measurements of pressures in the freely-beating right ventricle,[13] and later adapted by Chilian et al. for measurements of microvascular pressures in the beating, but restrained left ventricular epicardium.[35] To move a micropipet in synchrony with a beating heart for pressure measurements, the pipet is mounted in an electromechanical micromanipulator that consists of three electromagnets arranged in a perpendicular array. Movement of the micropipet is accomplished by changing the strength and polarity of the current driving each electromagnet, which corresponds to x, y, and z axes. Therefore, the position of the micropipet is determined by three electronic control signals that are derived from a joystick. An integral component of this system involves a modification of the stroboscopic illumination system described previously in this section for measurements of microvascular diameters. As discussed in the section on diameter measurements in the beating heart, a strobo-scopic light source is flashed once per cardiac cycle at the same point in successive cycles to create the illusion that the heart is motionless. For the pressure measurements, the strobe is initially flashed in the above-referenced mode, to create the illusion that the heart is not moving and under these conditions, the micropipet is positioned above the surface of the heart. Using the appropriate software, a computer can then be instructed to incrementally in-crease the delay between the original trigger point on the left ven-tricular dP/dt pulse and the strobe flash; thus, creating the illusion that the microvessel is moving in slow motion. Under these slow motion conditions, an operator can track the microvessel with the joystick, in other words, maintaining the same spatial relationship between the microvessel and the pipet during the entire cardiac cycle by moving the pipet with the electromagnets. Each cardiac cycle is divided into a defined set of intervals (100 or 200) and each interval is defined by a set of x, y, and z coordinates (currents driv-ing each magnet that are detected and stored by the computer). After the vessel has been tracked over a few composite cardiac cycles, the computer replicates the full cycle of movement by us-ing the electromechanical micromanipulator to automatically move the micropipet. Micropuncture of the vessel is performed by low-ering and advancing the electromechanical micromanipulator, which is positioned on a separate micromanipulator, and pressure within the microvessel is measured by the servo-null technique.

Caution must be exercised with such measurements of micro-vascular pressure in the beating heart, because it is difficult to ob-tain intraluminal pressures free from artifacts. In order to ensure that intravascular pressures of microvessels are measured, several criteria must be accepted. First, the phasic characteristics of the microvascular pressure waveform should be in the same frequency as aortic pressure. Second, increasing the gain of the servo-null pres-sure measuring system causes high frequency oscillations super-imposed on the pressure trace. If the micropipet is not in the vessel, an increase in gain does not produce oscillations, rather gross changes in mean pressures occur. Third, microvascular pressure must track modest changes in aortic pressure. Aortic pressure can be changed by gently snaring the abdominal aorta to increase pressure or by gently snaring the inferior vena cava to decrease pressure. Microvascular pressure must follow the change in aortic pressure, although the magnitude of the pressure change is expected to be less. Fourth, using microscopic analysis, there must be visual confirmation that the pipet is situated in the vessel. Once these criteria are satisfied, a measurement can be accepted as truly indicating microvascular pressures.

Conclusions: Pressure Measurements

Measurements of microvascular pressures are difficult, but yield important information regarding microvascular dynamics. Serial measurements of microvascular pressure along a network demonstrate the sites of major resistances, and following experimental interventions, dominant regulatory mechanisms for these different resistances. The area of greatest resistance is the one in which the pressure drop is most significant in the direction of blood flow. Traditionally, the site of major resistance in the vasculature has been thought to be located exclusively in arterioles, but data from many organ systems, such as skeletal muscle,[17] heart,[13,35] and brain[17] have provided a different prospective: a significant portion of the pressure dissipation occurs in small arteries. This concept was introduced following measurements of microvascular pressures using the servo-null technique; thus, these results have contributed significantly to our knowledge concerning microvascular dynamics in a variety of organ systems.

Summary

Measurements of pressures, diameters, and flow velocities of resistance arteries provide information about the regulation of resistance within a microvascular network. This information is critical to determine basic regulatory mechanisms involved in the control of oxygen delivery to an organ system under physiological and pathophysiological conditions. Unfortunately, no single measurement can provide unequivocal information regarding control mechanisms. Usually, measurements of at least two of the variables provide corroborative information from which to draw definitive conclusions. Another problem that hampers conclusions made from all of the above-mentioned methodologies regards the problem of microvascular heterogeneity. It has been verified in a number of microvascular networks that considerable non-uniformity of the distributions of resistance occurs, i.e., vessels at a certain vascular level or diameter size do not necessarily have the same pressure, flow, and resistance profiles. This, of course, complicates our understanding of microvascular regulatory mechanisms, but *may* offer a challenge to future investigators. Hopefully, in the near future, convergence of various disciplines, such as engineering, chemistry, physiology, and medicine, will enable to continue evolution of our approaches toward understanding the intricacies of microvascular physiology.

Acknowledgments

The author's studies and results presented here were partially supported by the National Heart, Lung, and Blood Institute of National Institutes of Health Grants HL32788 and HL01570.

References

[1] Landis, E. M. (1982) *Circulation of the Blood: Men and Ideas,* (Fishman, A. P. and Richards, D. W., eds.), American Physiology Society, Bethesda, MD, pp. 355–406.

[2] Landis, E. M. (1926) *Am. J. Physiol.* **75,** 548–570.

[3] Baker, C. H. and Nastuk, W. L. (1986) *Microcirculatory Technology,* Academic, Orlando, FL.

[4] Slaaf, D. W., Jongsma, F. H. M., Tangelder, G. J., and Reneman, R. S. (1986) in *Microcirculatory Technology,* (Baker, C. H. and Nastuk, W. L., eds.), Academic, Orlando, FL, pp. 211–228.

[5] Wolf, E. and Born, M. (1987) *Principles of Optics Electromagnetic Theory of Propagation Interference and Defraction of Light,* 6th Ed., Pergamon, New York.

[6] Baez, S. (1966) *J. Appl. Physiol.* **211,** 299–305.

[7] Intaglietta, M. and Tompkins, W. R. (1973) *Microvasc. Res.* **5,** 309–312.

[8] Goodman, A. H. (1988) *Innov. Tech. Biol. Med.* **9,** 350–356.

[9] Chilian, W. M., Layne, S. M., Eastham, C. L., and Marcus, M. L. (1987) *Circ. Res.* **61(Suppl. II),** II-47–II-53.

[10] Wiederhielm, C. A. (1963) *J. Appl. Physiol.* **18,** 1041,1042.

[11] Johnson, P. C. (1967) *J. Appl. Physiol.* **23**, 593–596.

[12] Martini, J. and Honig, C. (1969) *Microvasc. Res.* **1**, 244–256.

[13] Nellis, S. H., Liedtke, A. J., and Whitesell, L. (1981) *Circ. Res.* **48**, 342–353.

[14] Chilian, W. M., Layne, S. M., Eastham, C. L, and Marcus, M. L. (1989) *Circ. Res.* **64**, 376–378.

[15] Chilian, W. M. (1990) *Basic Res. Cardiol.*, **85**, 111–120.

[16] Chilian, W. M. and Layne, S. M. (1990) *Circ. Res.* **66**, 1227–1238.

[17] Renkin E. M. (1984) in *Handbook of Physiology, Section 2: The Cardiovascular System, vol. 4: Microcirculation* (Renkin, E. M. and Michel, C. C., eds.), American Physiology Society, Bethesda, MD, pp. 627–687.

[18] Hammersen, F. and Hammersen, E. (1984) *J. Cardiovasc. Pharmacol.* **6**, S289–S303.

[19] Lipowsky, H. H. and Zweifach, B. W. (1978) *Microvasc. Res.* **15**, 93–101.

[20] Davis, M. J. (1987) *Microvasc. Res.* **34**, 223–230.

[21] Wayland, H. and Johnson, P. C. (1967) *J. Appl. Physiol.* **22**, 333–337.

[22] Baker, M. and Wayland, H. (1974) *Microvasc. Res.* **7**, 131–143.

[23] Intaglietta, M., Silverman, N. R., and Tompkins, W. R. (1975) *Microvasc. Res.* **10**, 165–179.

[24] Jeurens, T. J. M., Arts, T., Reneman, R. S., and Slaaf, D. W. (1984) *Med. Biol. Eng. Comput.* **22**, 521–528.

[25] Fleming, B. P., Klitzman, B., and Johnson, W. O. (1985) *Am. J. Physiol.* **249**, H899–H905.

[26] Slaaf, D. W., Rood, J. P. S. M., Tangelder, G. J., Jeurens, T. J. M., Alewijnse, R., Reneman, R. S., and Arts, T. (1981) *Microvasc. Res.* **22**, 110–122.

[27] Borders, J. L. and Granger, H. J. (1984) *Microvasc. Res.* **27**, 117–127.

[28] Goodman, A. H. (1989) *FASEB J.* **3**, A 1387.

[29] Tillmanns, H., Lienberger, H., Thederan, H., Steinhausen, M., and Kübler, W. (1981) *Bibl. Anat.* **20**, 44–47.

[30] Steinhausen, M., Tillmanns, H., and Thederan, H. (1978) *Pflügers Arch.* **378**, 9–14.

[31] Ashikawa, K., Kanatsuka, H., Suzuki, T., and Takishima, T. (1984) *Microvasc. Res.* **28**, 387–394.

[32] Nellis, S. H. and Carroll, K. L. (1989) *Microvascular Mechanics: Hemodynamics of Systemic and Pulmonary Microcirculations* (Lee, J.-S. and Skalak, T. C., eds.), Springer-Verlag, New York, pp. 28–38.

[33] Fox, J. R. and Wiederhielm, C. A. (1973) *Microvasc. Res.* **5**, 324–335.

[34] Intaglietta, M. and Tompkins, W. R. (1971) *Microvasc. Res.* **3**, 211–214.

[35] Chilian, W. M., Eastham, C. L., and Marcus, M. L. (1986) *Am. J. Physiol.* **25**, H779–H788

Chapter 5

Chronic In Vivo Microcirculatory and Hemodynamic Techniques

Phillip M. Hutchins, Colleen D. Lynch, and Janie Maultsby

Introduction

This chapter will present the advantages and disadvantages of chronic in vivo preparations and describe the state of the art for some of these preparations. Investigations employing chronic in vivo techniques offer many advantages. For example, tumor necrosis factor alpha (TNFα) has been shown to inhibit the proliferation of endothelial cells in vitro, and to stimulate angiogenesis in vivo.[1] The investigation of the interaction of all physiological systems is not possible with in vitro techniques. On the other hand, it is difficult to isolate individual systems and study precise mechanisms of action in vivo. Extrapolation of results acquired from in vivo studies to human medicine is also more appropriate. Knowing the in vivo effects of pharmacological interventions or physiological disturbances on undisturbed microvascular beds is a definite benefit to the interpretation of the true effect of the agent. Ideally, we need to know the response of all vascular beds to a particular perturbation. However, windows into the internal vascular beds (such as

From *The Resistance Vasculature*, J. A. Bevan et al., eds. ©1991 Humana Press

kidney, heart, liver) are difficult to attain on a chronic basis. Therefore, representative vascular beds (such as skeletal muscle and brain) are used until newer techniques allow the further investigation of the internal microcirculations. Skeletal muscle composes approx one-half of the body mass and has been shown to contain virtually every vascular receptor investigated. The cerebral circulation is also important in many cardiovascular diseases and aging. Chronic, in vivo preparations of these two microvascular beds have been developed. Agents may be applied locally or systemically, depending on whether the investigation focuses on specific mechanisms of action or on systemic response and interaction.

Chronic Microvascular Preparations

Skeletal Muscle

This preparation was developed in our laboratory[2] and selected by NASA for an investigation of the effects of weightlessness on the microcirculation. The preparation has been considerably improved and refined in more recent studies,[3,4] and the most recent modifications are described below. While the microvascular dimensions and patterns are evaluated by video and microphotographic means the animal is unanesthetized and only lightly restrained. We have found that this approach minimizes the "stress of measurement" and allows measurements to be made at normal heart rates and blood pressures.

Dorsal microcirculatory chambers are implanted in rats weighing 130–140 g. The rat is given an injection of trimethoprim (40 mg/mL)/sulfadiazine (200 mg/mL) (Di-Trim[R]) two hours before surgery at a dose of 0.1 mL/100 g body wt. Animals are anesthetized with a 1/1 mixture of ketamine hydrochloride (100 mg/mL) and Rompun® (20 mg/mL) at a dose of 0.1 mL/l00 g body wt, injected intramuscularly. Supplemental anesthesia at a dose of 0.03 mL/100 g body wt is given as needed. All surgery is performed under sterile conditions.

The removal of hair from the surgical field is accomplished by means of animal clippers and a chemical depilatory agent (Nair®, Carter-Wallace, NY). Any residue from the depilatory agent is removed with a wet gauze pad. The area is then cleansed thoroughly with an antimicrobial agent.

The rat is placed, dorsal-side up, on the surgery board. Four 3-0 silk sutures are inserted through the skin and underlying skeletal muscle along the length of the vertebral column. An outline for the right-window incision is patterned by a circular dermal punch. Skin and muscle are dissected away from the skeletal muscle and then cut, leaving a hole approx 1.5 cm in diameter. The rat is then covered with a sterile surgical drape to prevent contamination of the surgical field. Placing the rat left-side down, the sutures are tied to a supporting D-shaped frame extended over the surgery board to expose at least 3 cm of skin on the right implant site. Any connective tissue left is dissected and removed, leaving exposed the left-side cutaneous maximus muscle. Trauma is minimized by moistening the tissue with a 0.9% sodium chloride solution. The right side of the microcirculatory chamber (current model: a 15-mm-diameter disk with a 1.5-mm flange) is then positioned in the hole in he skin, and holes for securing pins are made in the muscle. Sterile gauze is placed over the right side of the chamber and, while the body and chamber are supported, the animal is flipped over on the axis of the supporting D-frame to expose its left side. The rat is repositioned on the board. Openings for the pins on the left side are cut and the pins are pushed through. The window opening on the left side is marked as it was on the right side. A small incision is made to expose the underlying cutaneous maximus muscle. This muscle is carefully dissected from the skin. Exposed tissue is kept moist with a 0.9% sodium chloride solution. After freeing the muscle layer from its connective tissue on the underside of the skin, a cut is made along the edge of the outline and the skin is discarded. The stainless steel pins of the right window are punched through the cutaneous maximus with an 18-gage Luer stub adapter and the left window is fitted and adhered to the right window with cyanoacrylic adhesive. Air is eliminated by injecting 0.9% sodium chloride solution into the incision site. The holding sutures are removed from the D-frame and the animal is placed ventral-side down on the board. A chamber clamp is used to keep the chamber stationary until the two halves are securely fastened together. A small amount of cyanoacrylate glue is applied to the chamber-skin interface to seal the chamber against microbiological invasion. A petroleum-based antibacterial ointment is applied topically to the skin where the holding sutures had been placed. A second injection of Di-Trim[R] is given 24 h after the first.

Pharmacological agents may be applied topically to this preparation. A short length of thin (0.004 in. id × 0.008 in. od) stainless steel tubing is introduced into the backflap from the dorsal aspect and positioned between the muscle and cover glass. Solutions are pumped at low pressure through the chamber where the active agent then diffuses across the microvasculature of the skeletal muscle. In this manner, agents may be continuously applied to a chronic microvascular preparation without incurring a systemic response.

Brain

To our knowledge, this is the only chronic *rat* preparation of the cerebral microcirculation. We developed this preparation to test our hypothesis that "long-term vasodilation" is, in fact, vasoproliferation. The application of this technique is detailed in our recent studies with the well-known cerebral "vasodilator" nimodipine[5] and demonstraties the effect of hypertension on cerebral-vessel vasomotion.[6] This preparation allows the photographic registration of microvascular dimensions and patterns in the resting, unrestrained, unanesthetized animal. The clearly visible "pial" microcirculation travels over the cerebral cortex before diving into the brain parenchyma. The pial microcirculation has been shown to mimic the responses of the total cerebral circulation.[7] More rigorous and detailed studies of vasomotion and structural alterations may be obtained using maximal vasodilatation induced with 10% CO_2 in the lightly sedated animal connected to a continuous hemodynamic monitoring apparatus *(see below)*.

Prior to surgery, rats are anesthetized with a 1/1 mixture of ketamine hydrochloride (100 mg/mL) and xylazine (Rompun®) (20 mg/mL) at a dose of 0.1 mL/100 g body wt, injected intramuscularly. Supplemental anesthesia at a dose of 0.03 mL/100 g body wt is given as needed. All surgery is performed under sterile conditions.

The scalp is shaved and cleaned with Phisohex® and providone iodine scrub, and then disinfected with a Betadine solution. The rat's head is secured in a stereotaxic-type head restraint to hold it in place and minimize movement during removal of the skull. A midsagittal incision is made over the frontal and parietal portions of the scalp, and the soft tissues retracted laterally. The skull is dried and a 7-mm circle is scribed on the bone. An air-powered turbine dental drill (Star model 430-K, Den-Tal-Ez, Inc., Lancaster, PA) is used to cut through the skull along the circular outline. Care must be taken not to drill through the bone and into the underlying tissue and not to cause thermal injuries to these tissues with the heat generated by the drill. The skull and exposed tissues are kept moist and cool by repeated applications of artificial CSF (NaCl, 124 mM; KCl, 5 mM; $NaH_2PO_4-H_2O$, 1.24 mM; $MgSO_4$, 1.3 mM; $CaCl_2$, 2.5 mM; $NaHCO_3$, 26 mM; D-glucose, 10 mM). A 1-mm-wide, approx 0.5-mm deep shelf is drilled along the outer edge of the hole for the cover slip. The shelf is drilled so that the cover slip will lie flat and in the same plane. Once the hole and the shelf are drilled, the skull is removed by separating it from the dura. Bone dust and chips are removed from the dural surface by flushing with artificial CSF. The dura is then cut away, exposing the pial membrane. These tissues are also kept moist with artificial CSF that has been filtered through a 0.22 μm Millex GS antibacterial filter (Millipore Corp., Bedford, MA). Any bleeding resulting from drilling or removal of the dura can be stopped by dripping artificial CSF on the blood vessel until bleeding stops.

A 9-mm window made from a Thermonox™ tissue-culture cover slip (Miles Scientific, Naperville, IL) or glass for reflected-light Nomarski microscopy is placed on the shelf. The window is covered with an 8-mm circle of tape, exposing only the edges. This serves to prevent clouding of the window during the gluing process. Using wooden sticks, cyanoacrylate glue is applied around the edges of the window and to the edge of the skull on each side. When the glue is dry (about 3 min) the protective tape is removed from the window using forceps. Care must be taken not to scratch the surface with the forceps, and not to pull too hard and lift the window.

As with the dorsal microcirculatory chamber, pharmacological agents may be applied topically to the pial preparation. A short length of thin (0.004 in. id × 0.008 in. od) stainless steel tubing is introduced into the chamber from the lateral aspect and positioned between the pia and cover glass. Solutions are pumped at low pressure through the chamber, where the active agent then diffuses across the pial microvasculature. In this manner, agents may be continuously applied to a chronic microvascular preparation without incurring a systemic response.

The animals are allowed to recover for one week to assure both complete recovery from surgical trauma and tissue adaptation to the chamber.

Cheek-Pouch Transplants

The pioneering work of Joyner and others has led to the development of techniques for neonatal tissue transplants into the hamster cheek pouch for subsequent microcirculatory observation.[8–10] These procedures have opened windows into internal-organ microvascular beds (e.g., kidney, lung, and brain parenchyma), which have been previously unavailable for study. These techniques are particularly useful in the study of the developing vasculature and its interaction with surrounding anatomical structures.

Chick Chorioallantoic Membrane

Although the chick chorioallantoic membrane (CAM) is functionally different from the skeletal muscle and brain vascular beds, it may be desirable to investigate fully the angiogenic capacity of certain agents with the industry-standard CAM preparation.[11] Hatchery-purchased eggs (Hubbard Farms, Statesville, NC) are incubated at 38°C and 65% humidity. On d 3 of incubation, using a sterile technique, a 1 × 1-cm window is placed over the developing CAM. The window is then sealed with transparent tape. On d 7, putative angiogenic agents are pelleted and transplanted onto the CAM, the window reclosed, and the egg placed back in the incubator. The CAM is evaluated each day for viability and time to angiogenic initiation. In addition, the angiogenic responses are semiquantitatively assessed using a scale of 1+ to 4+. On d 7, 10, or

14 after pellet implantation (typically the last day for an experiment), the CAM is formalin-fixed *in situ* and removed for subsequent quantitation of the vascular response. The same microvascular dimensional and pattern analysis procedures as used in the skeletal muscle and pial microcirculations may be applied to this preparation *(see below)*.

Analytical Techniques

Evaluation of Long-Term Response

In the past we have categorized and dimensionally analyzed the microvasculature by its location along the vascular tree and by the resting and maximally-dilated size of vessels. In skeletal muscle, it is important to compare the number of arterioles open for blood flow to the total number that exist, since different levels of vascular tone are reflected by the percentage of total vessels that are open. Pattern analysis also recognizes tortuosity differences and growth patterns.[3] In microvascular preparations, in which the ends of vessel segments are somewhat tethered as a result of the restrictive nature of the chamber (e.g., skeletal-muscle backflap and pial surface vessels), changes in length are expressed as a more tortuous vessel. Recently, the use of fractals in the analysis of microcirculatory patterns and dimensions has proven to be very effective.[12,13]

We have recently changed our methods for assessing microvascular diameter. We have found that the pial arteriole dilatory response to breathing 10% CO_2 was not observed in 100 of the arterioles when assessed by digitizing "before" and "after" photomicrographs. This was worrisome to us, because it was inconsistent with the literature. In a recent study to readdress this problem, we found that, with the photomicrographic technique, we were occasionally obtaining the control photograph at the peak of a vasomotion cycle and the 10%-CO_2 photograph in the trough of the vasomotion cycle. Breathing 10% CO_2 did not always eliminate the vasomotion waves, but did always cause substantial vasodilatation when the vasomotion waves were averaged. In preparations that exhibit the beginning signs of inflammation, the arterioles do not always respond to 10% CO_2. Those preparations should be eliminated from analysis. The method of analyzing diameter therefore should include 3- to 4-min averages of vasomotion readings before and after pharmacological intervention. According to the approach suggested by Slaaf and colleagues,[14] the diameter should be integrated as a direct function, since diameter is directly related to flow (although inversely related to resistance). As Slaaf has shown, it is appropriate to integrate flow but not resistance.

Evaluation of Acute Response

Since chronic, long-term experiments require weeks to months to complete, the mechanisms of acute, or short-term, cardiovascular regulation and their correlation with long-term regulatory mechanisms may be investigated as well. The inherent fluctuations in vessel diameter, blood pressure, and heart rate periodically may be observed. How these oscillations are modified by long-term interventions may also be noted. Prony spectral-line frequency-analysis techniques[6,15–17] may be used to correlate microvascular vasomotion and systemic spontaneous fluctuations in heart rate and blood pressure. These techniques allow the simultaneous investigation of the effect of interventions on the normal minute-to-minute regulation of the macro- and microcirculations. Administration of pharmacological agents also permit the dissection of the physiological mechanisms operant in this regulation of blood pressure, pulse rate interval, and vasomotion.

Hemodynamic Monitoring

The simultaneous observation of the systemic hemodynamic and local microvascular effects of physiological or pharmacological intervention is highly desirable. Without such observation, systemic effects may not be clearly separated from alterations induced locally. Local and systemic interactions may be deduced more appropriately by the continuous monitoring of systemic hemodynamic responses. In addition, the blood pressure–heart rate relationship may be used to determine "on-line" baroreceptor function[18] and subjected to chaotic waveform analysis[19] to test further for systemic adjustments.

Surgery

Chronic abdominal aorta catheters are placed in rats. For surgery the animals are anesthetized with a 1/1 mixture of ketamine hydrochloride (100 mg/mL) and xylazine (Rompun®) (20 mL/mg) at a dose of 0.1 mL/100 g body wt, injected intramuscularly. All surgery is performed under sterile conditions. The catheter is inserted into the abdominal aorta according to the following procedure. Before surgery, the abdomen and the area from the hips to the tail on the dorsal side of the rat are shaven. These areas are cleaned with Betadine scrub and swabbed with Betadine solution. A small incision is made on the dorsal side near the tail. Using a pair of curved strabissmus scissors, the skin is dissected away from the muscle for placement of a silicone–Dacron™ implant (which protects the catheter from destruction by the rat). A ventral midline incision is made into the abdomen. The intestines are retracted with wet gauze sponges. The catheter is exteriorized at the back using a trochar. The aorta (between the renal arteries and the iliac bifurcation) is then dissected free from connective tissue. The catheter is placed in position with two pieces of Dacron™ mesh glued with Nexaband® and to the back muscle above and below the mesenteric vessels. A hole is made in the aorta using a bent 26-gage needle. The catheter (flexible Teflon™ and Microrenathane®) is inserted approx 4–5 mm upstream. The catheter is checked for patency and viability. The intestines are then repositioned and the muscle and subcutaneous tissue sutured with Vicryl™. Skin closure is accomplished using Nexaband®. The exteriorized portion of the catheter is threaded through a molded implant of silicone and Dacron™ mesh that is placed under the dorsal skin at the base of the tail. The catheter is threaded through stainless steel spring stock, which is then attached to the implant. The muscle and subcutaneous tissue is sutured closed using 5-0 Vicryl™ and the skin closed with Nexaband®. The rat is injected SC with Di-Trim^R (0.1 mL/100 g body wt) to prevent infection. The catheter is connected to a flow-through swivel, a pressure transducer, an amplifier, and a computer that will continuously monitor and measure average systolic and diastolic pressure, mean arterial pressure, heart rate, and indices of spontaneous baroreceptor control.

Computer Software

In our program, arterial pressure is sampled at a rate of approx 15,000 samples/min/rat, or 120,000 samples/min for eight rats. A software signal is generated to start a clock for each rat at the precise time the arterial pressure rises above the average pressure for the previous heart beat. The clock is then stopped, read, and restarted for the next signal. In this manner, a beat-to-beat registration of pulse interval and mean arterial pressure (along with the maximum—systolic—and the minimum—diastolic—pressure) is obtained with each beat. Whenever arterial pressure changes more than 6 mmHg over a 9-beat period, a calculation of baroreceptor reflex sensitivity (BRS) is triggered. The subsequent change in pulse interval, also over nine beats, and delayed by nine beats, is divided by the change in pressure over the previous nine beats to obtain one calculation of BRS. We reject pressure changes of <6 mmHg to eliminate small, unbuffered fluctuations in arterial pressure. The average for each of these variables is output to a monitor each minute, and the hourly averages along with the average differences in beat-to-beat and minute-to-minute values are printed out each hour. These indices of variation are used instead of the more usual standard deviation or coefficient of variation, since the average difference more appropriately indicates the variation in the sampled values.[20,21]

Future Directions

As pixel arrays become smaller and interpixel distances approach the resolving power of standard intravital microscopes, a totally implanted "microscope" with continuous monitoring of the microcirculation becomes more of a reality. A hypodermic, fiber-optic microscope has already been developed, with limited success. Smaller laser Doppler probes will allow the on-line registration of individual microvessel diameter and flow without the necessity of microscopic imaging. The computerized scanning techniques (e.g., PET, MRI, and the like) are approaching the resolution necessary for microvessel analysis. Time-allocation ("slow") scanning and new image-processing techniques will allow further refinement of these techniques. Rapid progress is being made in telemetry methods for continuous monitoring of hemodynamic variables. Advancements in this area will also spill over into the telemetering of microvascular images.

References

[1] Leibovich, S. J. and Wiseman, D. M. (1988) *Prog. Clin. Biol Res.* **266,** 131–145.

[2] Smith, T. L., Osborne, S. W., and Hutchins, P. M, (1985) *Microvasc. Res.* **29,** 360–370.

[3] Hutchins, P. M., Marshburn, T. H., Maultsby, S. J., Lynch, C. D., Smith, T. L., and Dusseau, J. W. (1988a) *Hypertension* **12,** 74–79.

[4] Hutchins, P. M., Marshburn, T. H., Smith, T. L., Osborne, S. W., Lynch, C. D., and Maultsby, J. (1988b) *Acta Aeronautical* **17,** 253–256.

[5] Yuan, X. Q., Smith, T. L., Prough, D. S., DeWitt, D. S., Dusseau, J. W., Lynch, C. D., Fulton, J. M., and Hutchins, P. M. (1990) *Am. J. Physiol.* **285,** H1395–H1401.

[6] Lefer, D. J., Lynch, C. D., Lapinski, K. C., and Hutchins, P. M. (1990) *Microvasc. Res.* **39,** 129–139.

[7] Rosenblum, W. I. and Kontos, H. A. (1974) *Stroke* **5,** 425–428.

[8] Shepard, J. M., Joyner, W. L, and Gilmore, J. P. (1987) *Circ. Res.* **61,** 228–235.

[9] Joyner, W. L., Mohama, R. E., Myers, T. O., and Gilmor, J. P. (1988a) *Microvasc. Res.* **35,** 122–131.

[10] Joyner, W. L., Young, R., Blank, D., Eccleston-Joyner, C. A., and Gilmore, J. P. (1988b) *Circ. Res.* **63,** 758–766.

[11] Dusseau, J. W. and Hutchins, P. M. (1989) *Microvasc. Res.* **37,** 138–147.

[12] Lefevre, J. (1983) *J. Theor. Biol.* **102,** 225–248.

[13] van Beek, J. H. G. M., Roger, S. A., and Bassingthwaighte, J. B. (1989) *Am. J. Physiol.* **257,** H1670–H1680.

[14] Slaaf, D. W., Oude Vrielink, H. H. E., Tangelder, G.-J., and Reneman, R. S. (1988) *Am. J. Physiol.* **255,** H1240–H1243.

[15] Burkhardt, P. (1983) PhD dissertation, University of California, San Diego, CA.

[16] Colantuoni, A., Bertuglia, S., and Intaglietta, M. (1984) *Am. J. Physiol.* *(Heart Circ.)* **15,** H508–H517.

[17] Meyer, J. U. and Intaglietta, M. (1986) *Ann. Biomed. Eng.* **14,** 109–117.

[18] Furlan, R., Guzzetti, S., Crivellaro, W., Dazzi, S., Tinelli, M., Baselli, G., Cerruti, S., Lombardi, F., Pagani, M., and Malliani, A. (1990) *Circulation* **81,** 537–547.

[19] Hutchins, P. M., Jansen, B. J. A., and Stryker-Boudier, H. A. J. (1990) *FASEB J.* **4,** A701.

[20] Mann, S., Millar-Craig, M. W., Gould, B. A., Altman, D. G., and Raftery, E. B. (1982) *Proc. Fourth Inter. Symp. Ambulatory Monitoring* (Stott, F. D., Raftery, E. B., Clement, D. L., and Wright, S. L., eds.), Academic, London, pp. 572–582.

[21] Von Montfrans, G. A. (1984) PhD thesis, University of Amsterdam.

Chapter 6

The Ultrastructure of Arterioles

Susan E. Luff

Introduction

The aim of this review is to outline the main ultrastructural features of the artery wall and the particular characteristics of the arterioles. The discussion will focus on more recent research developments and, in particular, the structural aspects of the vascular innervation. For further detailed reading, I refer you to the following classic papers and reviews.[1-14]

Basic Structure of Arterioles

Arterioles are generally considered to be the primary site of vascular resistance to blood flow in the peripheral vasculature. The terminology used in the literature for the different arteriole vessels can be confusing. Vessels 15–100 µm od are generally referred to as arterioles, and those ≤50 µm are also called terminal arterioles. However, the last segment that leads into the capillary network is referred to variously as precapillary arteriole, metarteriole, precapillary segment, precapillary arteriole, or precapillary sphincter. It is important to appreciate which terminology any particular

From *The Resistance Vasculature*, J. A. Bevan et al., eds. ©1991 Humana Press

author is using. I will use the terms *arteriole* for vessels 50–100 μm od, *terminal arteriole* for vessels with diameters <50 μm, and *precapillary arteriole* to describe the most distal segment.

Structure of the Arteriole Wall

All vessels other than capillaries have the same basic structure. The wall consists of three distinct layers that, from the lumen outwards, are termed the intima, media, and adventitia. The intima mainly consists of a single layer of endothelial cells, the media of contractile smooth muscle cells, and the adventitia is a layer of connective tissue containing predominantly fibroblasts and bundles of unmyelinated axons (Fig. 1).

Intima

The endothelial cells in the relaxed state are polygonal and elongated, being oriented longitudinally to the vessel axis.[10] In all branches of the arteriolar network, they are approx 2 μm in width at the level of the nucleus with a minimum width of approx 0.15 μm.

The nucleus is also elongated and oriented longitudinally to the vessel axis. In arterioles, they form a continuous layer lining the vessel lumen, whereas in capillaries they form a fenestrated or discontinuous layer. In arterioles, the luminal surface is relatively smooth (Fig. 1), having only occasional microvilli. The endothelial cells have a longitudinally oriented ridged contour on both their abluminal and luminal surfaces[15] and produce a layer of basal lamina, which coats the outer surface of the cell. The cells are intimately connected by junctions that in most arterioles are of the tight (occluding) type.[16] At these points of contact, there is no intervening basal lamina. These junctions are impermeable to molecules >2 nm in size[16] and hence transport of macromolecules occurs via transendothelial mechanisms. In the kidney the interendothelial junctions are complex structures that vary in different vessels. In the interlobular artery and afferent arterioles they are "occluding" type junctions, but in the efferent arteriole, there are regions that could be "leaky" to macromolecules.[17]

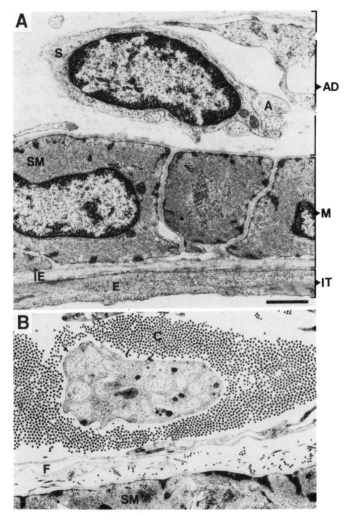

Fig. 1. Electron micrographs of the arteriole wall cut in longitudinal section showing; (**A**) the different regions of the arteriole wall and a terminal axon bundle containing four axons from the kidney cortex of the rabbit; (**B**) a preterminal axon bundle from a guinea-pig submucous arteriole. AD, adventitia; M, media; IT, intima; A, axon, of which there are four in this terminal axon bundle; E, endothelial cell; IE, internal elastic lamina; S, Schwann cell, which wraps around the axons; SM, smooth muscle cell; F, fibroblast processes; C, collagen. Arrows = mesaxons (in **B**). Mag. bar = 1 μm in **A** and **B**.

A striking feature of endothelial cells is the numerous plas-malemmal vesicles occurring on all cell surfaces (Fig. 2). They are uniform in size (60–70 nm in diameter), and have been referred to as pinocytotic vesicles. However, they are now thought to be per-manent structures similar to caveolae in smooth muscle. There are also numerous intracellular vesicles that sometimes appear fused with each other and with the plasmalemmal vesicles. Both types of vesicles (plasmalemmal and intracellular) are believed to be involved with transendothelial transport mechanisms.[18,19] Open channels from one side of the cell to the other have been reported in endothelial cells, and they are believed to be formed by the fusion of plasmalemmal and intracellular vesicles.[10] However, other work-ers have observed a more complex system of intracellular vesicles that are connected to each other and to abluminal and luminal caveolae forming "racemose" invaginations of the cell surface (Fig. 2B). They believe that transient connections occur between such structures to form complex channels across the cell through which macromolecules could diffuse.[18,19] Cytoplasmic vesicles are also concentrated in the perinuclear region, where they are thought to be involved in enzymatic activities and to act as a membrane re-serve for changes in vessel diameter.[18]

Endothelial cells contain the usual complement of organelles, such as endoplasmic reticulum, ribosomes, Golgi complex, mito-chondria, and glycogen. They also contain filaments of two differ-ent sizes, actin (5 nm) and intermediate (10 nm) filaments. The intermediate filaments are situated centrally, particularly in the region of the nucleus, and are believed to have a cytoskeletal func-tion, in particular, maintaining the position of the nucleus. The actin filaments are arranged in bundles (stress fibers) oriented longitu-dinally to the vessel axis, and predominantly lying close to the abluminal cell surface.[11] They are particularly numerous in the ter-minal arterioles and can also be observed close to endothelial and myoendothelial junctions[2] (Fig. 2A). It is not clear whether they are involved in contraction or if they have a purely cytoskeletal role.[20] Myosin has been identified in endothelial cells using immunofluo-rescence, although it is difficult to identify in electron micrographs. Bands of myosin have been reported within bundles of stress fibers, which the authors suggest could be indicative of the exist-

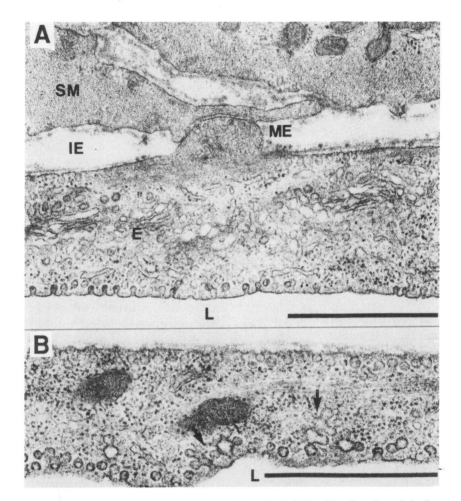

Fig. 2. Longitudinal sections through endothelial cells of an intralobular arteriole of a rabbit kidney; (**A**) a myoendothelial junction; (**B**) plasmalemma and intracellular vesicles showing the typical racimose formations of connecting vesicles (arrows). E, endothelial cell; IE, internal elastic lamina; L, lumen of the vessel; ME, myoendothelial junction; SM, smooth muscle cell.

ence of a sliding filament apparatus similar to that found in skeletal muscle.[11] It is known that stress fibers increase in number in endothelial cells of vessels subjected to high transmural pressure (e.g., in hypertensive conditions) and also in cells undergoing regeneration (*see* 11,12 for refs.).

The luminal surface of the endothelium is covered by a layer of basal lamina that is in turn overlaid by a glycocalyx (10–30 nm thick). The glycocalx is thought to function as an antithrombotic surface, and also has an ionic charge that may be involved in repelling blood cells. It is also proposed that this coat has receptors for molecules to be transported across endothelial cells.[10]

Media

The media in arterioles (≤100 μm od) consists of a single layer of smooth muscle cells (Fig. 1) that vary in size and shape in different arteriolar vessels. Most arteriole smooth muscle cells are uninucleate, the nucleus being ellipsoid, situated centrally and longitudinally to the cell axis. The cells are usually spindle-shaped, ranging in length from 30–60 μm, but with a relatively constant diameter.

The orientation of smooth muscle cells in different sized vessels has been controversial. Most early reports suggested a helical arrangement with varying pitch for different vessels.[1,2,21] More recent studies now show that vascular smooth muscle cells are generally circumferential.[15,22,23] Walmsley et al.,[24] in their work on hamster cheek pouch vessels, found that the cells were generally arranged circumferentially, both in contracted and relaxed states. They occasionally found small regions where they were helically arranged with a low pitch, which they believe could account for some of the earlier findings. The smooth muscle cells in most small vessels are arranged in such a way as to involve very little overlap.[2] However, in arterioles in skeletal muscle of the hamster, each smooth muscle cell is wrapped twice around the vessel in a single layer.[23]

Our knowledge of the shape and orientation of smooth muscle cells has increased significantly in recent years with the use of scanning electron microscopy and resin casting techniques. These studies have shown that there is considerable variation in the shape of smooth muscle cells in some small terminal and precapillary arterioles. They may have a central bulge containing the nucleus, from which arise a number of circumferentially oriented fingerlike projections.[22,23,25] The spindle-shaped smooth muscle cells of the larger arterioles have been shown to have small rod or nodular processes that appear to connect with adjacent smooth muscle cells.[22, 25] Overlying these smooth muscle cells at the medioadventitial border are occasional, irregularly shaped smooth muscle cells called asteroid cells. These are not present in precapillary arterioles.[25]

Caveolae and Endoplasmic Reticulum

Caveolae are characteristic ultrastructural features of the plasmalemma of smooth muscle cells. They are flask-shaped vesicles open to the extracellular space and are present on all cell surfaces (Fig. 3A). They are similar to the plasmalemmal vesicles in endothelial cells and are uniform in size (70–90 nm in diameter with 30–40 nm openings), which form permanent contacts with the extracellular space. In surface view, they are seen arranged in rows or narrow patches oriented longitudinally to the cell axis interdigitated with dense bands (Fig. 3). Freeze-fracture studies have shown that within the vesicle there is no evidence of intramembranous particles. However, such particles are present in the cell surface membrane surrounding the opening.[6] The caveolae are more densely distributed on the adventitial compared with the luminal surface and they are particularly abundant at the tapering ends of smooth muscle cells where they abut other smooth muscle cells. They are closely associated with endoplasmic reticulum (ER) (Fig. 3) and contain calcium in concentrations equivalent to the extracellular space. The ER is also rich in calcium, and hence, together they are thought to be involved with calcium activation of contraction. ER is also found in the cytoplasm in the perinuclear region of the cell and is often seen associated with mitochondria. These organelles also concentrate calcium, and again, are thought to be involved with activation of contraction. There are thought to be membrane surface connections between the ER and the caveolae forming a similar system to the connections between the sarcoplasmic reticulum and the transverse tubular system in skeletal muscle. For a recent review of the biochemistry of the contraction process in smooth muscle, *see* Hartshorne.[13]

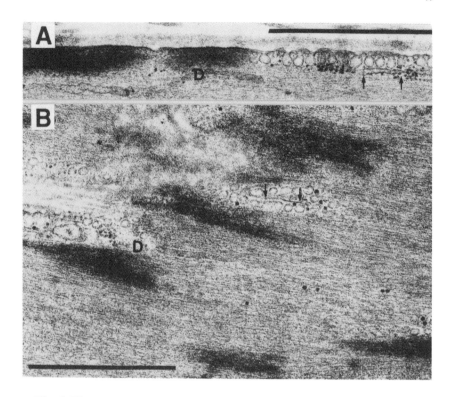

Fig. 3. Electron micrographs of the surface of smooth muscle cells of rabbit kidney intralobular arteriole, showing alternating regions of dense bands and caveolae. The caveolae are also seen closely associated with endoplasmic reticulum; (**A**) in longitudinal section; (**B**) in an oblique glancing section through the cell surface. D, dense band. Arrows, endoplasmic reticulum associated with caveolae. Mag. bar = 1 μm.

Dense Bands and Dense Bodies

Dense bands and dense bodies are opaque electron-dense regions in smooth muscle cells. They are similar in appearance but differ in their anatomical location within the cell. Dense bands are located on the plasmalemma, whereas the dense bodies lie in the cytoplasm (Fig. 1). The dense bands are thought to be attached to the plasmalemma rather than to be membrane thickenings. Both dense bands and dense bodies contain α-actin and have been equated functionally to Z-discs in skeletal muscle.[6] Dense bands and dense bodies are usually considered as separate and distinct organelles, as they differ in their protein content. Only dense bands contain viniculin. Both are associated with the contractile filaments and are involved in the contraction process (Fig. 3).

Dense bands range in width from 200 to 500 nm. In vascular smooth muscle, they typically have long projections (up to 1 μm, Fig. 4) that extend into the cytoplasm. They occur on all surfaces of the cell and are particularly abundant on the adventitial surface in arterioles[7] (Fig. 1). Dense bodies are less numerous in vascular smooth muscle than in visceral smooth muscle cells, and are arranged longitudinally to the cell axis. They are also associated with the contractile filaments (*see below*).

Filaments

Three types of filaments have been identified in vascular smooth muscle cells, thin actin filaments (7 nm), thick myosin filaments (14–16 nm), and intermediate filaments (10 nm).

Intermediate filaments consist of desmin and possibly vimentin. They are associated with dense bands and dense bodies, and are particularly numerous in developing and cultured cells. They also increase in number in response to cell injury (*see* 6 for refs.). They are not considered to be actively involved in the contractile process of the cell but are thought to be indirectly involved by acting as a cytoskeleton, and may be responsible for the elastic properties of smooth muscle cells.

The actin filaments are arranged longitudinally to the cell axis when relaxed and penetrate the dense bands and bodies (Fig. 3) at a very acute angle, whereas myosin filaments do not seem to penetrate them at all. In electron micrographs, the actin filaments are arranged either in bundles, or encircling myosin filaments. Their length still has to be determined.[12]

The large myosin filaments eluded electron microscopists for many years, probably owing to the labile nature of smooth muscle myosin. This hindered the resolution of the structural arrangement of these filaments and our understanding of the mechanism of contraction for many years. However, research in this area has advanced significantly with improved fixation methods[26] and more advanced biochemical techniques. It now seems probable that the

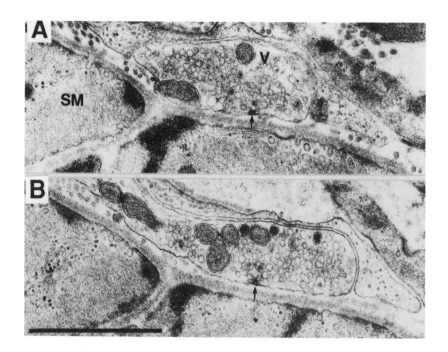

Fig. 4. Two serial sections (300 nm apart) through a varicosity of a single axon forming a neuromuscular junction with a smooth muscle cell of an intralobular arteriole showing a presynaptic membrane specialization present in both sections. The Schwann cell is wrapped around the axon on its adventitial surface. Basal lamina is seen as a single layer within the synaptic cleft. SM, smooth muscle cell; V, varicosity. Mag. bar = 1 μm.

mechanism of contraction in smooth muscle is similar to the sliding filament mechanism in skeletal muscle although there are some differences. The arrangement and structure of the filaments in smooth muscle cells are still not completely resolved. There are very detailed recent reviews on this topic.[7,12,13]

Extracellular Matrix

The space between smooth muscle cells in small arterial vessels tends to be smaller (approx 100 nm) than in larger vessels. The smooth muscle cells produce basal lamina that coats their outer surface. In electron micrographs, the basal lamina is seen as an amorphous electron-dense layer separated from the plasmalemma by an electron-translucent layer (lamina lucida) (Figs. 1 and 4). It is sometimes associated with fine filaments that connect it to collagen fibers in the interstitial space. It is well developed on the adventitial and lateral surfaces of the cell but appears attenuated on the endothelial surface. Between the lateral cell surfaces of arteriolar smooth muscle cells, the basal lamina generally appears as a single layer being shared by adjacent cells (Fig. 1). At the tapered ends of the cells, particularly with irregular cells, the basal lamina becomes a more extensive layer containing microfibrils; these are of two types, thin (3–5 nm) and thicker (11 nm). The latter sometimes appear to be attached to the electron-dense layer of the basal lamina and extend into the extracellular space. In this region of the interstitial space, the collagen fibers are more numerous, whereas they are very sparsely distributed between the lateral cell surfaces in arterioles (Fig. 1). The precise structure of basal lamina is not known. However, it is known to contain type IV collagen, laminin, entactin, and heparan sulfate proteoglycan.

The collagen fibers within the media are small (approx 30–35 nm in diameter) compared with those in the adventitia. In relaxed cells, they tend to lie almost parallel to the longitudinal axis of the smooth muscle cell and appear to be either embedded in the basal lamina or attached to it via microfibrils. They are never seen attached to the plasmalemma in vascular smooth muscle. Collagen fibers are thought to act as microtendons and to provide a skeletal framework within the vessel wall.[7]

Smooth muscle cells in some arterioles produce a fenestrated sheet of elastin (internal elastic lamina) that separates them from the intima (Fig. 1). The fenestrations are slots overlaid by basal lamina. The endothelial cells may project up through the slots and the overlying basal lamina and form "tight junctions" with smooth muscle cells (myoendothelial junctions)[15] (Fig. 2A). The internal elastic lamina is absent in the smaller terminal and precapillary arterioles except in the kidney.

Intercellular Junctions

Vascular smooth muscle cells are electrically coupled.[27] The structures that are assumed to be involved in this coupling are gap junctions (or nexuses) formed between adjacent cells. Some authors have reported a high density of gap junctions in small vessels and that the density of junctions increases with decreasing vessel diameter.[2,3] The existence of true nexuses in small arteries and arterioles has been disputed. Henderson[5] suggests that the predominant type of junction in these vessels is the "simple apposition" type that has a wider gap (i.e., ~10 nm cf 3 nm) and does not have associated membrane electron-densities. If true nexuses do not exist in small arteries and arterioles, then further studies are needed to clarify the structural arrangement involved in electrical coupling of the smooth muscle cells.

Adventitia

The thickness of this layer is very variable in different vessels and different animals. In the small arterial vessels, it can range from being either almost nonexistent in the smallest vessels to several microns thick, depending on the vessel bed (Fig. 1). It contains a number of cellular components, which include fibroblasts, macrophages, unmyelinated axons, and Schwann cells, all of which are bound up in an extracellular matrix and a loose network of collagen fibers (Fig. 1). The larger arterioles have an attenuated and fragmented elastic lamina at the medioadventitial border, which is usually absent in terminal and precapillary arterioles. The fibroblasts typically have an ellipsoid nucleus and often long fine tapering cell processes. Such cells are sometimes referred to as *veil cells* (Fig. 1B). Fibroblasts typically do not have a basal lamina around them. They are involved in producing the extracellular matrix and collagen fibers and are important in growth and repair. The collagen fibers are larger than those in the media having a diameter of approx 70 nm with a banding periodicity of 50–60 nm.

Innervation

Most small arteries and arterioles are predominantly innervated by mainly noradrenergic sympathetic postganglionic axons that originate in the prevertebral and paravertebral ganglia. The majority of axons are vasoconstrictor. There is a small proportion of vasodilatory sympathetic fibers that are cholinergic, which mainly supply the larger arterial vessels to skeletal muscle. Some vessels are also innervated by parasympathetic and sensory axons, e.g., those supplying the vessels to the head.[28-30]

In the last 10 years, an ever increasing number of neuropeptides has been demonstrated in different axons innervating blood vessels. Various functions have been ascribed to them, including neurotransmission, either as true transmitters, cotransmitters, or neuromodulators, or having influences in development, growth, and aging. Their role in these areas is still unclear. This subject has recently been reviewed by a number of authors.[30,31]

Histological Methods

A number of histological methods for visualizing catecholaminergic axons in the light microscope are available; for example, the aldehyde induced fluorescent method[32] and the glycoxylic acid method, which is probably more sensitive.[33]

The more recently developed immunohistochemical methods have greatly improved our ability to identify and distinguish between different axon types both in light and electron microscopy. This has involved specific antibodies to label catecholaminergic axons by their content of the synthesizing enzymes tyrosine hydroxylase (TH) and dopamine β-hydroxylase and peptides. The peptide that is predominantly found in axons surrounding blood vessels is neuropeptide-Y, which is usually colocalized with noradrenaline. The other peptides found in perivascular axons include vasoactive intestinal polypeptide, substance P, and calcitonin gene related peptide.

The axon terminals innervating blood vessels contain small (approx 50 nm) vesicles that are seen in electron micrographs to be either granular or agranular. The presence of small granular vesicles in axon terminals has been considered to be indicative of an adrenergic terminal, whereas those containing small agranular vesicles are considered to be cholinergic. In the latter case, this is an unreliable criteria as the preservation of granules in vesicles in electron microscopy is significantly affected by fixation. For a fixation method to preserve the granular nature of catecholaminergic terminals, *see* Transer and Richards.[34] The size and number of vesicles is similarly affected by fixation; glutaraldehyde, for example, tends to produce larger vesicle profiles than osmium. Other methods for identifying adrenergic axon terminals are those using 5- and 6-hydroxydopamine and immunohistochemical methods. As yet, there are no reliable methods for positively identifying peripheral cholinergic terminals. Some earlier studies used the existence of cholinesterase as being indicative of cholinergic terminals. However, this is a false indicator as this enzyme is present in noncholinergic nervous tissues.[4]

Arrangement of Axons

Most blood vessels are innervated by a plexus of branching, unmyelinated axons that is situated in the adventitia close to the medioadventitial border. This plexus is particularly dense in small arteries and arterioles. In some beds, the terminal and precapillary arterioles are either not innervated at all or receive a sparse innervation that is sometimes restricted to the vessel branching points; for example, guinea-pig submucous arterioles.[35,36]

The sympathetic postganglionic axons consist of two regions; a proximal nonvaricose region and a distal varicose region. The proximal region contains predominantly neurotubules and filaments and occasional mitochondria (Fig. 1B). Some of these axon profiles are seen to be full of large dense-cored vesicles[37] but rarely small synaptic vesicles. The axons, although not varicose, vary in diameter along their length.[36] They are contained in discrete bundles in the adventitia a little distant from the media, and are surrounded by Schwann cells that also wrap around each individual axon forming mesaxons. They often also have fibroblast processes and collagen fibers around them (Fig. 1B). These bundles can contain up to 50 axons, which are destined to innervate more distal regions of the vasculature, and in some cases, other tissue.

The distal segment of the axon has typical swellings or varicosities arranged in series along the length of the axon and separated by narrow intervaricose regions. The varicosities typically contain many small synaptic vesicles, some large dense-cored vesicles, and several mitochondria. They also contain neurofilaments and tubules, but in general, these do not extend through the varicosity.

The length of the terminal segment of individual axons is not known; neither is the extent of the region of vessel that they innervate. However, it is obvious that there is overlap of the area innervated by individual axons. There is evidence in the rat tail artery that axons innervate at the most a few millimeters of the vessel.[38] During development, in the rat mesentery, axons initially preferentially innervate the more distant segments of the vasculature, the more proximal region being innervated at a later time by the development of subsequent axons and not by collaterals from the initial innervation.[39]

The varicose axon segments also occur in discrete bundles that (although they have been described several microns away from the medioadventitial border in some vessels) are always located close to the medioadventitial border in arterioles (Figs. 1 and 4). Again, there is considerable variation in the number of axons per bundle and from our observations in the guinea-pig submucous arterioles and the rabbit kidney, bundles can contain 1–15 axons. Single axons (Fig. 4) tend to be more numerous in the terminal and precapillary arterioles, but this again varies in different vascular beds. These bundles are also enveloped by Schwann cells, but these often do not wrap completely around each axon (Fig. 1A) to form mesaxons. The Schwann cells form a mesh-like encasement around the axon bundles and hence, there are regions where the axons are exposed to the extracellular space, which can either be toward or away from the media. The bare regions of axon plasmalemma are usually covered by basal lamina (Fig. 1A), which is produced by the Schwann cell, but only when it is in contact with the axon.[40]

Varicosities are frequently seen in a cluster within a bundle. This is an important structural feature to consider when attempting to quantify varicosities from light micrographs. Also, as discussed above, the nonterminal axon bundles do not innervate the section of vessel in which they are situated and may not innervate vascular tissue at all. Hence, the identification in light micrographs of axons containing substances such as peptides in the axon plexus around a particular vessel is not necessarily indicative that such axons are innervating it. EM immunohistochemistry is required to confirm this aspect.

Structure of Axon Terminals

The size and shape of the varicosities varies considerably (cross-sectional diameter: 0.25–3 µm; length: 0.5–3 µm). The intervaricosity diameter is not constant along an individual segment and can range from 0.05 to 0.5 µm, but is more frequently 0.1–0.2 µm. The length of the intervaricosity segments varies from 0.2 to >7 µm.[36,41]

Three types of vesicles are found in varicosities innervating arterioles, two types of small vesicles (granular and agranular, 30–70 nm in diameter) and larger granular vesicles (dense-core vesicles, 80–100 nm in diameter) (Fig. 4). The number of vesicles contained in a terminal is also variable and affected by fixation procedures, although the number of vesicles tends to be directly related to the size of the varicosity.[41]

The neurotransmitter in sympathetic terminals is predominantly located in the small vesicles and is thought to be released from the axon terminal by exocytosis.

Noradrenalin is also stored in the large, dense-cored vesicles but it is not thought to be released from these during neurotransmission. There is also some extravesicular noradrenalin.[42] It has been postulated that large granular vesicles are manufactured in the perikaryon and transported via the axoplasm to the terminals.[43] The peptides are thought to be manufactured in the neuron soma and transported down the axon to the terminals where they are stored in large granular vesicles.

There has been considerable debate concerning the distance varicosities lie from their target smooth muscle cells. Most estimates of minimum and mean distance for different vessels have been measured from random thin sections. The minimum neuromuscular distance in blood vessels is of the order of 50–100 nm but more frequently, 80–100 nm. Such close appositions have been reported in many muscular arteries and are more frequently seen in smaller arteries.[37,44] Some authors have reported presynaptic membrane specializations in some of these close contacts as well as a cleft containing a single layer of basal lamina.[6,36] Axon terminals forming such close association with smooth muscle cells are reported to occur infrequently, i.e., <4% in random sections,[45] and several microns is thought to be the minimum separation of terminals from smooth muscle cells in some vessels.[44]

Recent serial section studies have now shown that almost all varicosities in guinea-pig submucous arterioles (50 μm od) form close contacts with smooth muscle cells.[41] These studies also showed that the varicosities form specialized structures with many of the characteristics of skeletal neuromuscular junctions (Fig. 4). A high frequency of close contacts has also been found in rat kidney arterioles[46] and in rat cerebral vessels.[47] Such junctions are also frequently observed in most muscular arteries.[37] The arrangement of terminals has been reported to change during contraction and dilation of the vessel, the terminals being closer to the smooth muscle cells during maximum dilation.[41,48]

Structure of the Neuromuscular Junction

In submucous arterioles, the synaptic cleft of the neuromuscular junction ranges from 35 to 100 nm, and always contains a single layer of basal lamina. Small synaptic vesicles are clustered at the presynaptic membrane (Fig. 4). The varicosities frequently show structural modifications associated with junctions; for example, the varicosities can flange out to form an extended surface contacting the sarcolemma. Varicosities that are situated some distance away from the sarcolemma sometimes extend a foot-like projection from the main body of the varicosity to form a junction. Some junctions (15%) also have a small region of the presynaptic membrane that appears thickened and is electron-dense, at which vesicles are clustered. Such a structure can always be seen in two or more successive sections[36] (Fig. 4). Vascular smooth muscle cells do not appear to exhibit any structural postsynaptic specialization. There has only been one freeze-fracture study on intramembrane ultrastructure[49] and the authors found "no clear modification of the smooth muscle membrane at the site of neuromuscular contact." These studies have also provided detailed information on varicosity size, shape, contact area, and presynaptic and membrane specialization area. There is considerable variation in all these parameters in different junctions. The contact areas vary considerably (0.03–2.97 μm^2) and are usually related to the size of the varicosity. The mean junctional area is also smaller in the smaller terminal arterioles (i.e., od ≤25 µm compared to 50 µm od vessels).

It is notable that presynaptic membrane specializations appear to be restricted to larger varicosities, i.e., those >1 µm in diameter. These specializations occupy <20% of the total junctional area. There is a smaller proportion of varicosities forming junctions in the smaller terminal arterioles compared to the 50 µm od vessels in the same vascular bed (i.e., 63% compared with 85%). It is difficult to assess the functional significance of the distribution of presynaptic membrane specializations and the differences in junctional areas. It is known that the probability of quantal release from these terminals is very low (<0.01),[50] so it is possible that the larger varicosities with presynaptic specializations are those that participate in release. This has been shown to be the case in skeletal neuromuscular junctions.[51]

Studies on the guinea-pig submucous arterioles have also demonstrated that the junctions occur in clusters. Hence, a single smooth muscle cell can have more than one junction from adjacent axons. Also a large proportion (45%) of junctions occur at the boundary of two adjacent smooth muscle cells.

It has generally been assumed that not all smooth muscle cells receive direct innervation but that they operate as a syncytium via electrical coupling of adjacent cells.[27] There is very little structural data from serial section studies on the frequency of junctions on significant lengths of vessels. However, from our own observations it seems likely that in arterioles, all smooth muscle cells may receive at least one junction.

Neuromuscular Junctions and Neurotransmission

The value of the mean cleft width as measured in random sections has frequently been used to correlate the structural arrangement of the innervation with the physiological response to nerve stimulation. It is also used frequently in studies investigating the possible neurohumoral role in vessel wall development and growth. There is a good correlation in many vessels of narrow mean cleft width with lower threshold and greater contractile response to small increases in nerve stimulation.[52] Consequently, on the basis of this knowledge and the infrequent observations of close neuro-muscular appositions in earlier studies, a model for neurotransmission was developed that involved the release of neurotransmitter from varicosities at variable distances from smooth muscle cells.[52] However, the correlation between narrow neuromuscular clefts with increased response to neural stimulation is not good for all vessels.[53]

There seems to be strong structural evidence in arterioles for us to consider that neuromuscular transmission occurs exclusively via junctions at least in arterioles, and possibly, also in larger muscular arteries as relatively high frequencies of neuromuscular junctions have also been found in these vessels.[37] There is also physiological evidence for this in that quantal release of transmitter has been demonstrated in both arterioles and larger arteries.[54,55] In addition, excitatory junction potentials occur only if noradrenalin is ionophoresed close to where the axon terminals are located in guinea-pig submucous arterioles.[56]

Acknowledgments

Most of the author's work on the innervation of blood vessels included in this chapter was supported by the National Heart Foundation of Australia. I wish to thank E. M. McLachlan and A. R. Luff for constructive comments and assistance in the preparation of this chapter.

References

[1] Rhodin, J. A. G. (1962) *Physiol. Rev.* **42,** 48–81.

[2] Rhodin, J. A. G. (1967) *J. Ultrastruct. Res.* **18,** 181–223.

[3] Somlyo, A. P. and Somlyo, A. V. (1968) *Pharmacol. Rev.* **20,** 197–272.

[4] Burnstock, G. and Costa, M. (1975) *Adrenergic Neurons.* Chapman and Hall, London.

[5] Henderson, R. M. (1975) *Methods in Pharmacology, vol. 3: Smooth Muscle,* (Daniel, E. E. and Paton, D. M., eds.), Plenum, New York, pp. 47–77.

[6] Gabella, G. (1981) *Smooth Muscle. An Assessment of Current Knowledge,* (Bülbring, E., Brading, A. F., Jones, A. W., and Tomita, T., eds.), Edward Arnold, London, pp. 1–46.

[7] Gabella, G. (1984) *Physiol. Rev.* **64,** 455–477.

[8] Forbes, M. S. (1982) *The Coronary Artery* (Kalsner, S., ed.) Oxford University Press, New York , pp. 3–60.

[9] Goldberg, B. and Rabinovitch, M. (1983) *Histology. Cell and Tissue Biology*, (Weiss, L., ed.), Macmillan, London, UK, pp. 139–177.

[10] Simionescu, N. and Simionescu, M. (1983) *Histology. Cell and Tissue Biology*, (Weiss, L., ed.), Macmillan, London, UK, pp. 371–433.

[11] Hammersen, F. and Hammersen, E. (1984) *J. Cardiovasc. Pharmacol.* **6**, S289–S303.

[12] Bagby, R. (1986) *Int. Rev. Cytol.* **105**, 67–128.

[13] Hartshorne, D. J. (1987) *Physiology of the Gastrointestinal Tract* (Johnson, L. R., ed.), Raven, New York, pp. 432–482.

[14] Herman, I. M. (1987) *Tissue Cell* **19**, 1–19.

[15] Carlson, E. C., Burrows, M. E., and Johnson, P. C. (1982) *Microvasc. Res.* **24**, 123–141.

[16] Simionescu, N., Simionescu, M., and Palade, G. E. (1978) *Microvasc. Res.* **15**, 17–36.

[17] Mink, D., Schiller, A., Kriz, W., and Taugner, R. (1984) *Cell Tissue Res.* **236**, 567–576.

[18] Noguchi, Y., Yamamoto, T., and Shibata, Y. (1986) *Cell Tissue Res.* **246**, 487–494.

[19] Bundgaard, M., Hagman, P., and Crone, C. (1983) *Microvasc. Res.* **25**, 358–368.

[20] Hammersen F. (1980) *Vascular Endothelium and Basement Membranes. Advances in Microcirculation*, vol. 9 (Altura, B. M., ed.), Karger, Basel, pp. 95–134.

[21] Phelps, P. C. and Luft, J. H. (1969) *Am. J. Anat.* **125**, 399–428.

[22] Fujiwara, T. and Uehara, Y. (1984) *Am. J. Anat.* **170**, 39–54.

[23] Komuro, T., Desaki, J., and Uehara, Y. (1982) *Cell Tissue Res.* **227**, 429–437.

[24] Walmsley, J. G., Gore, R. W., Dacey, R. G. Jr., Damon, D. N., and Duling, B. R. (1982) *Microvasc. Res.* **24**, 249–271.

[25] Shiraishi, T., Sakaki, S., and Uehara, Y. (1986) *Cell Tissue Res.* **243**, 329–335.

[26] Somlyo, A. P. and Somlyo, A. V. (1975) *Methods in Pharmacology*, vol. 3 (Daniel, E. E. and Paton, D. M., eds.), Plenum, New York, pp. 3–45.

[27] Bennett, M. R. (1972) *Monograph of the Physiological Society*, No. 30, Cambridge University Press, Cambridge, UK.

[28] Gabella, G. (1976) *Structure of the Autonomic Nervous System.* Chapman and Hall, London.

[29] Gibbins, I. L., Brayden, J. E., and Bevan, J. A. (1984) *Neuroscience* **13**, 1327–1346.

[30] Owman, C. (1988) *Handbook of Chemical Neuroanatomy, vol. 6: The Peripheral Nervous System* (Bjorklund, A., Hokfelt, T., and Owman, C., eds.),

Elsevier Science, Amsterdam, pp. 327–389.

31 Burnstock, G. and Griffith, S. G., (eds.) (1988) *Nonadrenergic Innervation of Blood Vessels, vol. 2: Regional Innervation*, CRC, Boca Raton, FL.

32 Falck, B., Hillarp, N. A., Thieme, G., and Torp, A. (1962) *J. Histochem. Cytochem.* **10**, 348–354.

33 Furness, J. B. and Costa, M. (1975) *Histochemistry* **41**, 335–352.

34 Tranzer, J. P. and Richards, J. G. (1976) *J. Histochem. Cytochem.* **24**, 1178–1193.

35 Furness, J. B. (1973) *J. Anat.* **115**, 347–364.

36 Luff, S. E. and McLachlan, E. M. (1988) *J. Neurocytol.* **17**, 451–463.

37 Luff, S. E. and McLachlan, E. M. (1989) *Blood Vessels* **26**, 95–106.

38 Sittiracha, T., McLachlan, E. M., and Bell, C. (1987) *Neuroscience* **21**, 647–659.

39 Hill, C. E., Hirst, G. D. S., and van Helden, D. F. (1983) *J. Physiol.* **338**, 129–147.

40 Bunge, R. P., Bunge, M. M., and Eldridge, C. F. (1986) *Annu. Rev. Neurosci.* **9**, 305–328.

41 Luff, S. E., McLachlan, E. M., and Hirst, G. D. S. (1987) *J. Comp. Neurol.* **257**, 578–594.

42 Tranzer, J. P. (1972) *Nature (New Biol.)* **237**, 57,58.

43 Geffen, L. B. and Ostberg, A. (1969) *J. Physiol.* **204**, 583–592.

44 Rowan, R. A. and Bevan, J. A. (1987) *Blood Vessels* **24**, 181–191.

45 Devine, C. E. and Simpson, F. O. (1967) *Am. J. Anat.* **121**, 153–176.

46 Barajas, L. and Müller, I. (1973) *J. Ultrastruct. Res.* **43**, 107–132.

47 Matsuyama, T., Shiosaka, S., Wanaka, A., Yoneda, S., Kimura, K., Hayakawa, T., Emson, P. C., and Tohyama, M. (1985) *J. Comp. Neurol.* **235**, 268–276.

48 Govyrin, V. A. (1976) *Physiology of Smooth Muscle*, (Bülbring, E. and Shuba, M. F., eds.), Raven, New York, pp. 279–285.

49 Devine, C. E., Simpson, F. O., and Bertaud, W. S. (1971) *J. Cell Sci.* **9**, 411–425.

50 Hirst, G. D. S., de Gleria, S., and van Helden, D. F. (1985) *Experientia* **41**, 874–879.

51 Herrera, A. A., Grinnell, A. D., and Wolowske, B. (1985) *J. Neurocytol.* **14**, 193–202.

52 Bevan, J. A. (1979) *Circ. Res.* **45**, 161–171.

53 Rowan, R. A. and Bevan, J. A. (1983) *4th International Symposium on Vascular Neuroeffector Mechanisms*, (Bevan, J. A., ed.), Raven, New York, pp. 75–83.

54 Hirst, G. D. S. and Neild, T. O. (1980) *J. Physiol.* **303**, 43–60.

55 Astrand, P. and Stjärne, L. (1989) *J. Physiol.* **409**, 207–220.

56 Hirst, G. D. S. and Neild, T. O. (1981) *J. Physiol.* **313**, 343–350.

Chapter 7

Geometry, Structure, and Mechanics of Resistance Arteries

Michael J. Mulvany

Introduction

The function of resistance arteries is to control the flow of blood. The more proximal resistance arteries (normally termed small arteries) appear to control the extent of perfusion of the different organs of the body. The role of the more distal resistance arteries (normally termed arterioles; that is, those arteries having not more than one complete layer of smooth-muscle cells) is to regulate the supply of blood within organs. In both cases, function is mediated through control of artery diameter and thus resistance. Resistance is determined by the lumen/pressure relationship, a relation often referred to as the "distensibility" of the artery.[1] This relationship is thus perhaps the most important characteristic of a small artery. However, it is complex: it is dependent on the amount, arrangement, and characteristics of the connective tissue and smooth-muscle cells, and is also determined by the activation level of the smooth-muscle cells. This is influenced by the intravascular pressure. This chapter will consider the role of resistance-artery

From *The Resistance Vasculature*, J. A. Bevan et al., eds. ©1991 Humana Press

geometry, morphology, and vascular mechanics in the ability of resistance arteries to exert their function.

Mechanics

Hydrodynamic Resistance

The resistance (R) presented by a resistance artery to the blood flowing through (*hydrodynamic resistance*) can, in principle, be calculated from the Poiseuille relation:

$$R = (8/\pi) \bullet n \bullet L \bullet (1/r_i^4) \tag{1}$$

where n is the viscosity of the perfusing liquid, L is the length of the artery, and r_i is the internal radius. Thus, resistance is exquisitely sensitive to lumen diameter; indeed, it has been demonstrated that, as would be expected from the Poiseuille relation and the principle of minimum work,[2] the flow of blood through the microvasculature depends on the cube of the lumen diameter.[3]

Tension–Length Relationship

The lumen diameter of a particular resistance artery will, as mentioned above, be determined in part by the properties of the passive components of the vascular wall and in part by the properties of the active components and the degree of their activation. These characteristics are best characterized by the so-called passive and active tension–length relations of the vascular wall. These have been examined in detail in mesenteric small arteries of the rat.[4] In these small arteries the active wall tension/internal circumference relation has a maximum at an internal circumference equal to about 90% of the internal circumference of the relaxed vessel when subjected to a transmural pressure of 100 mmHg (Fig. 1). The passive wall tension/internal circumference relationship is approx exponential, the wall tension doubling for a 6%[5] or 12%[6] increase in internal circumference.

Force–Velocity Relationship

Under in vivo conditions, most resistance arteries show vasomotion, that is, rhythmic changes in diameter. The speed and

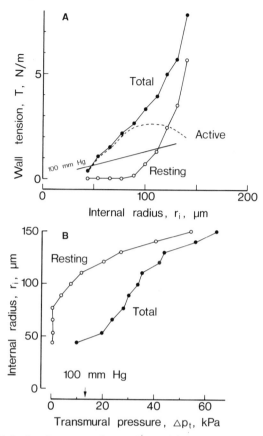

Fig. 1. (**A**) Relation between dependence of passive wall tension and total wall tension on internal radius, r_i, of a mesenteric small artery determined on a wire myograph.[16] The passive wall tension was determined in calcium-free physiological salt solution to eliminate tone. Total wall tension shows the wall tension attained when the vessel was activated with a cocktail of noradrenaline (10 μM) in 125 mM potassium solution. The difference between total and passive wall tensions is the active wall tension (stippled line). Internal radius is equal to (internal circumference)/(2•π). The straight line is an isobar corresponding, on the basis of the Laplace relation (Eq. 2), to 100 mmHg. The point at which this isobar intersects with the passive wall tension characteristic provides an estimate of the internal radius the vessel would have *in situ* when relaxed and exposed to a transmural pressure of 100 mmHg. (**B**) Calculated relations between transmural pressure and internal radius for the characteristic shown in the upper panel, under relaxed and under activated conditions. Transmural pressure calculated on the basis of the Laplace relation (Eq. 2).

extent of these changes are a function of the shortening velocity of the vessels.[7] The unloaded steady velocity of shortening
of mesenteric small arteries of the rat is estimated to be about
0.13 lengths/s (at 37°C[8]), which is similar to that seen in larger
vessels.[9] Lower shortening velocities occur when the vessel is
loaded according to the classic hyperbolic force/velocity relation.[10] Slaaf and colleagues[11] have demonstrated how the resistance of a small artery is affected by vasomotion.

Contractile Apparatus

The available evidence suggests that the contractile apparatus of resistance arteries is qualitatively similar to that of larger
arteries, both regarding its ability to produce force and in its
biochemical composition. When the force developed by small
arteries is expressed as the force developed per cross-sectional
area of the cell (*cell active stress*), the force development is remarkably similar within different resistance arteries and large
arteries (Table 1, *see* pp. 120,121). The contractile protein content of resistance arteries has been determined by Brayden and
colleagues.[12] In mesenteric small arteries of the rat, the actin
and myosin contents were found to be, respectively, 8.5 and 3.8
mg/g tissue (wet wt), which, given that the media composes
about 66% of the wall,[5] corresponds to ca. 13 and 6 mg/g media. This compares to values of 30.2 and 9.7 mg/g media, respectively, found in the carotid artery of the hog.[13] The quantitative
difference may be methodological, given the obvious difficulties in making such biochemical measurements in the minute
resistance arteries. Furthermore, the biochemical evidence suggests that connective-tissue content, especially elastin, may be
lower in resistance arteries, in which tissue concentrations of
67 mg/g wet wt (collagen) and 15 mg/g (elastin) have been
reported for mesenteric small arteries of the rat,[12] compared to
124 mg/g (collagen) and 111 mg/g (elastin) in rat carotid artery.[14] Although it is difficult to compare the methodology, these
data concerning resistance arteries are in qualitative agreement
with the ultrastructural data of Lee and coworkers.[15]

Wall/Lumen and Media/Lumen Ratios

Since, as indicated, the properties of the individual smooth-muscle cells are remarkably similar in different resistance arteries, regardless of arterial size, the mechanical function of the resistance artery is primarily determined by the morphological arrangement of the cells. Based on the Laplace equation, the wall tension, T, in a vessel of internal radius r_i that is subjected to a transmural pressure p_t ($= p_{in} - p_{out}$) is given by

$$T = p_t \bullet r_i \qquad (2)$$

For wall thickness, w, the force per unit wall area (or wall stress), s, is given by

$$s = T/w \qquad (3)$$

Combining Eqs. 2 and 3 then gives

$$p_t = s \bullet (w/r_i)$$
$$= 2 \bullet s \bullet (w/D_i) \qquad (4)$$

where D_i is the lumen (internal diameter). Given that the material of which vessels are made is similar in vessels of different sizes, it is likely that the wall stress is comparable along the vascular tree. Equation 4 thus implies that the wall/lumen ratio will be proportional to the transmural pressure and that this is an indication of the pressure that a vessel can withstand.

Similar considerations apply regarding the pressure against which a vessel can contract. The increase in wall tension, δT, resulting from an increase in transmural pressure, δp_t, will be given by

$$\delta T = \delta p_t \bullet r_i \qquad (5)$$

Since the vascular smooth-muscle cells are contained within the tunica media, it is the force developed by the smooth muscle per media area (or *active media stress*), $\delta \sigma$, that is of interest. This is given by

$$\delta \sigma = \delta T/m \qquad (6)$$

where m is the media thickness.

Combining Eqs. 5 and 6 then gives

$$\delta p_t = \delta\sigma \bullet (m/r_i)$$

$$\delta p_t = 2 \bullet \delta\sigma \bullet (m/D_i) \qquad\qquad (7)$$

The term δp_t is known as the *active pressure*.

The active media stress developed in resistance vessels in response to maximal activation is similar to that seen in larger vessels (Table 1). Thus, Eq. 7 shows that, in principle, the active pressure will be a function of the media/lumen ratio. In vitro data from wire-mounted preparations show (Table 1) that the pressures against which small arteries can contract ("effective pressure") appear to be well in excess of 150 mmHg (20 kPa), suggesting that, in vivo, small arteries are rarely, if ever, fully activated.

In practice, however, although calculation of effective pressure is a useful means of determining in vitro vessel viability,[16] the Laplace equation may not be appropriate to the resistance vasculature under in vivo conditions. Greensmith and Duling[17] have demonstrated that, in the constricted state, the luminal surface of the vessel is not smooth, but corrugated. These authors concluded that, if the Laplace formula is to be used to calculate the load on the vascular smooth muscle, the calculation should be based on an inner radius that extends from the center of the lumen to the base of the ridges formed by the internal elastic lamina. Clearly this radius will be greater than the mean radius presented by the artery to the blood flowing through it. Thus a radius calculated on the basis of the hydrodynamic resistance will not necessarily provide information that can be used to calculate wall tension and wall stress. The corrugation of the internal surface has the further consequence that some of the actomyosin filaments have a radial orientation. Nevertheless, Sleek and Duling[18] have shown that vasoconstriction is associated with a movement of the contractile filaments away from the corrugations, in such a manner that the large majority of the filaments are circumferentially oriented.

Determination of Vascular Morphology

Vessel Dimensions

Until recently, most quantitative morphological data concerning the resistance vasculature was obtained by examination of perfusion-fixed material. Such observations, however, cannot determine to what extent the morphological characteristics studied had been affected by the vasculature contracting more strongly in response to the fixation process. Furuyama[19] attempted to circumvent this problem by estimating vascular diameter from the total length of the internal elastic lamina (which, as mentioned above, corrugates during vascular contraction). The validity of the method is, though, dependent on the conditions under which the internal elastic lamina becomes smooth.[20] More recently, the morphological characteristics of resistance arteries have been quantified under in vitro conditions. With both wire[5,21] and pressure[22] myographs, direct microscopy has been used for determination of vessel and wall dimensions, including measurements of adventitial, medial, and intimal thicknesses. With the pressure myograph, this morphological data can be readily normalized by relating it to a standardized transmural pressure. With the wire myograph, this is not so straightforward, but dimensions may be normalized on the basis of the resting wall tension/internal circumference characteristic. From this characteristic, the Laplace relation may be used to determine the dimensions corresponding to a specific transmural pressure.[16]

Cell Size

Although Nomarski interference microscopy permits observation of individual, living smooth-muscle cells in mounted arteries,[4,21,23] definition is poor. Thus, quantitative assessment of the constituent parts of the vascular wall still requires conventional histological investigation.

As a means for comparing the cellular content of different tissues, a number of investigators have determined the ratio of cell volume to cell surface area using standard morphometric

Table 1
Mechanics of Arteries

Artery	Activator	Lumen diameter (μm)	Media thickness (μm)	Cell vol density (%)	Active tension (N/m)	Active pressure (kPa)	Active stress media (kPa)	Active stress cell (kPa)	Ref.
Conduit arteries									
Hog, carotid	5HT–	–	60	–	–	–	370	33	34
Rat, tail	NA + K	561	–	–	3.5	12.8	–	–	
Small arteries									
Rat									
Mesenteric	NA + K	210	13.3	71	2.7	32.2	221	311	5
Mesenteric	NA	200	–	–	2.3	23	–	–	35
Mesenteric	ADH	229	–	–	2.6	22.7	–	–	36
Mesenteric	NA + K	193	9.5	–	3.2	33.0	344	–	37
Cerebral	K	178	14.6	73	1.6	18	111	154	12
Coronary	5HT + K	203	10.9	81	2.5	24	234	289	38

Rabbit									
Ear	NA/5HT	210	9.8	72	2.2	21	229	318	39
Ear	K5O	300	–	–	2.7	18	–	–	40
Human									
Omental	K	199	9.6	75	2.1	21	209	279	41
Subcutaneous	ADH	180	13.4	–	2.2	24	160	–	42
Subcutaneous	NA + K	–	–	–	2.1	–	–	–	43

[a]Lumen diameter, D_i, and corresponding media thickness, m, were measured on a wire myograph after normalization. Cell volume density within media, V_C, was measured in histological sections. Active tension, δT, is force measured on a wire myograph in response to a given activator divided by wall length. Active pressure, δp, is the pressure against which vessels could contract, based on the Laplace relation, $\delta p = \delta T/(D_i/2)$. Active stress, $\delta\sigma$, is force per unit media area, $\delta\sigma_{media} = \delta T/m$, or per unit cell area, $\delta\sigma_{cell} = \delta\sigma_{media}/V_C$, see ref. 16. All individuals were normotensive. Table adapted from table shown in ref. 1.

[b]Abbreviations: NA, noradrenaline in supramaximal concentration; K, 125 mM potassium; K50, 50 mM potassium; 5-HT, supramaximal concentration of serotonin; ADH, supramaximal concentration of vasopressin.

techniques.[24] Based on the assumption that cells in different tissues have the same form, the greater the value of this ratio, the larger the cell. However, cell form in different tissues clearly need not be similar, and other methods are required. Measurement of cell volume, hence of cell number, has been attempted in resistance arteries using two approaches: First, scanning electron microscopy has been used. This technique has shown that, in cerebral resistance arteries of the dog, for example, smooth-muscle-cell volume ranged from over 1000 μm^3 in proximal small arteries to under 500 μm^3 in distal small arteries.[25] Similar findings concerning the intestinal microvasculature[25] have also been reported. A second approach is to determine cell volume using a "disector."[27] Here the number of "downward-pointing" nuclear ends within a section are determined, and the ratio of section volume to number of nuclear ends then gives (after multiplication by cell volume density within the media) an unbiased estimate of the cell volume. This method has given a value of 1256 μm^3 for the volume of smooth-muscle cells in mesenteric small arteries of the rat.[28]

Sampling

Although the methods described above can provide information about the dimensions and cellular composition of individual vessels, there is an inherent difficulty in using such methods for comparing the vasculature of various groups of individuals. For example, even though there is abundant evidence that the media/lumen ratio of resistance vessels of hypertensive individuals is increased,[1] one cannot be sure that the "same" vessel is being compared. Commonly, vessels have been categorized on the basis of the branching pattern (e.g., *see* ref. 16), but this presupposes that the branching pattern is identical in all groups, an assumption that needs to be proved. Furthermore, random selection of vessels has an inbuilt bias, since larger vessels are more likely to be selected than smaller vessels. A new approach is therefore needed, in which the resistance vasculature in the various organs is considered as a whole. In principle, the total number of vessels and the branching pattern need to be established and the length, lumen diameter, and wall dimensions of each vessel determined under carefully defined conditions, in much the same way as is being done with arteriolar networks.[29-31] By avoiding the need to include every vessel in the analysis, new sampling methods[32] raise the possibility of making unbiased estimates of the mean values of many of these parameters more easily.

Conclusion

The morphological and mechanical properties of the individual smooth-muscle cells in vessels of different sizes are remarkably similar. The arrangement and number of the cells, however, varies widely in the various vessels. For example, cell arrangements that give large values of media/lumen ratio enable the vessels to contract against high pressures; arrangements in which large numbers of cells are arranged around a large lumen permit high flow rates. Thus, although, in the short term the lumen diameter of a resistance vessel is dependent on the properties of its connective tissue and on the level of activation of the smooth-muscle cells, in the longer term the geometrical arrangement and amount of the wall material plays a crucial, if not dominant, role.

References

[1] Mulvany, M. J. and Aalkjaer, C. (1990) *Physiol. Rev.* (in press).

[2] Murray, C. D. (1926) *Proc. Natl. Acad. Sci. USA* **12**, 207–213.

[3] Mayrovitz, H. N. and Roy, J. (1983) *Am. J. Physiol.* **245**, H1031–H1038.

[4] Mulvany, M. J. and Warshaw, D. M. (1979) *J. Gen. Physiol.* **74**, 85–104.

[5] Mulvany, M. J, Hansen, P. K., and Aalkjaer, C. (1978) *Circ. Res.* **43**, 854–864.

[6] Freslon, J. L. and Giudicelli, J. F. (1983) *Br. J. Pharmacol.* **80**, 533–543.

[7] Mulvany, M. J. (1983) *Prog. Appl. Microcirc.* **3**, 4–18.

[8] Mulvany, M. J. (1979) *Biophys. J.* **26**, 401–413.

[9] Murphy, R. A. (1980) *Handbook of Physiology, Section 2: The Cardiovascular System, vol. 2: Vascular Smooth Muscle* (Bohr, D. F., Somlyo, A. P., and Sparks, H. V. Jr., eds.), American Physiological Society, Bethesda, MD, pp. 325–351.

[10] Hill , A. V. (1938) *Proc. R. Soc.* **126**, 136–195.

[11] Slaff, D. W., Vrielink , H. H. E. O., Tangelder G. J., and Reneman, R. S. (1988) *Am. J. Physiol.* **255**, H1240–H1243.

[12] Brayden, J. E., Halpern, W., and Brann, L. R. (1983) *Hypertension* **5**, 17–25.

[13] Murphy, R. A., Herlihy, J. T., and Mergerman, J. (1974) *J. Gen. Physiol.* **64**, 691–705.

[14] Cox, R. H. (1978) *Am J. Physiol.* **234**, H280–H288.

[15] Lee, R. M. K. W., Forrest, J. B., Garfield, R. E., and Daniel, E. E. (1983) *Blood Vessels* **20**, 72–91.

[16] Mulvany, M. J. and Halpern, W. (1977) *Circ. Res.* **41**, 19–26.

[17] Greensmith, J. E. and Duling, B. R. (1984) *Am. J. Physiol.* **247**, H687–H698.

[18] Sleek, G. E. and Duling, B. R. (1986) *Circ. Res.* **59**, 620–627.

[19] Furuyama, M. (1962) *Tohoku J. Exp. Med.* **76**, 388–414.

[20] Lee, R. M. K. W., Forrest, J. B., Garfield, R. E., and Daniel, E. E. (1983) *Blood Vessels* **20**, 245–254.

[21] Mulvany, M. J. and Halpern, W. (1976) *Nature* **260**, 617–619.

[22] Osol, G. and Halpern, W. (1985) *Am. J. Physiol.* **249**, H914–H921.

[23] Mulvany, M. J. and Warshaw, D. M. (1981) *J. Physiol.* **314**, 321–330.

[24] Weibel, E. R. (1979) *Stereological Methods, Practical Methods for Biological Morphometry* (Academic, London).

[25] Shiraishi, T., Sakaki, S., and Uehara, Y. (1986) *Cell Tissue Res.* **243**, 329–335.

[26] Gattone, V. H., Miller, B. G., and Evan, A. P. (1986) *Anat. Rec.* **216**, 443–447.

[27] Baandrup, U., Gundersen, H. J. G., Mulvany, M. J. (1985) *Adv. Appl. Microcirc.* **8**, 122–128.

[28] Korsgaard, N. and Mulvany, M. J. (1988) *Hypertension* **12**, 162–167.

[29] Engelson, E. T., Schmid-Schönbein, G. W., and Zweifach, B. W. (1986) *Microvasc. Res.* **31**, 356–374.

[30] Koller, A., Dawant, B., Liu, A., Popel, A. S., and Johnson, P. C. (1987) *Am. J. Physiol.* **253**, H154–H164.

[31] Schmid-Schönbein G. W., Firestone, G., and Zweifach, B. W. (1986) *Blood Vessels* **23**, 34–49.

[32] Gundersen, H. J. G. (1986) *J. Microsc.* **143**, 3–45.

[33] Herlihy, J. T. and Murphy, R. A. (1973) *Circ. Res.* **33**, 275–283.

[34] Mulvany, M. J., Nilsson, H., Nyborg, N., and Mikkelsen, E. (1982) *Acta Physiol. Scand.* **116**, 275–283.

[35] Nilsson, H. (1984) *Acta Physiol. Scand.* **121**, 353–361.

[36] Bund, S. J., Heagerty, A. M., Aalkjaer, C., and Swales, J. D. (1988) *Clin. Sci.* **75**, 449–453.

[37] Bund, S. J., Aalkjaer, C., Heagerty, A. M., Leckie, B., and Lever, A. F. (1989) *J. Hypertens.* **7**, 741–746.

[38] Nyborg, N. C. B., Baandrup, U., Mikkelsen, E. O., and Mulvany, M. J. (1987) *Pflugers Arch.* **410**, 664–670.

[39] Walmsley, J. G., Owen, M. P., and Bevan, J. A. (1983) *Am J. Physiol.* **245**, H840–H848.

[40] Hwa, J. J. and Bevan, J. A. (1986) *Am J. Physiol.* **251**, H182–H189.

[41] Aalkjaer, C. and Mulvany, M. J. (1981) *Blood Vessels* **18**, 233–244.

[42] Aalkjaer, C., Heagerty A. M., Petersen K. K., Swales J. D., and Mulvany, M. J. (1987) *Circ. Res.* **61**, 181–186.

[43] Richards, N. T., Poston, L., and Hilton, P. J. (1989) *J. Hypertens.* **7**, 1–3.

Chapter 8

Growth and Modification in Number of Resistance Vessels

Russell L. Prewitt, Donna H. Wang, Tetsuya Nakamura, and Edward G. Smith

"Blood vessels are, of course, organs. In this instance the primary tissue is endothelium. All else is auxiliary tissue, which, in vessels larger than capillaries, becomes added so as to provide support, resistance, flexibility, elasticity or contractility" (L. B. Arey, 1963).[1]

Normal and Abnormal Growth of Resistance Vessels

Normal Postnatal Growth

The mechanisms for the growth of existing resistance vessels are not completely understood. However, three of the histomechanical laws formulated by Thoma[2] in 1896 are applicable: (1) the increase in the size of the lumen of a vessel depends on the rate of blood flow, (2) growth in thickness of the vessel wall depends on its tension, and (3) the change in the length of a vessel is dependent on the tension exerted in the longitudinal direction by tissues

From *The Resistance Vasculature*, J. A. Bevan et al., eds. ©1991 Humana Press

outside it. How these mechanical stimuli, which are the likely initiating events, are transduced into growth of a vessel is of intense scientific interest and is the subject of this chapter.

The peripheral resistance vessels develop from dominant capillaries in the proximal portion of a growing capillary network. As new capillaries form by sprouting in the distal portion, the proximal capillaries with lower rates of flow degenerate, leaving the preferential channels with higher rates of blood flow to become arterioles and venules.[3] Pericytes and fibroblasts around the capillary endothelial cells give rise to vascular smooth muscle cells,[1] thus forming terminal arterioles. As early as 4–5 wk of age in the rat, most of the resistance vessels are formed, down to the level just before the precapillary arterioles, in the vascular beds of intestine[4] and skeletal muscles.[5] The vascular network appears as a miniature version of the mature vascular bed (Fig. 1). As the animal grows, most of the vessels enlarge through increases in length, diameter, and thickness of the vascular wall, whereas only the terminal arterioles[5] and capillaries[6,7] increase in number rather than caliber, the terminal arterioles most likely developing from preferential capillary channels as new capillaries are added.

Incorporation of ^{3}H-thymidine in the mesenteric artery of 80–90 g rats was found to be little higher than in the adult.[8] Thus, DNA replication and cellular hyperplasia cease at an early age in this major artery. The age at which hyperplasia ceases in the smaller arteries and arterioles has not been established, but is of importance for studies of the effects of growth factors on vascular structure.

Blood vessels in the normal adult individual are quiescent as determined by turnover of the capillary endothelium,[9] except for the cyclical changes in the female reproductive organs,[10] during chronic exercise in skeletal muscle,[11] and during repair of injury.[3,12] The capillary density and capillary to fiber ratio increases in skeletal muscles of endurance trained men,[11] and this is undoubtedly accompanied by enlargement of resistance vessels to handle the increased blood flow, but little is known about the latter changes. Wound healing angiogenesis primarily consists of capillary sprouting, but as already noted, when the capillary network grows, the feeding capillaries take on a smooth muscle coat to form new terminal arterioles.[3,12]

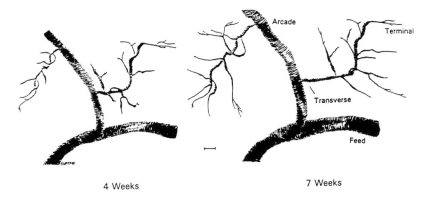

Fig. 1. Illustration of the growth of arterioles in the rat cremaster muscle from 4–7 wk of age. The feed, arcade, and transverse arterioles increased in lumen size, wall thickness, and length, whereas only the terminal arterioles increased in number. The calibration bar represents 100 μm.

Vascular Growth in Hypertension

Hypertrophy of the resistance arteries in hypertension has been known for over 100 years.[13] Growth of the vessel wall is widespread from the aorta down to small arteries of 200 μm.[14] Hypertrophy in the aorta of spontaneously hypertensive rats (SHR)[15] and renal hypertensive rats[16] is caused by cellular hypertrophy rather than hyperplasia, but a true cellular hyperplasia is found in the resistance arteries of SHR.[17] The stimulus for hypertrophy of the vascular wall is thought to be the increased wall stress associated with the rise in blood pressure. Hypertrophy reduces wall stress and leads to a mechanical advantage that amplifies vasoconstrictor responses in the hypertensive individual.[18] Cultured vascular smooth muscle cells from SHR show an enhanced growth and growth factor responsiveness that may indicate a genetic predisposition for vascular hypertrophy.[19]

The vascular response is related to the nature of the stimulus as well as the vessel, because hyperplasia is also found in the aorta during aortic ligation hypertension, which may present a greater stimulus.[20,21] Thus, a mild stimulus may initiate cellular hypertrophy and the associated polyploidy, whereas a greater stimulus may initiate events leading to cell division.[22]

Below 150 μm, the arteries and arterioles have a very different response to hypertension. Except for the pial circulation,[23] no dimensional changes are found in SHR arterioles.[24–26] In renal hypertension, the arterioles remodel so that they have a reduced lumen, increased wall-to-lumen ratio, and increased wall thickness, but either no increase[27] or an actual decrease[28,29] in wall cross-sectional area. Even the pial arterioles, which hypertrophy to a certain extent in the SHR, also remodel to achieve a higher wall-to-lumen ratio.[30] Interestingly, in human hypertension, intestinal arterioles undergo similar structural changes to achieve an increase in wall-to-lumen ratio without hypertrophy.[31] The reason for the different responses among the various forms of hypertension and vessel sizes is not known. However, the structural changes in the renal models appear to be related to the increase in pressure, because they do not occur in the normotensive cremaster muscle of rats with aortic coarctation hypertension.[32]

Atherosclerosis

Although mainly confined to the larger arteries, atherosclerosis involves growth of vascular smooth muscle cells and has been the stimulus for much of the research on vascular growth factors. The major hypothesis explaining smooth muscle replication in the vessel wall is the response-to-injury hypothesis of Ross.[33] In brief, the hypothesis states that smooth muscle cells in the wall normally exist in a quiescent state. When the endothelium is injured, platelets release a factor or factors that stimulate smooth muscle migration and replication within the arterial intima.

Rarefaction and Proliferation

Long–Term Autoregulation

The number of terminal arterioles in many tissues is not constant, but decreases with chronic elevations of blood pressure and increases when blood pressure and flow decrease following ligation of a feeding artery. Decreases in the number of arterioles, known as rarefaction, occurs in skeletal muscle,[25–27,34,35] skin,[36] intestine,[37] and mesentery[38] of hypertensive animals, and the intestine[31] and conjunctiva[39,40] of hypertensive patients. Small arterioles in the lung also rarefy in pulmonary hypertension.[41] Structural arteriolar rarefaction is preceded by a period of functional rarefaction, when the arterioles are not perfused under resting conditions, but will open to flow on application of a vasodilator.[25-27] It was thought that the nonperfused vessels degenerate leaving a structurally rarefied bed, but direct evidence has been lacking until recently when degenerating arterioles were found in the cremaster muscle of rats with reduced renal mass hypertension[42] at the time that rarefaction occurs in this model.[43]

The number of small arterioles proliferates in the cremaster muscle during the 3 wk following ligation of the external spermatic artery,[44] one of the two feed vessels. In addition, arteriolar rarefaction does not occur in the normotensive cremaster muscle of the rat with aortic coarctation hypertension.[32] This close relationship between arteriolar density and perfusion pressure suggests that changes in arteriolar number may be a mechanism for long-term autoregulation of blood flow.

Vasoconstrictors

Spontaneous hypertension[26,35] (SHR) and the initial stages of renal hypertension[27,45] are characterized by strong active vasoconstriction in the small arterioles related to increased activity of vasoconstrictor systems[46] and/or increased sensitivity to vasoconstrictor agents.[35] This time frame is coincident with functional rarefaction and suggests that vasoconstrictor mechanisms, such as the sympathetic nerves, the renin-angiotensin system, and/or vasopressin, may contribute to arteriolar closure. Studies by Bohlen[35] show that sensitivity to norepinephrine is greatly increased in cremaster arterioles of the SHR and acute denervation results in an increase in the number of perfused arterioles. On the other hand, treatment with the converting enzyme inhibitor, captopril, exacerbates rarefaction in the one-kidney, one-clip hypertensive rat.[47] Thus, angiotensin II appears to be more important as a growth factor than a vasoconstrictor in this regard. Elevated intracellular free calcium is the final pathway for active vasoconstriction of any origin and the calcium channel blocker, nitrendipine, decreases arteriolar rarefaction in the one-kidney, one-clip hypertensive rat.[48] Thus, vasoconstriction to the point of arteriolar closure, whatever the cause, appears to be a prerequisite for structural rarefaction.

Factors Leading to Growth of the Vascular Wall

Physical Factors

Pressure

It is well established that an increase in transmural pressure, through alterations in vascular wall stress, stimulates vascular wall growth. Vascular wall hypertrophy occurs in the hypertensive, but not the normotensive regions of animals with aortic coarctation hypertension.[32,49] Vascular beds protected from pressure rises by ligation do not hypertrophy in the SHR[50] or the DOCA-salt model of hypertension.[51] Antihypertensive treatment reverses vascular wall hypertrophy along with blood pressure.[28,52] Even isolated arteries cultured under increased levels of wall stress increase their incorporation of ^{3}H-thymidine, indicative of replication of DNA[53] and cyclical stretching of cultured vascular smooth muscle cells stimulates the production of matrix components and hypertrophy.[54]

As pressure rises during the onset of hypertension, wall stress increases in arteries, stimulating an increase in wall thickness, thereby decreasing wall stress. In large arteries, wall stress sometimes returns to control levels after the structural changes have taken place,[28] and other times it remains elevated.[52] In arterioles, however, enhanced vasoconstriction can prevent rises in wall stress in the face of increased intraluminal pressure because of the Laplace relationship, which states that wall stress is equal to the transmural pressure times the radius divided by the wall thickness. This may explain, at least in part, why vascular wall hypertrophy usually does not occur in resistance vessels smaller than 150 μm.[14,22,24–29]

Flow

Changes in blood flow influence the size of blood vessels independently of changes in blood pressure. This has been demonstrated in the rabbit by Langille and O'Donnell,[55] who ligates the internal carotid artery to decrease blood flow in the common carotid without altering intraluminal pressure. The response appears to be mediated by, or at least require the presence of, the endothelium that is exposed to the shear stress of the flowing blood. When flow is decreased, the initial response is an increase in tone to decrease in the size of the lumen. This reduction can be reversed with a vasodilator, but after one month, the changes become structural. In 6-wk-old rabbits, vascular wall growth is inhibited and the wall cross-sectional area is reduced along with the lumen.[56] In adult rabbits, however, the vessel is remodeled to a smaller lumen with the same wall cross-sectional area, at least within the one month time-scale of this experiment.

As organs grow in a young animal, the capillary bed proliferates, flow increases, and the resistance vessels enlarge. Whether the stimulus for growth of existing resistance vessels in this case is the increase in flow, as proposed in 1896 by Thoma,[2] has not been established, but it is a likely possibility. As tissue grows and the capillary bed enlarges, an increasing demand for blood flow through the resistance vessels will result in vasodilation. The terminal resistance vessels will be exposed to vasodilator metabolites, while the upstream vessels will dilate in response to the increase in flow or wall shear stress,[57] probably through the release of endothelial-derived relaxing factor (EDRF). Vasodilation will increase wall stress, which, as previously discussed, may stimulate growth of the vessel, or the endothelial cells may decrease secretion of growth inhibitors or release an as yet undetermined growth factor that acts in a paracrine fashion to stimulate growth of the adjacent smooth muscle cells. Definite evidence to support either modulation of wall stress or endothelial-derived growth factors and inhibitors as the mechanism for this response is presently lacking.

Humoral Factors

Sympathetic Nerves

The work of Bevan et al.[58] has demonstrated the trophic influence of the sympathetic nerves on wall structure of small peripheral arteries, especially in young, growing rabbits. Sympathetic innervation contributes to development of hypertrophy in cerebral resistance vessels[59] and protects against stroke and hypertensive encephalopathy in stroke-prone spontaneously hypertensive rats.[60] Yamori et al.[61] have demonstrated the induction of polyploidy in vascular smooth muscle cells by norepinephrine and epinephrine in vivo and in tissue culture. Some evidence showing an increased sympathetic innervation of selected vascular beds in SHR[62] and increased sympathetic nerve activity[46] support the hypothesis that the sympathetic nerves are responsible for initiating hypertension, and a predisposition for the ensuing hypertrophic response in the arteries.

The mechanism for the influence of the sympathetic nerves on vascular structure is not clear, but there are three components that could be responsible, (1) impulse rate, (2) norepinephrine itself, or (3) the secretion of an unidentified trophic factor. Catecholamines, including norepinephrine, stimulate growth of cultured rat aortic vascular smooth muscle cells[63] and an hypertrophic response from cultured cardiac muscle cells mediated through the α_1-adrenergic receptor.[64] The same receptor may mediate the response in vascular smooth muscle cells, because the in vitro growth response to epinephrine was blocked by the α_1-adrenergic receptor antagonist, phentolamine.[63]

Sympathetic innervation can have an inhibitory effect on vascular growth under some circumstances. Sympathetic denervation stimulated hypertrophy in the rabbit saphenous vein, ear artery,[65] and cerebral artery,[66] and it increased collagen synthesis in the aorta.[67] An inhibitory trophic effect was demonstrated by the addition of a sympathetic ganglion extract to cultured vascular smooth muscles,[68] whereas the depletion of norepinephrine from nerve endings does not produce the same result as denervation.[65] These data support the hypothesis that a trophic factor from sympathetic nerves prevents dedifferentiation of the vascular smooth muscle cells to a synthetic phenotype.[68]

Vasoactive Agents

Many agents primarily associated with vasoconstriction have recently been shown to stimulate vascular growth (Table 1), and conversely, some of the growth factors have been shown to have vasoactivity.[69,70] Apparently both functions share signaling path-

Table 1
Vasoactive Growth Modulators

Norepinephrine	Endothelin
Angiotensin II	Prostaglandins (+/-)
Vasopressin	Bradykinin
Serotonin	Adenosine

Growth Factors

Platelet-derived growth factor (PDGF)
Basic-fibroblast growth factor (B-FGF)
Insulin-like growth factor-1 (IGF-1)
Vascular endothelial growth factor (VEGF, VPF)
Epidermal growth factor (EGF)
Interleukin-1 (IL-1)
Transforming growth factor-β (TGF-β) (+/-)

(+/-) Indicates growth stimulation and inhibition.

ways, such as the activation of the phosphatidylinositol second messenger system[71] and intracellular Ca^{2+} mobilization.[72] How the two responses are differentiated is presently unknown, but it may depend on the length of receptor occupancy or require the activation of more than one signal transduction pathway to allow a vasoconstrictor response to proceed to a growth response.[73]

Angiotensin II (AII), which stimulates cellular hypertrophy in cultured aortic smooth muscle cells[74] and hypertrophy and hyperplasia in cultured smooth muscle cells from mesenteric resistance arteries,[75] appears to be necessary for normal growth of blood vessels in vivo.[47] Treatment of rats with the converting enzyme inhibitor, captopril, decreased wall mass of the aorta and arterioles, as well as the number of terminal arterioles in normotensive and hypertensive rats, without decreasing blood pressure in the latter.[47] Although AII circulates in the plasma, it is more likely to function in an autocrine or intracrine manner to regulate vascular growth.[76] In addition to activation of the inositol lipids, AII stimulates the expression of protooncogenes *c-fos*[77] and *c-myc* and the production of platelet-derived growth factor (PDGF) A-chain in rat aortic smooth muscle cells.[78] AII also increases the number of receptors for the BB homodimer of PDGF and potentiates PDGF-BB stimulated DNA synthesis in vascular smooth muscle.[79] This correlates very well with the ability of converting enzyme inhibitors to decrease growth of carotid lesions caused by balloon injury.[80]

Arginine vasopressin (AVP) induces proliferation of a number of cell lines[81,82] and enhances the mitogenic response of Swiss 3T3 cells to epidermal growth factor (EGF).[81] Others have shown that AVP causes hypertrophy, but not hyperplasia or even inhibits proliferation of cultured rat or human aortic smooth muscle cells.[83,84] AVP and AII share the same phosphatidylinositol second messenger system leading to activation of protein kinase C and Ca^{2+} mobilization, which are thought to result in growth.[85]

Serotonin, a vasoconstrictor released from the α granules of blood platelets, increases the uptake of ^{3}H-thymidine into DNA by cultured aortic smooth muscle cells.[86] Release of serotonin, when platelets are exposed to subendothelial connective tissue, could potentiate the response to PDGF and exacerbate the development of atherosclerotic lesions.[33]

The prostaglandins appear to either promote[87,88] or inhibit[89-91] cell proliferation depending on the time and the concentration of prostaglandin added to, or generated by, the cells in tissue culture. In vivo, the results also seem to vary, but PGE_1 and PGE_2 have been shown to promote angiogenesis.[92,93]

Bradykinin and related kinins exert a variety of pharmacological effects by stimulating or inhibiting smooth muscle tone in various species and vascular beds. Bradykinin, however, is also mitogenic for lymphocytes,[94] and stimulates Na^+ influx and DNA synthesis in human diploid fibroblasts.[95]

Endothelin stimulates DNA synthesis in vascular smooth muscle cells and induces the expression of protooncogenes *c-fos* and *c-myc* in cultured rat aortic smooth muscle cells and causes proliferation.[96] Endothelin also stimulates mitogenesis through inositol lipid turnover in Swiss 3T3 fibroblasts.[97]

Adenosine, a vasodilator produced in response to tissue hypoxia, stimulates angiogenesis in chick chorioallantoic membrane[98,99] and growth of cultured vascular endothelial cells.[100]

Growth Factors

Physical and vasoactive modulators of vascular growth must use a signaling mechanism to activate the intracellular machinery for hypertrophy and/or replication. A variety of growth factors listed in Table 1, identified primarily from tissue culture work, are candidates for this role. Much work is dedicated to illuminating the cellular events required for these factors to promote growth. Activation of competence factors, the protooncogenes, *c-myc* and *c-fos*, are first steps in the initiation of replication.[76] Regulatory signals, such as tyrosine protein kinase activation, leading to phosphoinositide hydrolysis, Ca^{2+} mobilization, Na^+/H^+ antiport activation, and protein kinase C activation, also are thought to be involved in progression through the cell cycle.[101] It is not certain which of these regulatory signals are required for DNA synthesis and cell proliferation, as it has recently been shown that PDGF-induced activation of phospholipase C, the initial event for activation of the phosphatidylinositol second messenger system, is not required for induction of DNA synthesis in fibroblasts.[102] This is a major cellular mechanism for growth induction and the one that growth stimulating vasoconstrictors and growth factors have in common.

Besides PDGF and serotonin, platelets also contain EGF and transformin growth factor-β (TGF-β).[103] EGF has been found to both promote growth of rat vascular smooth muscle cells and induce contraction in helical strips from rat aortas.[70] TGF-β inhibits proliferation of endothelial cells in vitro,[104] however, it can stimulate or inhibit growth of certain nonendothelial cells, depending on whether the cells are anchored or not and on the presence of EGF.[105] On the basis of these results, it has been suggested that TGF-β acts as a bidirectional regulator of cell growth in vitro.[103]

Vascular endothelial growth factor (VEGF), which may be the same as vascular permeability factor (VPF),[106] shows a similarity to the B chain of PDGF and is secreted by endothelial cells.[107] It could function as an angiogenesis factor, especially in growing tumors.

The endogenous role of these growth factors for normal development of resistance vessels has not been defined, but they are postulated to mediate vascular growth. Recently, Morrow et al.[108]

reported an increased expression of fibroblast growth factors (FGF) in rabbit skeletal muscle subjected to chronic motor nerve stimulation, a model of exercise conditioning known to exhibit vascular growth. FGFs must be transported out of the skeletal muscle cells to initiate angiogenesis, but there is presently no evidence of this. FGFs are thought to be sequestered inside the cells of origin.[109]

Endothelial cells have been shown to release a substance with PDGF-like activity preferentially into the basal subendothelial compartment,[110] an ideal location for paracrine induction of vascular wall growth. Paracrine and autocrine activity by growth factors in the vascular wall, initiated by blood flow or stretch, are the most likely mechanisms by which resistance vessels could adapt to needs of a distal growing vascular network, whereas the diffusible angiogenesis factors may function to stimulate capillary growth. The origin, mechanisms of action, and interaction of these growth factors and growth inhibitors in vivo are extremely complicated. Much more information will be required before their role in normal development of the vascular wall will be understood.

References

[1] Arey, L. B. (1963) *The Peripheral Blood Vessels* (Orbison, J. L. and Smith, D. E., eds.), Williams and Wilkins, Baltimore, MD, pp. 1–16.

[2] Thoma R. (1896) *Textbook of General Pathology*. Translated by A. Bruce, Adam and Charles Black, London, UK, vol. 1.

[3] Sandison, J. C. (1928) *Am. J. Anat.* **41**, 475–496.

[4] Unthank, J. L. and Bohlen, H. G. (1987) *Circ. Res.* **61**, 616–624.

[5] Wang D. H. and Prewitt, R. L. (1990) *FASEB J.* **4**, A722.

[6] Hudlicka, O. and Tyler, K. R. (1986) *Angiogenesis*. Academic, London, UK.

[7] Banchero N. (1982) *The Physiologist* **25**, 385–389.

[8] Crane, W. A. J. and Dutta, L. P. (1969) *J. Pathol. Bacteriol.* **88**, 291–301.

[9] Engerman, R. L., Pfaffenbach, D., and Davis, M. D. (1967) *Lab. Invest.* **17**, 738–743.

[10] Reynolds, S. R. M. (1963) *Handbook of Physiology Section 2: Circulation*. (Hamilton, W. F. and Dow, P., eds.), American Physiological Society, Washington, DC, **II**, 1585–1618.

[11] Brodal, P., Ingjer, F., and Hermansen, L. (1977) *Am. J. Physiol.* **232**, H705–H712.

[12] Clark, E. R. and Clark, E. L. (1940) *Am. J. Anatomy* **66**, 1–49.

[13] Pickering, G. (1982) *Circulation of the Blood Men and Ideas* (Fishman, A. P. and Richards, D. W., eds.), American Physiological Society, Bethesda, MD, pp. 487–541.

[14] Furuyama, M. (1962) *Tohuko J. Exp. Med.* **76**, 388–414.

[15] Owens, G. and Schwartz, S. (1982) *Circ. Res.* **51**, 280–289.

[16] Owens, G. and Schwartz, S. (1983) *Circ. Res.* **53**, 491–501.

[17] Mulvany, M., Baandrup, U., and Gundersen, H. (1985) *Circ. Res.* **57**, 794–800.

[18] Folkow, B., Hallback, M., Lundgren, Y., Sivertsson, R., and Weiss, L. (1973) *Circ. Res.* **23(Suppl. 1)**, 2–16.

[19] Scott–Burden, T., Resink, T., and Buhler, F. R. (1989) *J. Cardiovasc. Pharmacol.* **14(Suppl. 6)**, S16–S21.

[20] Bevan, R., Marthens, E., and Bevan, J. (1976) *Circ. Res.* **38(Suppl. II)**, 58–62.

[21] Owens, G. and Reidy, M. (1985) *Circ. Res.* **57**, 695–705.

[22] Owens, G. K. (1989) *Am. J. Physiol.* **257**, H1755–H1765.

[23] Harper, S. L. and Bohlen, H. G. (1984) *Hypertension* **6**, 408–419.

[24] Bohlen, H. G. and Lobach, D. (1978) *Blood Vessels* **15**, 322–330.

[25] Chen, I. I. H., Prewitt, R. L., and Dowell, R. (1981) *Am. J. Physiol.* **241**, H306–H310.

[26] Prewitt, R. L., Chen, I. I. H., and Dowell, R. (1982) *Am. J. Physiol.* **243**, H243–H251.

[27] Hashimoto, H., Prewitt, R. L., and Efaw, C. W. (1987) *Am. J. Physiol.* **253**, H933–H940.

[28] Stacy, D. L. and Prewitt, R. L. (1989) *Circ. Res.* **65**, 869–879

[29] Ono, Z., Prewitt, R. L., and Stacy, D. L. (1989) *Hypertension* **14**, 36–43

[30] Baumbach, G. L. and Heistad, D. D. (1989) *Hypertension* **13**, 968–972

[31] Short, D. S. and Thomson, A. D. (1959) *J. Pathol. Bacteriol.* **78**, 321–334.

[32] Stacy, D. L. and Prewitt, R. L. (1989) *Am. J. Physiol.* **256**, H213–H221.

[33] Ross, R. (1986) *N. Engl. J. Med.* **314**, 488–500.

[34] Hutchins, P. M. and Darnell, A. E. (1974) *Circ. Res.* **34(Suppl.)**, I-161–I-165.

[35] Bohlen, H. G. (1979) *Am. J. Physiol.* **236**, H157–H164.

[36] Haack, D. W., Schaffer, J. J., and Simpson, J. G. (1980) *Proc. Soc. Exp. Biol. Med.* **164**, 453–458.

[37] Bohlen, H. G. (1983) *Hypertension* **5**, 739–745.

[38] Henrich, H., Hertel, R., and Assman, R. (1978) *Pflugers Arch.* **375**, 153–159.

[39] Harper, R. N., Moore, M. A., Marr, M. C., Watts, L. E., and Hutchins, P. M. (1978) *Microvasc. Res.* **16**, 369–372.

[40] Sullivan, J. M., Prewitt, R. L., and Josephs, J. A. (1983) *Hypertension* **5**, 844–851.

[41] Reid, L. M. (1979) *Am. Rev. Respir. Dis.* **119**, 531–546.

[42] Hansen-Smith, F., Greene, A. S., Cowley, A. W. Jr., and Lombard, J. H. (1990) *Hypertension* **6(Part 2),** 922–928.

[43] Greene, A. S., Lombard, J. H., Cowley, A. W. Jr., and Hansen-Smith, F. M. (1990) *Hypertension* **6(Part 2),** 779–783.

[44] Hogan, R. D. and Hirschmann, L. (1984) *Microvasc. Res.* **27,** 290–296.

[45] Meininger, G. A., Lubrano, V. M., and Granger, H. J. (1984) *Circ. Res.* **55,** 609–622.

[46] Judy, W. V., Watanabe, A. M., Henry, D. P., Vesch, H. R. Jr, Murphy, W. R., and Hockel, G. M. (1976) *Circ. Res.* **38(Suppl. II),** II-21–II-29.

[47] Wang, D. H. and Prewitt, R. L. (1990)*Hypertension* **15,** 68–77.

[48] Smith, E. G. and Prewitt, R. L. (1990) *FASEB J.* **4,** A1271.

[49] Bevan, R. D. (1976) *Blood Vessels* **13,** 100–128.

[50] Hansen, T. R., Abrams, G. D. and Bohr, D. F. (1974) *Circ. Res.* **34/35 (Suppl I.),** I-101–I-107.

[51] Berecek, K. H and Bohr, D. F. (1977) *Circ. Res.* **40(Suppl. I),** I-146–I-152.

[52] Wolinsky, H. (1971) *Circ. Res.* **28,** 622–637.

[53] Hume, W. R. (1980) *Hypertension* **2,** 738–743.

[54] Leung, D. Y. M., Glagov, S., and Mathews, M. B. (1976) *Science* **191,** 475–477.

[55] Langille, B. L. and O'Donnell, F. (1986) *Science* **231,** 405–407.

[56] Landille, B. L., Bendeck, M. P., and Deeley, F. W. (1989) *Am. J. Physiol.* **256,** H931–H939.

[57] Smiesko, V., Kozik, J., and Dolezel, S. (1989) *Blood Vessels* **22,** 247–251.

[58] Bevan, R. D. (1984) *Hypertension* **6(Suppl. III),** III-19–III-26.

[59] Hart, M. N., Heistad, D. D., and Brody, M. J. (1980) *Hypertension* **2,** 419–423.

[60] Sadoshima, S., and Heistad, D. D. (1982) *Hypertension* **4,** 904–907.

[61] Yamori, Y., Mano, M., Nara, Y., and Horie, R. (1987) *Circulation* **75 (Suppl. I),** I-92–I-95.

[62] Head, R. J. (1989) *Blood Vessels* **26,** 1–20.

[63] Blaes, N. and Boissel, J-P. (1983) *J. Cell Physiol.* **116,** 167–172.

[64] Simpson, P. (1985) *Circ. Res.* **56,** 884–894.

[65] Branco, D., Albino-Teixeira, A., Azevedo, I., and Osswald, W. (1984) *Arch. Pharmacol.* **326,** 302–312.

[66] Dimitriadou, V., Aubineau. P., Taxi, J., and Seylaz, J. (1988) *Blood Vessels* **25,** 122–143.

[67] Fronek, K. (1983) *Ann. Biomed. Eng.* **11,** 607–615.

[68] Chamley, J. H. and Campbell, G. R. (1975) *Cell Tissue Res.* **161,** 497–510.

[69] Berk, B. C., Alexander, R. W., Brock, T. A., Gimbrone, M. A., and Webb, R. C. (1986) *Science* **232,** 87–90.

[70] Berk, B. C., Brock, T. A., Webb, R. C, Taubman, M. B., Atkinson, W. J.,

Gimbrone, M. A., and Alexander, R. W. (1985) *J. Clin. Invest.* **75,** 1083–1086.

[71] Habenicht, A. J. R., Glomset, J. A., King, W. C., Nist, C., Mitchell, C. D., and Ross, R. (1981) *J. Biol. Chem.* **256,** 12329–12335.

[72] Berridge, M. J. and Irvine, R. F. (1984) *Nature* **312,** 315–321.

[73] Rozengurt, E. (1986) *Science* **234,** 161–166.

[74] Geisterfer, A. A. T., Peach, M. J., and Owens, G. K. (1988) *Circ. Res.* **62,** 749–756.

[75] Lyall, F., Morton, J. J., Lever, A. F., and Gragoe, E. J. (1988) *J. Hypertension* **6(Suppl. 4),** 438–441.

[76] Dzau, V. J. and Gibbons, G. H. (1988) *Am. J. Cardiol.* **62,** 30G–35G.

[77] Naftilan, A. J., Pratt, R. E., Eldridge, C. S., Lin, H. L., and Dzau, V. J. (1989) *Hypertension* **13(6 pt 2),** 706–711.

[78] Naftilan, A. J., Pratt, R. E., and Dzau, V. J. (1989) *J. Clin. Invest.* **83,** 1419–1424.

[79] Bobik, A., Grinpukel, S., Little, P. J., Grooms, A., and Jackman, G. (1990) *Biochem. Biophy. Res. Comm.* **166,** 580–588.

[80] Powell, J. S., Clozel, J–P., Muller, R. K. M., Kuhn, H., Hefti, F., Hosang, M., and Baumgartner, H. R. (1989) *Science* **245,** 186–188.

[81] Rozengurt, E., Legg, A., and Pettican, P. (1979) *Proc. Natl. Acad. Sci.* **76,** 1284–1287.

[82] Payet, N., Deziel, Y., and Lehoux, J. G. (1984) *J. Steroid Biochem.* **20,** 449–454.

[83] Geisterfer, A. A. T. and Owens, G. K. (1989) *Hypertension* **14,** 413–420.

[84] Campbell-Boswell, M. and Robertso, A. L. (1981) *Exp. Mol. Pathol.* **35,** 265–276.

[85] Doyle, V. M. and Ruegg, U. T. (1985) *Biochem. Biophy. Res. Comm.***131(1),** 469–476.

[86] Nemecek, G. M., Coughlin, S. R., Handley, D. A., and Moskowitz, M. A. (1986) *Proc. Natl. Acad. Sci.* **83,** 674–678.

[87] Taylor, L. and Polgar, P. (1977) *FEBS Lett.* **79,** 69–72.

[88] Bettger, W. J. and Ham, R. G. (1981) *Prog. Lipid Res.* **20,** 265–268.

[89] Nilsson, J. and Olsson, A. G. (1984) *Atherosclerosis* **55,** 77–82.

[90] Loesberg, C., Van Wijk, R., Zandbergen, J., Van Aken, W. G., Van Mourik, J. A., and De Groot, P. H. G. (1985) *Exp. Cell Res.* **160,** 117–125.

[91] Morisaki, N., Kanzaki, T., Motoyama, N., Saito, Y., and Yoshida, S. (1988) *Atherosclerosis* **71,** 165–171.

[92] Ziche, M., Ruggiero, M., Pasquali, F., and Chiarugi, V. P. (1985) *Int. J. Cancer* **35,** 549–552.

[93] Form, D. M. and Auerbach, R. (1983) *Proc. Soc. Exp. Biol. Med.* **172,** 214–218.

[94] Perris, A. D. and Whitfield, J. F. (1969) *Pro. Soc. Exp. Biol. Med.* **130,** 1198–1202.

[95] Owen, N. E. and Villereal, M. L. (1983) *Cell* **32,** 979–985.

[96] Komuro, I., Kurihara, H., Sugiyama, T., Takaku, F., and Yazaki, Y. (1988) *FEBS Lett.* **238,** 249–252.

[97] Takuwa, N., Takuwa, Y., Yanagisawa, M., Yamashita, K., and Masaki, T. (1989) *J. Biol. Chem.* **15, 264,** 7856–7861.

[98] Dusseau, J. W., Hutchins, P. M., and Malbasa, D. S. (1986) *Circ. Res.* **59,** 163–170.

[99] Adair, T., Montani, J. P., Strick, D. M., and Guyton, A. C. (1989) *J. Physiol.* **256,** H240–H246.

[100] Meininger, C. J., Schelling, M. E., and Granger, H. J. (1988)*Am. J. Physiol.* **255,** H554–H562.

[101] Wasteson, A., Douglas, G., and Ljungquist, P. (1987) *Thrombosis and Haemostasis.* (Verstraete, M., Vermylen., Lijnen, H. R., and Arnout, J., eds.), Leuven University Press, Leuven, Belgium, pp. 281–300.

[102] Hill, T. D., Dean, N. M., Mordan, L. J., Lau, A. F., Kanemitsu, M. Y., and Boynton, A. L.(1990) *Science* **248,** 1660–1663.

[103] Folkman, J. and Klagsburn, M. (1987) *Science* **235,** 442–447.

[104] Frater-Schroder, M., Muller, G., Birchmeier, W., and Bohlen, P. (1986) *Biochem. Biophy. Res. Comm.* **137,** 295–302.

[105] Shipley, G. D., Tucker, R. F., and Moses, H. L. (1985) *Proc. Natl. Acad. Sci.* **82,** 4147–4151.

[106] Keck, P. J., Hauser, S. D., Krivi, G., Sanzo, K., Warren, T., Feder, J., and Connolly, D. T. (1989) *Science* **246,** 1309–1312.

[107] Leung, D. W., Cachianes, G., Kuang, W.-J., Goeddel, D. V., and Ferrara, N. (1989) *Science* **246,** 1306–1309.

[108] Morrow, N. G., Kraus, W. E., Moor, J. W., Williams, R. .S, and Swain, J. L. (1990) *J. Clin. Invest.* **85,** 1816–1820.

[109] Moscatelli, D., Presta, M., Joseph-Silverstein, J., and Rifkin, D. B. (1986) *J. Cell Physiol.* **129,** 273–276.

[110] Zerwes, J. G. and Risau, W. (1987) *J. Cell Biol.* **105,** 2037–2041.

Chapter 9

Myogenic Properties
of Blood Vessels In Vitro

George Osol

Introduction and Historical Perspective

It would be a disservice to begin a chapter on the myogenic behavior of blood vessels in vitro without describing the work of Sir William Bayliss almost 90 years ago. Using primitive recording techniques (a smoked drum), he removed a carotid artery from the neck of a dog three hours after asphyxiation, cannulated one end, tied off the other, and filled the segment with defibrinated blood. The preparation was connected to a mercury manometer and, when pressure was elevated, the vessel underwent a powerful contraction and "appeared to writhe like a worm." Decreasing the pressure produced "considerable relaxation." Unfortunately, the experimental detail is scant and we have no quantitative measure of the response because, in Bayliss' words, "I regret that the tracing was spoilt in varnishing, so that I am unable to reproduce it here."[1] Nonetheless, Bayliss' observation captures the quintessence of myogenic theory as it applies to vascular behavior—pressure or

From *The Resistance Vasculature,* J. A. Bevan et al., eds. ©1991 Humana Press stretch-induced activation of smooth muscle cells that leads to a decrease in arterial diameter.

Although Bayliss went on to other areas of research, his observations generated considerable debate in the ensuing years, and the 1902 paper is often cited in the physiological literature.[2] Subsequent experiments in vitro and in vivo led to a formulation of the myogenic hypothesis, described by Folkow in 1964.[3] Since then, several authors have attempted to unify the experimental and theoretical aspects of myogenic behavior in the form of predictive mathematical models.[4–6] Because of space limitations, the reader is referred to these papers and to an excellent review by Johnson[7] for a more complete analytical and historical perspective.

In this chapter, my goal is to review and summarize more recent findings, particularly those that provide insights into the cellular mechanisms that may underlie myogenic behavior. Although the word "myogenic" is used often and seems fairly straightforward, when it comes to discussing the myogenic properties of blood vessels, it is helpful to put the term into physiological perspective by first considering the role of myogenic behavior in blood-flow autoregulation and then discussing the definitions of, and differences between, myogenic tone and myogenic reactivity.

General Concepts and Definitions

Autoregulation

Most organs of the body possess some capability for blood-flow autoregulation, a process that ensures the maintenance of a relatively constant whole-organ blood-flow in spite of occasionally wide fluctuations in systemic pressure.[8] Stated most simply, the process of autoregulation involves arterial constriction to increased, and dilation to decreased perfusion pressure, so that whole-organ blood-flow remains relatively invariant.

This view is somewhat oversimplified because it ignores the existence of regional mechanisms (such as recruitment, adjustments in metabolic demand, efficiency of oxygen extraction, and so on) and of geometric differences (such as the pattern of branching and functional interrelationships between serial segments of the vasculature) that are important when considering the behavior of a particular organ or vascular bed. It serves our purpose, however, since most of these factors are not present in vitro and yet, as Bayliss and others have shown, isolated vessels often demonstrate the appropriate responses to stretch or pressure, suggesting that at least some portion of the autoregulatory response can be attributed to mechanisms that are intrinsic to the cells of the vascular wall.

Most often, myogenic behavior is seen in smaller, "resistance" vessels, although the definition of a resistance vessel varies from one vascular bed to another. For instance, in the cerebral circulation, large vessels contribute significantly to total resistance.[9] Conversely, in the splanchnic circulation, which is highly branched and of a very different geometry, the major portion of vascular resistance is localized in very small (<100 μm) arteries and arterioles.[10] It is also worth noting that myogenic behavior is not restricted to the arterial side of the circulation—a number of in vitro studies have been conducted on veins (for example, the rat portal vein[11] or rabbit facial vein[12]), as an experimental model of stretch-induced vascular tone.

Myogenic Tone vs Myogenic (?) Reactivity

For dilation to occur following a decrease in perfusion pressure, a vessel must already be in a state of partial constriction; in vitro, this state of sustained constriction in the absence of any chemical or electrical activation imposed by the investigator is defined as *basal tone*.[3] Other terms that are synonymous and have been used interchangeably are "spontaneous" or "intrinsic" tone.

An issue that has become controversial in the last few years is whether the source of this basal tone is truly myogenic, i.e., originating solely from the vascular smooth muscle cell, or "vasogenic"—a product of smooth-muscle interaction with other cell types, for example, the endothelium. The evidence available to date, presented below, overwhelmingly supports the former opinion.

There is another layer of complexity, however, in that basal, or myogenic tone is not necessarily synonymous with myogenic reactivity, which has been defined as the ability of the vascular smooth muscle to contract in response to a sudden stretch or an increase in transmural pressure. As pointed out by Johansson, although myogenic tone and myogenic reactivity are often associated, they are not necessarily one, and the latter may invoke mechanisms that are additional to and superimposed upon those that govern basal myogenic tone.[13] In this regard, the consensus on the role of the endothelium is much less clear, and the experimental evidence is sometimes contradictory.

Comparison of Methodologies Used for Studying Myogenic Activity In Vitro

The distinction between myogenic tone and reactivity becomes important in light of the fact that two very different methodologies (wire-mounted vs pressurized segments; *see* the chapter by Halpern, this volume) have been used to study myogenic activity in vitro, with little attempt to distinguish between the mechanisms that underlie the appearance of basal tone from those that may modulate the level of preexisting (basal) tone as a function of some physical force, such as intravascular pressure or stretch.

Because they affect the interpretation of data, some differences between wire-mounted and pressurized vessels are worth considering before reviewing what is known about the cellular mechanisms underlying myogenic activity.

Wire-Mounted Vessels

The wire method is isometric, and study of the myogenic response involves stretching a segment of vessel and measuring the level of force production as a function of the amplitude of stretch. In most studies, a vessel is mounted unstretched, equilibrated, and then stretched to some multiple of the unstretched diameter. Once the response has been recorded, the vessel is returned to the unstretched state and the protocol repeated.

Following the initial rapid increase in force, nonmyogenic vessels exhibit stress–relaxation only, a response that probably arises from a reorientation of connective tissue elements within the wall and is characterized by an exponential decrease in force to some new "passive" level.

Conversely, a wire-mounted myogenic vessel may exhibit some stress–relaxation that is quickly overshadowed by a gradual increase in force as a result of stretch-induced smooth-muscle activation. This response can be quantified, and various experimental manipulations (such as examining the effect of different rates of stretch,[14,15] selectivity of pharmacologic agonists,[16] measurement of calcium influx,[17] and the like) may be performed.

The vascular wall of a wire-mounted vessel experiences circumferential stretch, but not a transmural pressure, and the smooth-muscle cells cannot shorten unless so accomodated by the investigator. In general, the demonstration of myogenic tone in ring-mounted segments is unusual and, when present, is seen more frequently in small, rather than large, vessels. When present following a stretch, the tone may be sustained for several hours[18] or may be quite transient, disappearing within a few minutes.[15,19]

Pressurized Vessels

In the pressurized system, the artery is cannulated and may be studied in the presence or absence of intralumenal flow.[20-28] The preparation is allowed to equilibrate at some transmural pressure, and the experimental manipulation most often consists of changing the transmural pressure and observing the effect on vessel diameter.

As in wire-mounted vessels, myogenic tone is typically characteristic of smaller pressurized arteries, although one must again recall the work of Bayliss as a reminder about the dangers of generalization. The occurrence of tone is much more common in pressurized than in wire-mounted vessels, having been observed in arteries taken from the cerebral, renal, skeletal, splanchnic, cremaster, uterine, and coronary circulations, and in vessels obtained from a variety of animal species.[20-28] Interestingly, identical vessels taken from the same animal will often develop tone if pressurized, but not when mounted on wires (Halpern, personal communication).

Myogenic tone may appear suddenly during equilibration or, if a vessel is equilibrated at a low transmural pressure, immediately following an increase in transmural pressure above some threshold. Cerebral arteries from normotensive rats, for example, usually develop spontaneous tone at transmural pressures above 40 mmHg, at which time lumen diameter decreases by 30–35% in several minutes.[26] Once present, tone remains stable for hours and sometimes has a superimposed rhythmicity (vasomotion); the latter behavior is particularly true of vessels obtained from hypertensive animals.[29]

Under isobaric conditions, the development of tone is a result of shortening of vascular smooth muscle cells. Because the diameter of a cell increases with shortening, the reduction in lumen diameter is accompanied by an increase in both wall thickness and axial length. The endothelial cell layer and the underlying internal elastic lamina are therefore simultaneously buckled in a circumferential direction and extended axially, and the stresses on the passive connective-tissue matrix (collagen and elastin) are also altered by the active shortening of vascular smooth muscle.

Clearly, the spatial relationships assumed by the cells within the vascular wall of a pressurized vessel are not only extremely complex, but also completely different from those present in the same artery when it is stretched taut on wires, since the response of a pressurized vessel is not isometric. Hence, as the smooth muscle cells within the vascular wall shorten, reducing lumen diameter, the diameter of each cell increases, resulting in significant increases in wall thickness. Furthermore, the increase in cell diameter translates into an increase in the length of the vessel segment and an infolding of the internal elastic elamina. The combination of infolding and lengthening affects the endothelium, so that each endothelial cell must be buckled in a circumferential, and stretched in an axial, direction. Likewise, the extracellular matrix must undergo a rearrangement secondary to smooth muscle contraction. Unfortunately, without a precise understanding of matrix pattern and composition, it is impossible to predict just how force is redistributed within the wall of a vessel whose smooth muscle is in a state of partial activation and shortening (tone).

In a wire-mounted vessel segment, on the other hand, smooth muscle cells can produce force but cannot shorten (isometric response), and all the spatial changes described above, which occur in vivo, are not possible.

Furthermore, the forces that impinge upon a vessel are different as well, both initially (immediately following stretch or pressure change) and with time, since the vascular wall of a wire-mounted artery experiences only circumferential tension, not the radial compression and circumferential and longitudinal tension that are generated by a true transmural pressure.

Up to this point, we have discussed the appearance of myogenic tone only, not myogenic reactivity. To demonstrate the latter, the stretch on a wire-mounted vessel or the pressure inside a cannulated segment must be changed. In many studies with wire-mounted vessels, the phenomenon under study is the effect of some variable on the induction of tone, not the modulation of pre-existing tone.

Conversely, in most experiments with pressurized segments it is myogenic reactiviy, rather than the appearance of myogenic tone, that is often the subject of study. Changes in lumen diameter are measured as a function of transmural pressure, and may be expressed in absolute terms (μm or mm), or relative to the diameter at some specific pressure or state (% change per mmHg, % of maximal constriction, % of fully relaxed diameter, and so on). The pattern of response is opposite for increases vs decreases in pressure (constriciton vs dilation) and, at low transmural pressures, the artery may completely deactivate and behave passively. If transmural pressure is raised beyond a certain level, the force produced by the smooth muscle may be overcome, and the vessel will dilate out to a passive diameter (forced dilatation).

In summary, myogenic tone may be present in wire-mounted arteries or veins, although it is more common in pressurized vessel segments. Myogenic reactivity involves a modulation of the basal tone as a function of transmural pressure or stretch. If the pattern of response is such that arterial diameter decreases with increasing pressure, the potential for flow autoregulation is present.

Cellular Mechanisms
Underlying Myogenic Tone and Reactivity

Calcium

Myogenic tone, be it observed on wires or in a pressurized vessel, is dependent on the presence of extracellular calcium, and its development is associated with a measurable increase in calcium influx.[17] Incubation in a calcium-free solution or the administration of calcium-entry blockers, such as diltiazem,[16,18,30] most often leads to a complete loss of basal tone, although Nakayama demonstrated that contraction of the rabbit basilar artery produced by quick stretch depends not only on transmembrane supply of calcium, but also on the release of calcium from the inner surface of the plasma membrane.[31]

The mechanisms by which stretch is coupled to calcium influx are not known, but may involve calcium entry through a population of stretch-operated channels in the smooth-muscle-cell membrane, as suggested recently by Davis and Burch, thus far in abstract form only.[32] Bevan and colleagues made a convincing case for the existence of a distinct subpopulation of calcium channels operated by stretch.[16,18] Their conclusions were based on pharmacological interventions that distinguished stretch-induced tone from that attributable to receptor stimulation or membrane depolarization by potassium, such as sensitivity to calcium-channel blocking agents and the absence of a stretch-induced mechanism in vessels that develop other forms of tone. (Note: Additional comments about the relationship between calcium and myogenic tone may be found in the chapter by Laher and van Breemen, in this volume.)

Membrane Potential

An association between electrical activity and myogenic tone was described in two early studies which found an increased frequency of propagated electrical activity during the stretch response in subcutaneous arteries and in the portal vein.[33,34] Subsequent observations support a correlation between tone and membrane potential, with a good correlation between force or constriction and membrane depolarization reported in wire-mounted[35] and pressurized [36,37] vessels. The correlation is especially convincing for phasic activity, although the magnitude of rapid change in potential may be more important that the absolute value (note the correlation between spike activity and force, but not between actual potential value and force, in ref. 38).

Endothelium

Several lines of experimental evidence support the concept that myogenic tone may exist in the absence of an endothelium in vitro and in vivo.[28,39,40] Although the response of a wire-mounted vessel to sudden stretch may or may not be mediated by the same mechanisms as basal constriction in a pressurized segment, basal tone has been observed in both types of preparations in previously denuded vessels.[28,39,41,42] One exception is a study by Katusic et al., in which stretch was found to elicit an increase in force in wire-mounted segments of the dog basilar artery.[19] This response was dependent on calcium influx and could be inhibited by pretreatment with indomethacin or by endothelium removal, leading the authors to conclude that stretch induced the release of an endothelium-derived prostanoid vasoconstrictor. The tone was short-lived, however, disappearing within a few minutes, and its physiological role is uncertain.

A role for the endothelium in myogenic reactivity is more difficult to rule out. First, it is clear that the endothelium has the potential to at least influence basal tone through the release of any of a number of vasoactive substances, some of which have not been identified. Second, this versatile tissue is known to respond to physical forces directly by changing membrane ionic conductance[43] and secretory activities. For example, the release of prostacyclin increased significantly when the pattern of transmural pressure changed from static to pulsatile.[44,45] Shear stress augmented this response further.[45] Shear stress is also known to modulate the release of endothelin, a potent endothelial vasoconstrictor, in an inverse manner (increased shear stress → decreased endothelin release).[46] Thus, it would appear likely that, at least in vivo, some influence of the endothelium is superimposed on the basal tone of vascular smooth muscle.

It is worth noting that, in vivo, damaging the endothelium of mouse cerebral arterioles by a light-dye method did not affect the ambient diameter of small mouse pial arterioles, which normally operate in a state of partial constriction.[40] Although this observation supports the concept of a smooth-muscle tone that is independent of the endothelium, it does not definitively rule out an influence of the endothelium on myogenic reactivity, since the lack of diameter change could be a result of the removal of a balance of vasoconstrictor and vasodilator endothelial substances; a balance that may be shifted by a change in transmural pressure (Rosenblum, personal communication).

In isolated, pressurized vessels, the presence of myogenic tone in denuded arteries has been reported in several studies on arteries from the cerebral and coronary circulations of the rat, in which endothelium removal produced a slight but significant increase in the degree of basal constriction. When transmural pressure was changed, myogenic reactivity was preserved, although the pressure–diameter curve was shifted to smaller diameters in a parallel fashion.[28,41]

Conversely, removal of the endothelium from cerebral arteries of the cat by perfusion with a solution of collagenase and elastase completely abolished their responsiveness to changes in transmural pressure, leading Harder to conclude that this response is endothelium-dependent.[47] Furthermore, measurements of membrane potential showed that pressure-induced smooth-muscle-cell depolarization was lost following removal of endothelium, and that constriction was associated with the release of a constrictor substance into the lumen as well.[47,48] By perfusing two vessels in series (cerebral–cerebral[48] or carotid–coronary[49]), constriction could be observed in a denuded distal artery and was attributed to a pressure-induced release of a vasoconstrictor, as yet unidentified. Considering the evidence for the preservation of myogenic tone, however, it is somewhat puzzling that vessel diameters after endothelium removal were identical to those found when verapamil was used suggesting either a complete loss of tone, or an insensitivity to calcium-entry blockade.[47]

In our laboratory, perfusion with a dilute detergent (CHAPS) was found to be an effective tool for endothelium removal in small pressurized cerebral and mesenteric arteries.[42,50] After denuding rat cerebral arteries, there was a small but significant increase in basal tone, with well-preserved diameter responses to such constrictor or dilator agents as $PGF_{2\alpha}$ and papaverine. As in the aforementioned study,[47] the constriction to an increase in transmural pressure observed prior to denuding the arteries was lost.[42]

Endothelium removal was verified both functionally (loss of ACh dilation) and by electron microscopy. In transmission micrographs, the cell layer was completely removed and the smooth-muscle cells in the media appeared normal, with the exception of an occasional cell next to the internal elastic lamina. When additional vessels were examined, it became apparent that this damage was localized to and associated with areas of contact between the endothelium and smooth muscle. Most often, myoendothelial junctions were formed by the protrusion of an endothelial foot process through the internal elastic lamina and in close apposition to the cell membrane of an underlying smooth-muscle cell. The structural and permeability properties of this type of junction are not known, but removing endothelium using a detergent was clearly destroying not only the endothelium, but some portion of an underlying smooth muscle as well (the assumption of limited damage was based on the observation that constrictor responses to PGF_{2a} were not affected to any degree).

In an attempt to circumvent this problem, we developed a new technique to remove the endothelium mechanically in small arteries by using a human hair.[51] The benefit of a mechanical method is that it takes advantage of the inherent polarity of these junctions and of the relatively thick internal elastic lamina, which acts as a physical buffer zone between endothelium and smooth muscle. Using this technique, we found that removal of endothelium did not abolish the reflex constriction to an increase in transmural pressure,[52] in agreement with the findings of Kuo et al. in small coronary vessels from the rat.[28]

Mechanisms of Transduction

Possibly the most interesting question of all remains unresolved; that is, what are the cellular transduction mechanisms through which physical forces affect the production of a balanced counterforce by vascular smooth muscle?

One possibility is that membrane deformation leads directly to a deformation of channel proteins, thereby altering ion flux and producing calcium influx either directly or through membrane depolarization, which is known to increase calcium-channel activity. This view is supported by findings in isolated vascular smooth-muscle cells[32] and, in this case, the mechanism can be termed myogenic in the true sense of the word.

Alternatively, changes in the physical forces experienced by a cell may be transduced by mechanisms that involve another physical "sensor," such as the extracellular matrix or cytoskeleton, whose deformation induces the release of a chemical second messenger. In this regard, studies with cultured vascular smooth muscle from the pig coronary circulation[53] and with other cell types have demonstrated rapid changes in adenylate cyclase induced by mechanical deformation, such as stretch or hyposmotic swelling.

In comparison with the myriad signals that a cell is able to recognize at the membrane level, relatively few mechanisms have been identified by which information is conveyed into the interior of a cell. The principal second-messenger pathways are those involving cAMP or the phosphoinositide–diacylglycerol pathway that leads to the production of inositol triphosphate and the induction of protein kinase C (PKC). Pharmacological activation of the PKC pathway in segments of the rabbit facial vein or basilar artery selectively potentiated stretch-induced tone, possibly by increasing intracellular sensitivity to calcium.[55,56]

Very recently, we found that small cerebral arteries possessing basal myogenic tone in vitro could be almost fully constricted or relaxed by exposure to pharmacological agents that activate (indolactam) or inhibit (staurosporine, calphostin C) PKC activity in smooth muscle, respectively.[57]

Although the results of these in vitro studies await confirmation in vivo, the available evidence does suggest that vascular smooth-muscle cells can respond to physical forces directly by altering membrane ionic conductance[32] and intracellular enzyme activation.[53–57] The actual structure(s) that are involved in sensing and transducing physical forces into chemical messages, however, have remained curiously elusive, in spite of a fairly steady effort on the part of many investigators for nearly a century.

References

[1] Bayliss, W. M. (1902) *J. Physiol.* **28,** 220–231.

[2] Garfield, E. (1976) *Curr. Contents* **19,** 5.

[3] Folkow, F. (1964) *Circ. Res.* **15(Suppl. I),** 279–287.

[4] Koch, A. R. (1964) *Circ. Res.* **14(Suppl. I),** 269–277.

[5] Borgstrom, P., Grande, P. O., and Mellander, S. (1982) *Acta Physiol. Scand.* **116,** 363–376.

[6] Lush, D. L. and Fray, J. C. S. (1984) *Am. J. Physiol.* **247,** R89–R99.

[7] Johnson, P. C. (1980) *Handbook of Physiology, Section 2: The Cardiovascular System, vol. 2: Vascular Smooth Muscle,* (Bohr, D. F., Somlyo, A. P., and Sparks, H. V. J. Jr., eds.), American Physiological Society, Bethesda, MD, pp. 409–442.

[8] Johnson, P. C. (1986) *Circ. Res.* **59,** 483–495.

[9] Heistad, D. D. and Kontos, H. A. (1983) *Handbook of Physiology, Section 2: The Cardiovascular System, vol. 3, Peripheral Circulation and Organ Blood Flow, Part 2* (Shepherd, J. T. and Abboud, F. M., eds.), American Physiological Society, Bethesda, MD, pp. 137–182.

[10] Bohlen, H. G. and Gore, R. W. (1977) *Microvasc. Res.* **14,** 251–264.

[11] Johansson, B. and Mellander, S. (1975) *Circ. Res.* **36,** 76–83.

[12] Laher, I. and Bevan, J. A. (1989) *J. Hypertens.* **7(Suppl. 4),** S17–S20.

[13] Johansson, B. (1989) *J. Hypertens.* **(Suppl. 4),** S5–S8.

[14] Sparks, H. V. (1964) *Circ. Res.* **14 (Suppl. I),** 254–260.

[15] Nakayama, K. (1982) *Am. J. Physiol.* **242,** H760–H768.

[16] Bevan, J. A. (1982) *Am. J. Cardiol.* **49,** 519–524.

[17] Laher, I., van Breemen, C., and Bevan, J. A. (1988) *Circ. Res.* **63,** 669–672.

[18] Bevan, J. A. and Hwa, J. J. (1985) *Ann. Biomed. Eng.* **13,** 281–286.

[19] Katusic, Z. S., Shepherd, J. T., and Vanhoutte, P. M. (1987) *Am. J. Physiol.* **252,** H671–H673.

[20] Uchida, E., Bohr, D. F., and Hoobler, S. W. (1967) *Circ. Res.* **21,** 525–536.

[21] Duling, B. R., Gore, R. W., Dacey, R. G., and Damon, D. N. (1981) *Am. J. Physiol.* **241,** H108–H116.

[22] Davignon, J., Lorenz, R. R., and Shepherd, J. T. (1963) *Am. J. Physiol.* **209,** 51–59.

[23] Halpern, W., Osol, G., and Coy, G. S. (1984) *Ann. Biomed. Eng.* **12,** 463–479.

[24] Harder, D. R., Smeda, J., and Lombard, J. (1985) *Circ. Res.* **57**, 319–322.

[25] Uchida, E. and Bohr, D. F. (1969) *Circ. Res.* **25**, 549–555.

[26] Osol, G. and Halpern, W. (1985) *Am. J. Physiol.* **249**, H914–H921.

[27] Vinall, P. E. and Simeone, F. A. (1981) *Stroke* **12**, 640–642.

[28] Kuo, L., Chilian, W. M., and Davis, M. J. (1990) *Circ. Res.* **66**, 860–866.

[29] Osol, G. and Halpern, W. (1988) *Am. J. Physiol.* **254**, H28–H33.

[30] Osol, G., Osol, R., and Halpern, W. (1986) in *Essential Hypertension: Calcium Mechanisms and Treatment* (Aoki, K., ed.), Springer-Verlag, Tokyo, pp. 107–114.

[31] Nakayama, K., Suzuki, S., and Sugi, H. (1986) *Jpn. J. Pysiol.* **36**, 745–760.

[32] Davis, M. J. and Burch, L. M (1989) *FASEB J.* **3**, A1383.

[33] Johansson, B. and Bohr, D. F. (1966) *Am. J. Physiol.* **210**, 801–806.

[34] Holman, M. E., Dasby, C. B., Suthers, M. B., and Wilson, J. A. F. (1968) *J. Physiol. (Lond.)* **196**, 111–132.

[35] Nelson, M. T., Standen, N. B., Brayden, J. B., and Worley, J. F. (1988) *Nature* **336**, 382–385.

[36] Smeda, J. S., Lombard, J. H., Madden, J. A., and Harder, D. R. (1987) *Pflugers Arch.* **408**, 239–242.

[37] Harder, D. R. (1984) *Circ. Res.* **55**, 197–202.

[38] Harder, D. R., Brann, L., and Halpern, W. (1983) *Blood Vessels* **20**, 154–160.

[39] Hwa, J. J. and Bevan, J. A. (1986) *Am. J. Phusiol.* **250**, H87–H95.

[40] Rosenblum, W. I. and Nelson, G. H. (1988) *Stroke* **19**, 1379–1382.

[41] McCarron, G., Osol, R., and Halpern, W. (1989) *Blood Vessels* **26**, 315–319.

[42] Osol, G., Osol, R., and Halpern, W. (1988) *Resistance Arteries* (Halpern, W., Pegram, B. L., Brayden, J. E., Mackey, K., McLaughlin, M. K., and Osol, G., eds.), Perinatology, Ithaca, NY, pp. 162–169.

[43] Olesen, S.-P., Clapham, D. E., and Davies, P. F. (1988) *Nature* **311**, 168–170.

[44] Pohl, U., Forstermann, U., Busse, R., and Bassenge, E. (1985) *Prostaglandins and Other Eicosanoids in the Cardiovascular System. Proc. 2nd Int. Symp.* (Schor, ed.), Nurnberg-Furth, Karger, Basel, pp. 553–558.

[45] Frangos, J. A., Eskin, S. G., McIntire, L. V., and Ives, C. L. (1985) *Science* **227**, 1477,1478.

[46] Yanagisawa, M., Kurihara, H., Kimura, S., Tomobe, Y., Kobayashi, M., Mitsui, Y., Yazaki, Y., Goto, K., and Masaki, T. (1988) *Nature* **332**, 411–415.

[47] Harder, D. R. (1987) *Circ. Res.* **60**, 102–107.

[48] Kauser, K., Stekiel, W. J., Ruganyi, G., and Harder, D. R. (1989) *Circ. Res.* **65**, 199–204.

[49] Rubanyi, G. M. (1988) *Am. J. Physiol.* **255**, H783–H788.

[50] Tesfamariam, B., Halpern, W., and Osol, G. (1985) *Blood Vessels* **22,** 301–305.

[51] Osol, G., Cipolla, M., and Knutson, S. (1989) *Blood Vessels* **26,** 320–324.

[52] Osol, G., Cipolla, M., and Osol, R. (1990) *Blood Vessels* **27,** 51.

[53] Mills, I., Letsou, G., Sumpio, B., and Gewirtz, H. (1989) *Circulation* **80 (Suppl. II),** II-197.

[54] Watson, P. (1990) *J. Biol. Chem.* **265,** 6569–6575.

[55] Laher, I. and Bevan, J. A. (1987) *J. Pharmacol. Exp. Ther.* **242,** 566–572.

[56] Laher, I., Vorkapic, P., Dowd, A., and Bevan, J. A. (1989) *Biochem. Biophys. Res. Commun.* **165,** 312–318.

[57] Osol, G., Laher, I., and Cipolla, M. (1991) *Circ. Res.* **68,** 359–367.

Chapter 10

The Myogenic Response

In Vivo Studies

Paul C. Johnson

Historical Aspects

Much of the early evidence for the myogenic response was obtained from in vivo experiments on whole organs. Bayliss' classic studies of the myogenic response[1] included both in vivo and in vitro preparations, but data were presented only on in vivo measurements of organ volume. These measurements indicated that intravascular pressure reduction caused vasodilation. However, these studies provided only indirect evidence for a myogenic mechanism in the blood vessels, and later investigators pointed out that alternative explanations were at least as plausible as a myogenic response.

Relatively little work was done on the question of the myogenic response for several decades, until Folkow[2] showed in pressure-flow studies that vascular responses to changes in intravascular pressure could not be explained simply on the basis of neural reflexes or release of vasodilator metabolites. Many other whole-

From *The Resistance Vasculature*, J. A. Bevan et al., eds. ©1991 Humana Press

organ pressure-flow studies followed, on kidney, brain, intestine, and other organs.[3] These also indicated that a myogenic mechanism provided a suitable explanation for autoregulatory adjustments to arterial pressure changes. These studies involved calculation of vascular resistance, and for a time it was debated whether the resistance change was on the arterial or the venous side; this was resolved by a number of techniques, including the use of an indirect method of capillary-pressure measurement that showed conclusively that, except in unusual circumstances, the resistance change occurred on the arterial side of the vascular bed.[4] Thus the necessary participation of the "resistance vessels" was obvious. Further delineation of the response of large and small arteries and the various orders of arterioles to changes in intravascular pressure using fine catheters inserted retrograde into the small arteries[5] showed that vessels larger than 1 mm id appeared to provide a significant fraction of total vascular resistance. But the relative importance of these vessels, compared to the arterioles, in flow regulation appears to depend importantly on the vascular bed. In brain and certain other organs, including the kidney, a substantial fraction of total resistance is found in the arteries.[6,7] Since these organs also show very pronounced autoregulation, there is prima facie evidence that the arteries in these organs are resistance vessels and contribute importantly to local regulation. Interestingly, the myogenic response has long been implicated by whole-organ studies in the strong autoregulation in brain and kidney.[8]

Microcirculatory Studies

A limitation of whole-organ studies is that they cannot provide information on the response of the different generations of the arterial network to pressure changes. In vivo microscopy has provided a more direct assessment of the role of the various segments of the arterial network in the myogenic response. As noted by Renkin,[9] microcirculatory studies have often focused on the terminal portion of the arteriolar network at the expense of the larger arterioles and smaller arteries. There are practical reasons for this, since the larger vessels are often less accessible for viewing in the microcirculatory preparations commonly employed. Thus, the relative paucity of data on this question from such studies does not necessarily indicate that the larger vessels are less important.

Despite this limitation, in recent years, information has been obtained from microcirculatory studies on the large arterioles and small arteries. For example, in the rat cremaster muscle, Morff and Granger[10] studied responses of the large 1A arterioles (average diameter 80–90 μm) to arterial and venous pressure elevation and compared their behavior to that of 2A and 3A arterioles (average diameters 45–50 μm and 20–25 μm, respectively). Only the 3A vessels showed behavior consistent with the myogenic hypothesis under all circumstances. The 1A vessels generally showed only a passive narrowing to arterial pressure reduction and no significant change in arteriolar diameter with venous pressure elevation. However, recent studies, described in this chapter, have shown that under certain circumstances the myogenic response can be quite prominent in these vessels. Studies in the isolated cat mesentery by Gannon et al.[11] showed that the largest arterioles studied (which were slightly larger than 50 μm) dilated about 15% with arterial pressure reduction from 95 to 56 mmHg. There appeared to be a slightly greater increase in the percentage of dilation in the smaller arterioles with the same reduction in arterial pressure, but the difference was only marginally significant ($p < 0.10$). Evidence from a variety of studies suggests that this response is primarily myogenic. Studies on the perfused rabbit ear, reviewed later in this chapter provide evidence that arterial vessels of 70–1000 μm id may exhibit a myogenic response when flow-dependent vasodilation is abolished.

Coordination of Vascular Responses

The response of an arteriole to alteration of arterial or venous pressure will depend, to a degree, on the prevailing intravascular pressure in that vessel, which of course will differ in different regions of the network. The response will also depend on the proximity of the vessel to the site of pressure change. Thus, an elevation of venous pressure would be expected to produce a much greater rise in pressure in the distal arterioles than in the proximal vessels. Despite this, Burrows and Johnson[12] found no difference in magnitude of arteriolar constriction among mesenteric vessels experiencing wide differences in magnitude of pressure change caused by a step rise in venous pressure. Presumably the differences in pressure change reflected differences in network location. They deduced from this observation that the response of arterioles was not only attibutable to the local pressure change, but also was very likely influenced by conducted responses from other areas of the network. Since all the vessels they studied were localized to the central region of the mesenteric sector, such conduction could play a significant role.

This idea is not entirely new; it was proposed in 1964 by Folkow[13] in a somewhat different form, when he suggested that periodic vasomotion in terminal arterioles would be conducted upstream to larger vessels, which would then sum input from many daughter branches. Folkow based his hypothesis on evidence of conduction in arteries, but studies by Intaglietta and coworkers[14] have shown that vasomotion can be conducted from branching points in the arteriolar network to upstream vessels in an apparently decremental fashion. Moreover, recent studies by Segal, Damon, and Duling[15] provide strong evidence for conducted responses of both vasoconstrictor and vasodilator activity in the arteriolar network. The length constant for the conducted response to locally-applied vasoactive agents appears to be 1.5–2.0 mm and thus still allows for differential behavior within the arteriolar network over greater distances.

Series–Coupled Independent Effectors

It has been hypothesized that the myogenic response would cause a different pattern of dilation in the large arterioles, compared to the small arterioles, during arterial pressure reduction. According to the series-coupled, independent-effector hypothesis, the larger proximal arterioles would be expected to show a greater proportional change in diameter than the smaller distal vessels when arterial pressure is reduced modestly,[8] because the dilation of the proximal arterioles would tend to restore pressure in the distal arterioles. Aukland[16] has developed experimental evidence as well as a theoretical model from studies on the kidney and has proposed that the concept fits the behavior of the renal arterial vessels during arterial pressure reduction. According to this concept, the arteries, which have significant resistance in the kidney, play an important role in autoregulation through the myogenic mechanism. The degree of differential behavior that might be expected would obviously depend on the contribution of the conducted responses described above, which would tend to attenuate and mask local tendencies toward differences in response. Studies in the bat wing do not support this hypothesis, perhaps because of conducted responses and also because the gain is low in the larger arterioles, allowing pressure changes to be reflected to the smaller arterioles.[17]

Influence of Neural Factors

The magnitude of the myogenic response in the arterioles and in arteries may be importantly influenced by neural and hormonal factors. Meininger and Faber[18] have shown that denervated 1A arterioles in the rat cremaster muscle (120 μm id) are not myogenically active in vivo under normal circumstances, but constrict in response to transmural pressure rise when norepinephrine (3×10^{-8} to 1×10^{-5} M) is added to the superfusion bath. This observation has important implications for homeostasis, since it indicates that the myogenic response is probably modulated in vivo as sympathetic activity changes. The elevated vascular tone induced during alpha-2- and, to a lesser degree, alpha-1-mediated contraction can be partially reversed when arterial pressure is lowered.[19] This enhancement of the myogenic response may also cause the larger vessels to play a more important role in flow regulation when sympathetic activity is elevated, since they will now respond actively, rather than passively, to changes in arterial pressure. The importance of this modification in the response of the larger arterioles depends on the contribution of these vessels to total vascular resistance and on whether or not the more distal arterioles are undergoing a similar enhancement of the myogenic response.

Recent studies in our own laboratory provide further support for sympathetic enhancement of the myogenic response. In the exteriorized cat sartorius muscle, autoregulation was weak under normal conditions but it was significantly enhanced during 8-Hz sympathetic-nerve stimulation; in fact, flow actually increased during pressure reduction.[20] Subsequent study has shown that this superregulation of flow is caused by enhanced dilation of all the arterioles in the network during arterial presssure reduction. Under normal conditions first-order arterioles, which have an average diameter of about 80 μm in this muscle, dilated by only 4–5% with arterial pressure reduction from 110 mmHg to 80, 60, and 40 mmHg, which was not statistically significant. During sympathetic-nerve stimulation (8 Hz) these arterioles constricted by about 25%, when arterial pressure was reduced to 80, 60, and 40 mmHg they dilated by 11–22%. The autoregulation found in this muscle with normal vascular tone is abolished when an oxygen-rich suffusate is applied over the muscle, suggesting that a metabolically linked or tissue–oxygen-dependent mechanism is involved.[21] However, the enhanced autoregulation during sympathetic–nerve stimulation appears not to be abolished by an oxygen-rich suffusate (Ping and Johnson, unpublished observations), which favors the conclusion that superregulation is attributable to a myogenic response.

Influence of Flow–Dependent Vasodilation

As reported elsewhere in this volume (Chapter 11), the larger arterioles and small arteries possess a potent flow–dependent vasodilator mechanism. Smiesko, Lang, and Johnson[22] found that this extended down to the arteriolar level; they reported an average 65% dilation in arcading arterioles in rat mesentery when flow was greatly increased. The dilation appeared to be graded according to the magnitude of the flow increase, and the mechanism was capable of inducing maximal dilation. According to other studies, this dilation is a result of increased wall shear stress.[23] This mechanism might interact with a myogenic mechanism during autoregulation of blood flow in vivo. Griffith and Edwards[24] found that in the presence of hemoglobin or N–monomethyl–L–arginine (both of which inhibit EDRF production) flow-related vasoconstriction was enhanced in the rabbit ear perfused with a crystalloid solution. Under the conditions of the experiment, this constriction could be either flow- or pressure-mediated. One explanation of their findings is that EDRF opposes a myogenic response in this vascular bed.

Influence of Oxygen and Metabolic Factors

Alteration of intravascular pressure will alter vascular tone by the myogenic mechanism, and this will in turn change blood flow, perivascular, and tissue oxygen levels. Thus the change in vascular tone seen in vivo with a change in intravascular pressure will reflect changes in oxygen tension as well as in pressure. Meininger et al.[25] found that elevation of intravascular pressure in the cremaster muscle by the box technique led to a fall in blood flow, perivascular PO_2, and, presumably, tissue oxygen levels, but the third–order arterioles constricted nonetheless, indicating that myogenic mechanisms are dominant over oxygen-dependent mechanisms in these vessels. First- and second-order arterioles dilated slightly with the pressure elevation, indicating that any myogenic response present in these vessels was not sufficient to prevent passive dilation in the presence of a fall in oxygen tension. There is other evidence that a reduction in tissue oxygen levels in skeletal muscle attenuates the myogenic response. Morff and Granger[10] found that the constriction in third-order arterioles induced by venous pressure elevation was enhanced when an oxygen-rich suffusate was used over the cremaster muscle. This suggests either an effect of oxygen directly on the vascular smooth muscle or an indirect effect, through altered tissue metabolite levels, on the myogenic response. Studies by Lang and Johnson[26] showed that the dilation to pressure reduction in cat mesentery was unaffected by elevated ambient oxygen, which supports other evidence, from the same vascular bed, that the response is the result of myogenic, rather than metabolic, factors. Since metabolic activity is low in mesentery, this observation is consistent with the idea that the effect of oxygen on the myogenic response is through an effect on tissue metabolism.

Interactions Among Control Mechanisms

Flow-dependent dilation, O_2-dependent vasoconstriction, and the pressure-dependent myogenic response probably interact under normal circumstances, and the interaction may become more important during physiological perturbations. For example, during periods of increased functional demand, the distal arterioles would dilate as a result of increased metabolic demand, and the proximal arterioles and small arteries would dilate because of the increased flow. At the same time, pressure changes in the arterioles would occur and be likely to induce a change in vascular caliber through a myogenic mechanism, except for the fact that both flow-dependent dilation and increased metabolism would weaken the myogenic response. However, if sympathetic activity were increased at the same time, as is often the case in muscular exercise,[27] the myogenic response could be enhanced, rather than diminished. Thus, a complex pattern of response to changes in flow, intravascular pressure, neural activity, and vasoactive metabolites must take place in the arterioles and small arteries during functional hyperemia. Similarly, during other disturbances, such as hemorrhagic hypotension and shock, it can be expected that changes in flow, neural activity, and elevated blood levels of epinephrine and hormones influence the myogenic response of the arterioles to the change in intravascular pressure.

Summary

In vivo studies provide direct evidence that the myogenic response has a widespread distribution in the arterial network, especially in the various orders of the arterioles, but also extending to vessels commonly considered to be small arteries. The evidence suggests that the response is less prominent in the larger vessels. The constriction induced by the myogenic response may be conducted from vessel to vessel, amplifying the effect in vivo over that which might be seen in an isolated vessel. The myogenic response may be enhanced by catecholamines and by sympathetic-nerve activity. It is evident that a complex response of the arterioles may occur during physiological adjustments, such as functional dilation, hemorrhage, and other circumstances in which there are simultaneous changes in intravascular pressure, oxygen delivery, oxygen consumption, and sympathetic activity, as well as in circulating hormones and autacoids that have vascular effects.

These considerations point to the importance of in vivo studies in understanding the integration of the myogenic response with other control mechanisms. They also underscore the difficulty of determining the native behavior of the myogenic response from in vivo studies alone. From this standpoint both in vitro and in vivo studies may be expected to make essential contributions to our understanding of the mechanism of the myogenic response and its role in circulatory homeostasis.

Acknowledgments

The author gratefully acknowledges the aid of Peipei Ping in the preparation of this chapter. Original work referred to in this chapter was supported by a Grant-in-Aid FG 2–5–88 to Paul C. Johnson from the American Heart Association, Arizona Affiliate, and Grant HL AM 15390 from the National Institutes of Health.

References

[1] Bayliss, W. M. (1902) *J. Physiol. (Lond.)* **28**, 220–231.

[2] Folkow, B. (1949) *Acta Physiol. Scand.* **17**, 289–310.

[3] Johnson, P. C. (1964) *Circ. Res.* **14–15(Suppl. 1)**, 1291.

[4] Johnson, P. C. and Hanson, K. M. (1962) *J. Appl. Physiol.* **17**, 503–508.

[5] Haddy, F. J., Fleishman, M., and Emanuel, D. A. (1957) *Circ. Res.* **5**, 247–251.

[6] Lassen, N. (1978) *Peripheral Circulation*, vol. 12 (Johnson, P. C., ed.), Wiley, New York, pp. 337–358.

[7] Tonder, K. and Aukland, K. (1979) *Renal Physiol.* **2**, 214–221.

[8] Johnson, P. C. (1980) *Handbook of Physiology, Section 2: The Cardiovascular System, vol.2: Vascular Smooth Muscle* (Bohr, D. F., Somlyo, A. P., and Sparks, H. V. Jr., eds.), American Physiological Society, Bethesda, MD, pp. 409–442.

[9] Renkin, E (1984) *Handbook of Physiology, Section 2: The Cardiovascular System, vol. 4: The Microcirculation* (Renkin, E. M. and Michel, C. C., eds.), American Physiological Society, Bethesda, MD, pp 627–688.

[10] Morff, R. J. and Granger, H. J. (1982) *Circ. Res.* **51**, 43–55.

[11] Gannon, B. J., Rosenberger, S. M., Versluis, T. D., and Johnson, P. C. (1983) *Microvasc. Res.* **26**, 1–14.

[12] Burrows, M. E. and Johnson, P. C. (1983) *Am. J. Physiol.* **245**, H796–H807.

[13] Folkow, B. (1964) *Circulation Res.* **15**(Suppl. 1), 279–287.

[14] Colantuoni, A., Beruglia, S., and Intaglietta, M. (1984) *Am. J. Physiol.* **246**, H508–H517.

[15] Segal, S., Damon, D. S., and Duling, B. R.(1989) *Am. J. Physiol.* **256**, H832–H837.

[16] Aukland, K. (1989) *J. Hypertension* **7** (Suppl. 4), S71–S76.

[17] Davis, M. J. and Pinkston, P. M.(1987) *Fed. Proc.* **46**, 152A (abstract).

[18] Meininger, G. A. and Faber, J. E. (1989) *FASEB J.* **3**, A257 (abstract).

[19] Faber, J. E. and Meininger, G. A. (1989) *FASEB J.* **3**, A258 (abstract)

[20] Ping, P. and Johnson, P. C. (1989) *FASEB J.* **3**, A270 (abstract).

[21] Sullivan, S. M. and Johnson, P. C. (1981) *Am. J. Physiol.* **241**, H804–H815.

[22] Smiesko, V., Lang, D. J., and Johnson, P. C. (1989) *Am. J. Physiol.* **257**, *(Heart Circ. Physiol.* **26**), H1958–H1965.

[23] Melkumyants, A. M., Balashov, T. A., Smiesko, V., and Khayutin, V. M. (1986) *Bull. Exp. Biol. Med.* **101**, 568–570.

[24] Griffith, T. M. and Edwards, D. H. (1990) *Am. J. Physiol.* **258** *(Heart Circ. Physiol.* **27**), H1171–H1180.

[25] Meininger, G. A., Mack, C. A., Fehr, K. L., and Bohlen, H. G. (1987) *Circ. Res.* **60**, 861–870.

[26] Lang, D. G. and Johnson, P. C. (1988) *Am. J. Physiol.* **255**, H131–H137.

[27] Mark, A. L., Victor, R. G., Nerhed, C., and Wallin, B. G. (1985) *Circ. Res.* **57**, 461–469.

Chapter 11

Flow-Dependent Vascular Tone

John A. Bevan

Introduction

Driven by the heart, the blood flows along the arterial tree through vessels of low resistance and large diameter to those of high resistance and small diameter. This process influences the arterial wall through two forces, pressure and shear stress. The effects of pressure, which are exerted radially in all directions perpendicular to the artery surface, are to dilate the blood vessel. The forces acting on the blood vessel wall when flow is laminar, owing to the shearing or slipping of the blood, are exerted in its long axis. Most vascular smooth muscle cells are oriented in a low helix around the long axis of the blood vessel. Pressure by stretching these cells longitudinally, and possibly, also by influencing the endothelium initiates the myogenic response, a contraction. This topic is dealt with in Chapters 9 and 10. In this chapter, the response of the resistance artery to flow exerted through changes in shear stress normal to the long axis of the smooth muscle cells is discussed.

From *The Resistance Vasculature*, J. A. Bevan et al., eds. ©1991 Humana Press

When blood is flowing through a vessel, there is a thin fluid layer immediately in contact with the inner surface of the intima, which does not move. However, a layer of fluid in contact with this stationary layer must move in response to the pressure gradient and fluid viscosity. The effect of viscosity is to produce frictional drag between the concentric layers and, in particular, that attached to the vessel lumen. This drag is known as wall shear, which for laminar flow and neuwtonian viscosity is:

$$\tau = \mu j = 4\,\mu Q / \pi r^3$$

When τ is the shear stress, j wall shear rate, μ viscosity, Q flow rate, and r, radius.

In this chapter, two flow-related phenomena, contraction and relaxation, will be described. Provided viscosity does not change (and this is not necessarily a correct assumption, *see* Chapter 5), the increased shear rate associated with an increase in flow would be offset, at least in part, if flow caused an increase in diameter. On the other hand, an increase in shear rate related to flow-increase would only be accentuated if flow caused a decrease in diameter. I propose *(see below)* that these two effects of flow interact and bring the artery wall to an intermediate level of tone. There have been a number of proposals that the drag on the blood vessel wall regulate a number of vascular parameters, including its diameter, not only in the short-term, resulting in changes in active smooth muscle tone, but on a long-term basis leading to local adaptive structural change and thus influencing regulation of vessel size, diameter, length, and branching.[1-3]

Flow–Velocity Relations

Flow velocity measurements in vessels greater than 30 μm obtained by direct observation are summarized in Fig. 1. Most measurements are restricted to smaller translucent blood vessels because they are usually obtained by observation of erythrocyte or platelet movement. Variation in flow rate in vessels with diameters between 30–130 μm varies from 3–30 mm/s, and there is a correlation between flow velocity and diameter ($r = 0.66$; $p < 0.05$). Flow is pulsatile in small arteries and arterioles of this size.[4]

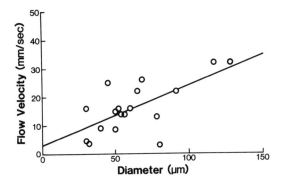

Fig. 1. Relationship between flow velocity and diameter observed in resistance arteries 30–130 μm diameter from a number of vascular beds and species (*see* refs. 4, 45, 51–59).

Intrinsic Tone

A number of studies involving a variety of approaches have established that the smooth muscle cells of most small arteries and arterioles in vivo are in a state of tonic contraction. This state of contraction is related to the topic of this chapter, in that flow-contraction may contribute to it, and because flow-dilation can only become manifest when it is present. In general, the relative level of contraction or intrinsic tone reported to be present in vivo increases as vascular diameter diminishes (*see below*). However, there appear to be many exceptions to this generalization. The strongest quantitative evidence for intrinsic tone comes from the local application in vivo of maximally effective concentrations of vasodilators. In Fig. 2, we have summarized published data on the level of active tone present in the resistance arteries within the diameter range 30–200 μm. The variability of this measurement even in vessels of the same size from the same vascular bed of the same species is considerable (*see* for example, Meininger[5]). The most common response to the application of a vasodilator was dilation, from almost no change to an increase of more than 90%. Occasionally, constriction was observed reflecting presumably passive changes caused by the action of the dilator on connecting vessels. The dominant response of first order arterioles, however, was constric-

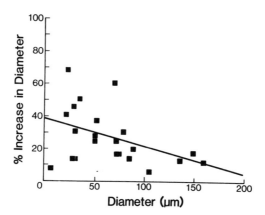

Fig. 2. Relationship between changes in diameter on exposure to maximum dilating concentrations of adenosine and observed initial diameter of resistance arteries (30–300 μm) from a number of vascular beds and species (*see* refs. 5,17,18,58,60–65. $r = 0.42$).

tion. It follows, therefore, that the observed diameter of a small artery in vivo may be very different from the "relaxed" diameter. This latter measurement presumably should be the common denominator for comparison of arterial size.

A number of factors (both constrictor and dilator) may influence the level of small artery tone. These include neural, both constrictor and dilator, circulating and locally produced vasoactive substances, flow, and pressure. There may be causes that are intramurally propagated (*see* for example, ref. 6). The tone resulting from the latter three would be considered intrinsic. One of the theses of this chapter is that intraluminal flow is a significant resistance artery tone-influencing and regulating factor and that it exerts its effect through the interaction of simultaneously elicited constrictor and dilator components.

Evidence For Flow-Dilation

A number of experimental studies (*see* Table 1, pp. 174,175) support the conclusion that changes in intraluminal flow can by a local action reduce the tone of the smooth muscle cells in the blood vessel wall. The observations will be presented in historical sequence. The first demonstration of flow-induced vasodilation was by Schretzenmayer,[7] who observed that hyperemia of the dog hind limb was accompanied by femoral artery dilation. Subsequently, flow-dilation in large arteries has been reported in a number of laboratories and in a variety of vessels from several species. Invariably in these experiments, when the endothelium was rubbed or otherwise destroyed, the flow-related changes in diameter disappeared.

The first observation of flow-dilation in a resistant artery was made by Johnson and Intaglietta[8] on cat mesenteric arcade arteries. These investigators concluded that both pressure and flow contributed to the active wall tone in small arteries, but of the two, pressure was the most important. Griffith et al.[9,10] studied the hydrodynamic properties of the rabbit ear artery network, measuring arterial dimensions with a microradiographic technique. Several orders of arteries dilated with increased whole bed flow. In their analysis, EDRF was implicated as the coordinating factor in the diameter changes. This conclusion was based on the assumption that hemoglobin is a specific inhibitor. They concluded that in this preparation, the tonic release of EDRF inhibited the myogenic response. That the entire effect of hemoglobin can be attributed to its inhibition of EDRF has been questioned.[11,12]

Bevan and Joyce,[13] and Bevan et al.[14] (Fig. 3) demonstrated flow-dilation in isometrically mounted vascular segments of resistance arteries from several vascular beds from the cat and rabbit. Dilation could still be demonstrated after endothelium-removal, but was significantly less. It still occurred in the presence of indomethacin. Methylene blue and hemoglobin modestly reduced it, which raises doubt about a role for EDRF or other endothelium-derived vasoactive factors that seem to have been implicated in the study of flow changes in large arteries.[15,16] Smiesko et al.[17] examined flow-induced dilation in the arcade arteries of the rat mesentery (Fig. 4). The extent of dilation correlated with red cell velocity and volume flow. Although intravascular pressure was not measured, an ingenious experimental design precluded an explanation of the observed dilation based on a fall in intravascular pressure and a reduction in myogenic tone. No attempt was made to determine the cellular origin of the flow-initiated effect. Koller and

Table 1
Experimental Studies in Flow Dilation

Reference	Artery/ diameter µm	Animal, anesthetic	Technique	Pressure, maintained or measured	Endothelium role	Indomethacin-sensitive
Dilation						
Johnson and Intaglietta[8]	Mesentery	Ketamine Cat α-Chlor-alose	Graded Occlusion	No	Not tested	—
Smiesko et al.[45]	Mesenteric arcade, 40–110	Pentobarbital Rat α-Chloralose Urethane	Single and graded paired occlusion	Role of changes excluded	Not tested	—
Griffith et al.[9,10]	Ear artery bed, 100–200	Rabbit —	Perfusion	No	Hemoglobin to to implicate EDRF	—
Bevan, Joyce[13] Bevan, Joyce Wellman[14]	Ear, 200–400 Brain Tongue, 200–400	Rabbit — Cat —	In vitro Isometric	Yes role excluded Yes role excluded	Possible Partial role	No No

				Parallel arterial occlusion		
Koller and Kaley[18]	Rat, Pento-barbital	Cremaster, 21.5 + 0.5		No	Implicated by light dye/endothelial damage	Yes
Garcia-Roldan, Bevan[19]	Rabbit —	Pial, Coronary, 200–600	Perfusion	Controlled	No role	No
Kuo et al.[21]	Pig—	Coronary, 40–80	Perfusion measured	Controlled	Implicated	—
Bognar, Gow[22]	Rat —	Abdominal aorta	Opened segment	Yes role excluded	Possible partial role	No
Constriction						
Bevan, Joyce[23,24]	Rabbit —	Pial, Ear Mesenteric	In vitro Isometric	Yes role excluded	No role	No
Garcia-Roldan Bevan[19,35,36]	Rabbit	Pial, Coronary, 200–600	In vitro Perfusion	Yes role excluded	No role	No
*Sipkema et al.[26]	Rabbit —	Femoral, 1500	In vitro	Yes role excluded	—	—

*Although this chapter is devoted to resistance arteries, this observation on a larger artery has been included because of the sparsity of data available on flow-induced constriction.

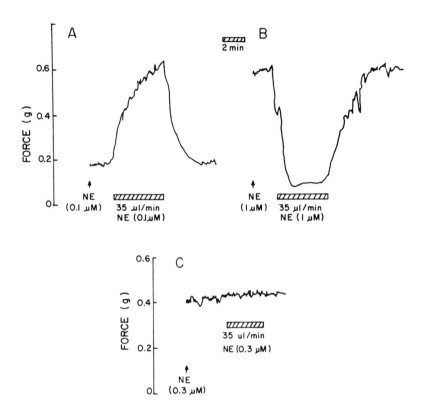

Fig. 3. Responses of paired adjacent isolated segments of rabbit ear resistance arteries (unstretched outer diameter 150 μm; ~200 μ od) mounted in a myograph. Concentration of norepinephrine (NE) in tissuebath is shown below arrows. Period of infusion is indicated by hatchbars and concentration of NE in infusate below hatchbars. A: intraluminal infusion of physiological saline solution (PSS) into artery with low level of tone initiated by NE (bath concentration, 0.1 μM); B: infusion into adjacent vessel with a high level of wall tone owing to PSS containing NE (1 μM); C: infusion into artery with an intermediate level of tone resulting from NE (0.3 μM).[46]

Kaley[18] observed arteries (21.5 ± 0.5 μm od) of the rat cremaster muscle. Utilizing parallel arteriolar occlusion, they showed that the increases in diameter could be correlated with increase in red blood cell velocity, which in turn could be related to volume flow. Presumed specific inactivation of the endothelium by a light dye treatment prevented this flow effect. Most recently flow-dilation

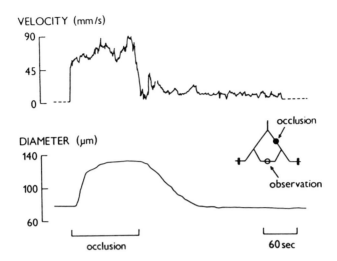

Fig. 4. Effect of occluding a rat mesenteric feed artery on dual slit red cell velocity and internal diameter of arcading arteriole.[45]

has been demonstrated in a perfused resistance artery[19] (Fig. 5) under conditions in which pressure and flow could be precisely controlled and vessel diameter registered automatically with a video system.[20] Subsequent experiments revealed that these effects still occurred after mechanical endothelium inactivation, and that the relationship between flow and diameter was similar whether the endothelium was present or not. Flow-induced inhibition of the myogenic tone occurred when intravascular pressure, and in consequence the myogenic response, was relatively low. Kuo and colleagues[21] demonstrated flow-mediated dilation in coronary arteries cannulated and connected to two independent reservoir systems. Flow was initiated by raising and lowering the two reservoirs by the same increments while maintaining the same pressure gradient. Intravascular pressure was monitored. The greatest magnitude of flow-dilation occurred at 60 cm H_2O and was attenuated at higher and lower levels. The dilation to flow, but not the myogenic response was abolished by endothelium denudation.

Finally, Bognar and Gow[22] demonstrated shear stress-induced dilation using a rotating disk to apply measured shear stress to an opened small arterial segment. An imposed stress

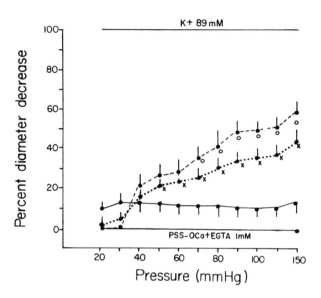

Fig. 5. Relationship between percentage diameter change and pressure of epicardial coronary arteries ($582 \pm 30\,\mu m$ od). Responses are expressed in relationship to diameter differences between O Ca^{2+} plus EGTA (1 mM) and 89 mM K^+. The latter represents approx 90% of tissue maximum. Vessel diameters were measured at 6 mmHg pressure at the beginning of the experiment. Intravascular flow decreased significantly the arterial diameter at low pressure (30 mmHg) and increased significantly the arterial diameter at high pressure (50–130 mmHg), over 70 mmHg the effect is rate-related. X: different from No flow; O: different from 20 μL/min. Flow rate: 100 μL/min (- -); 20 μL/min (••); No flow (—).

of 0.13 N.m^{-2} produced a mean relaxation of 72% of the contraction to norepinephrine (10^{-6}M). The response was not blocked by indomethacin. When the endothelium was removed by gentle rubbing, relaxation still occurred, although attenuated by 50%. The time-course of relaxation was slower than that described following infusion into the segment mounted in a myograph.

Evidence For Flow-Constriction

Flow-induced constriction was first demonstrated in a myograph mounted artery segment.[23,24] When held at optimum length, the contraction is a sigmoid function of flow rate, is relatively slow in developing in comparison with agonist tone, and abruptly reverses on cessation of flow. It is independent of the endothelium and unaffected by blockers of classical surface receptors and by indomethacin and is reversed by a vasodilator , such as papaverine. The contraction does not occur in the presence of Mn^{2+} and in zero calcium physiological saline solution plus EGTA (1 mM). It is observed either in the absence or in association with lower levels of agonist tone. Flow-constriction has been observed on perfusion of the rabbit pial artery *(see above)*[19] (Fig. 5), but only after the artery has myogenically responded to intravascular pressure greater than 50 mmHg. The pressure gradient required to achieve the increased flow through the perfused segment was responsible for only a fraction of the segments total response. There are a number of strong arguments for distinguishing flow from pressure-induced contraction.[25] Sipkema et al.[26] perfused rabbit femoral arteries after gossypol. All vessels constricted to flow, this constriction still occurred after hemoglobin was used to bind EDRF although the mean diameter of the artery was decreased. The effect of endothelium removal was not examined.

Models of Flow Sensing
and Coupling Mechanism
in Resistance Artery (Fig. 6)

All studies of flow dilation in larger arteries that involved destruction of the intima by balloon,[17,27] direct rubbing,[27-29] and H_2O_2[29] showed loss of flow dilation following the procedure. This occurred even if responses to acetylcholine and other agents presumed to cause endothelial-dependent vasodilation were not completely abolished. In several studies, the procedures used to destroy the intima did not change the level of intrinsic tone in the arterial segments being examined.

The hypotheses of flow-induced changes in tone center around the participation or lack of participation of the endothelium in the response. In view of the existence of endothelium-derived relaxing factor (EDRF) and other vasodilator and constrictor factors that

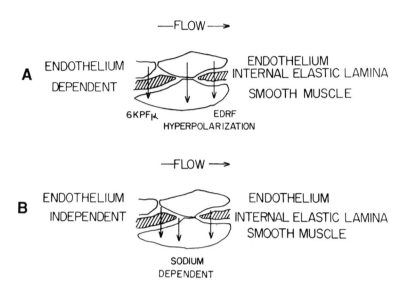

Fig. 6. Diagrammatic representation of two models of flow-induced change of tone in resistance artery. (**A**) Model based on the idea that the mechanoreceptor/transducing system is present in the endothelium. The shear stress resulting from intraluminal flow causes: (i) the release of an endothelial derived vasoactive substance-EDRF, 6-ketoprostaglandin $F_{1\alpha}$ and/or some other substance that diffuses into the tunica media to cause relaxation and/or: (ii) hyperpolarization resulting from increased conductance of K channel in the endothelium. The spreading hyperpolarization possibly through gap junctions results in relaxation of the tone of the vascular smooth muscle cell. (**B**) Model encompassing the idea that the mechanoreceptor and transducing system is in the extracellular matrix in the artery wall and the vascular smooth muscle cell. The shear stress-induced drag on the endothelial layer results in deformation of the subjacent tissues and changes that are sodium dependent.

can be released from the endothelium, it would seem most logical that the cells in contact with blood and that are directly affected by the shear of the flowing blood should participate in the blood vessel response to flow. This hypothesis is consistent with several studies that show that the procedures used to achieve "endothelium-removal" abolished the flow effect. However, there is evidence *(see below)* that flow-induced changes in tone can occur after demonstrated removal of this inner layer of cells. Techniques of endothelium-removal involve manipulations of stretch and wall distortion whose effects are not necessarily restricted to the intimal layer of cells, especially in smaller arteries where the internal elastic lamina is thin and when the subintima is only a few microns thick. Detergents applied to remove the tunica intima have been shown to cause a different functional effect on the blood vessel from carefully applied mechanical methods.[30,31] It would be very difficult to control their diffusion into the wall and avoid effects on the inner smooth muscle cells. If the flow sensors were associated with some structural complexity of the tissues beneath the endothelium, such as the extracellular matrix and/or the basement layer of the vascular smooth muscle cells, then chemical techniques and possible rubbing might injure them. Responsiveness of the smooth muscle cells to directly-acting constrictors and dilators, particularly in high dose and structural observations, do not necessarily provide the most sensitive indicies of the integrity of the entire spectrum of vascular smooth muscle cell function.

Endothelium-Dependent Flow-Dilation (Fig. 6)

The evidence in support of the endothelium being both the mechanosensor and transducer is as follows:

1. In all studies of large arteries in vivo[16,27-29] when this question was explored, there was unanimity that flow-dilation disappeared after rubbing or applying destroying chemicals to the inner surface of the artery wall.
2. Koller and Kaley[18] showed that light/dye treatment that appears to be selective for the endothelium cells in that constrictor and dilator drug-induced responses remained intact, and prevented the observed dilation.
3. Tissue culture studies have shown that shear stress can exert many effects on structure and function of endothelial cells. These include pinocytosis, production of vasoactive substances and metabolites, cellular alignment and changes in deformability (for summary, *see* Davis[32]).
4. Perfusion of dog femoral artery segments at 2 mL/min and higher caused a flow-dependent release of 6 ketoprostaglandin $F_{1\alpha}$ and a substance with a characteristic similar to that released by acetylcholine. Further details of the release mechanism were not studied.[15] That

resistance arteries show an acetylcholine-dependent endothelial dilation was first shown by Owen and Bevan.[33]

5. Patch clamp studies of single arterial endothelium cells exposed to laminar shear stress in a capillary tube, show a K^+-selective shear-stress activated ionic current. This could represent a flow-response coupling mechanism originating in the endothelium, which influences the tone of the underlying vascular smooth muscle cell. The latter step, however, has not been demonstrated.[34] These various studies appear to implicate two possible endothelial-based mechanisms of dilation: (a) the release of endothelial vasoactive chemicals as mediators of the flow effect or; (b) shear stress-induced hyperpolarization of endothelial cells by activation of K^+ channels leading probably by passive spread to direct vascular smooth muscle cell relaxation.

Endothelium–Independent Flow–Dilation (Fig. 6)

The evidence in support of the vascular smooth muscle or its surrounding extracellular matrix being the mechanosensor and transducer is as follows:

1. Flow-dependent dilation in the absence of the endothelium has been seen in three experimental systems.
   (a) In an isometric myograph using a ring mounted segment in which bath solution is infused through an intraluminal pipet. Flow-dilation still occurred after loss of all endothelial cells and loss of acetylcholine dilation.[14]
   (b) Flow over the inner surface of an opened-up vascular segment mounted in a myograph caused dilation before and after endothelium removal (Gow, personal communication). Using a rotating disk to apply measured shear stress, Bognar and Gow[22] found a similar effect.
   (c) Increased flow through isolated vascular segments[35,36] both from the brain and from the heart caused dilation, when intravascular pressure was less than 40–50 mmHg. This occurred after endothelium removal by rubbing and after treatment with 1-NMMA, an inhibitor of the metabolism of nitric oxide. The shear rate in these segments was calculated assuming laminar flow, to be within the physiological range.

2. Experimental flow-dilation is dependent on sodium in the extracellular space. Dilation was diminished when extracellular sodium was lowered by substitution by N-methyl-d-glucamine, choline chloride, or lithium chloride, and when intracellular sodium was raised indirectly. It was selectively blocked by amiloride (ID_{50} $1–3 \times 10^{-5}M$), but not by tetrodotoxin.[43]

Endothelium-Independent Flow-Constriction (Fig. 6)

This response has been observed in resistance arteries from a number of vascular beds using the same three tissue techniques used to demonstrate flow-dilation. The contraction is a function of the rate of flow and is unaltered by endothelium removal by physical methods.[23,24] Flow-induced contraction-like dilation is dependent on the presence of sodium in the extracellular space.[37] There is one paper describing this phenomenon in an elastic artery.[26] Presumably, like endothelial independent flow-induced dilation, the sensor/transducer for this effect must reside in the vascular smooth muscle cells and/or its surrounding extracellular tissue.

These studies seem to implicate a role for tissues subjacent to the endothelial cells—the subintima, internal elastic lamina, extracellular matrix of the media, basement membrane of the vascular smooth muscle cells, and the cell membranes of the cells themselves. The connective tissues surrounding vascular smooth muscle cells have a high capacity for the binding of cations, particularly the polyanionic proteoglycans,[38] which are known to bind large quantities of sodium. The anionic properties are derived from the presence of highly sulfated glycosamine glycan chains that are covalently bonded to a protein core. About 90% of sodium is distributed extracellularly and is presumably bound to such substances. Siegel et al.[39] have distinguished two fractions: one bound to connective tissue and a rapidly exchanging vesicle fraction. The transmembrane flux rate for sodium is high.[40,41] The extracellular sodium pool for this flux must reside to a great extent in these extracellular components of the vascular smooth muscle cells.

Presumably, shear stress of the artery wall could be mechanically communicated to subintimal tissues and, via the extensive extracellular matrix of the artery wall, to the connective tissue proteoglycans surrounding the inner smooth muscle cells. It might be speculated that shear stress might uncoil or distort, or in some way change the configuration of these molecules, thus influencing their anionic binding properties.[42]

The common characteristics of endothelium independence, absence of effects of indomethacin, and above all susceptibility to changes in extracellular sodium, suggest a common sodium dependent flow-sensor for constriction and dilation.[43] Stretch-dependent ion channels have been found in both the endothelium and vascular smooth muscle cells. These channels defined by patch studies are generally cationic selective, but poorly discriminate between biologically active ions. There is only one tissue in which both stretch-dependent myogenic and flow-dependent tone have been studied, a branch of the rabbit ear artery (*see* this chapter and ref. 44). These two effects can be distinguished on the basis of the force necessary for their involvement or activation and in their ionic dependence. In this blood vessel, both the myogenic and the flow response occur after endothelium removal.

Significance

Intrinsic Tone

Undoubtedly, many factors contribute both positively and negatively to the intrinsic/basal tone of resistance arteries, which is an essential feature of the peripheral vascular tree. It seems likely on the basis of in vitro studies that flow is one of these factors. The dilation that follows flow increase in vivo can be as great as the dilation caused by a vasodilator.[45] Flow-contraction has not been identified in vivo, and from what is known at the moment, it will be difficult to distinguish such an effect from other concomitant changes associated with the experimental procedures. It is too speculative to extrapolate quantitatively from the in vitro experiments to the in vivo state. What might be concluded, however, is that the capacity of resistance arteries to dilate or constrict with flow is large and could be responsible for significant changes in diameter.

There are persuasive arguments from in vitro studies that the response to flow is a result of the interaction of constrictor and dilator components[46] (Fig. 7). In vitro studies on a myograph show that flow-dilation is directly related and flow-constriction indirectly related to the active level of wall tone as a result of norepinephrine. These effects overlap in the middle range of wall force. In fact, flow tended to change or alter the level of wall force toward a null

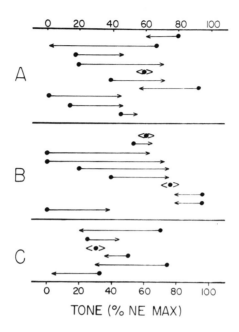

Fig. 7. Change in level of wall tone expressed as percent maximum developed to norepinephrine (NE), with infusion into lumen of isolated segment of rabbit ear resistance artery. Results of three different experiments, A, B, and C are shown. Wall tone level was altered by changing bath concentration of NE. <o>, balance (null) point of these experiments when changes with flow are <5%.[46] • : preflowtone level; →: direction and size of response to flow (35 μL/min).

point, which was characteristic for that particular vessel. At the null point, constrictor and dilator influences are in balance and flow causes little change.[46] The import is that flow independent of pressure might be responsible for the maintenance of intrinsic tone in a small artery.

Little is known about the interactions of flow with tone-influencing moieties other than pressure. Our own experiments suggest that flow-dilation and constriction can override myogenic tone—the exact effect depending on the level of intravascular pressure. At least under isometric in vitro conditions, sympathetic vasoconstriction appears to be additive with flow-induced tone,[47] but there is further complexity. If sympathetic activity were to change smooth muscle tone, this would be responsible for secondary changes in flow and pressure that would in turn influence the final response.

Integration of Tone Across the Vascular Bed

A systematic study has been made of the level of tone in successive branches of an entire vascular bed supplying the rat cremaster muscle. The changes in diameter that follow the local application of adenosine ($10^{-3}M$) were extremely variable in direction and in size for both the larger named vessels and their branches.[5] For example, the mean change in diameter of the external spermatic artery, diameter 136 ± 8 μm, was $12 \pm 4\%$. However, the range of change varied from +10% (constriction) to more than ⁻95% (dilation). These data show that even in the same vascular bed, presumably under standard conditions, arteries of the same diameter exhibit very different levels of tone. Among the many implications of this study is that intrinsic tone is not a simple function of artery diameter alone. The finding of flow-induced changes in small arteries as well as large, raises the possibility that intraluminal flow could be a common denominator contributing to the coordination of vascular tone across an entire vascular bed. There is presently no direct data that quantitatively supports such a concept. There are, however, observations consistent with a decisive role for flow in controlling tone in a small blood vessel. Mayrovitz and Roy[48] found a positive correlation between flow and in vivo diameter with a third power dependence of flow on diameter. It can be shown that this relationship suggests a constancy of wall shear throughout this vascular bed for arteries in the range of 6–108 μm. The relationship that these investigators found is so close that it is difficult to escape the conclusion that this is not an important regulated homeostatic characteristic.

Local increased tissue activity is known to be coupled with local increase in vascularity, resulting presumably from an increase in local metabolism. This increase in tissue flow would increase upstream flow in successively larger arterial segments which, it seems likely, would result in dilation. A long debated question is "how is this dilation accomplished?" It is possible that flow change could optimize these adaptive changes in diameter. Vessels of diameter greater than those directly influenced by metabolic activity are known to contribute to total vascular resistance, and thus an integrative response of the entire tree would be thermodynamically sound, minimizing power loss.

Independence of Vascular Adjustment to Pressure and Flow

One of the experimental criteria for establishing a flow-induced change in tone in an artery is that the change should not be accounted for by alteration in intravascular pressure. The weight of evidence is that flow and pressure effects are independently sensed and the responses are mediated by different cellular mechanisms.[25] The import is that the peripheral vasculature can respond independently to these forces.

It is not difficult to visualize physiological circumstances in which the inlet pressure to a particular regional vascular bed might increase, associated with either an increase or decrease in regional flow. The opposite combination of circumstances can also be imagined. The dissociation of the vascular response to pressure and to flow would provide a desirable flexibility in the homeostatic response.

Growth

Rodbard[1] summarized his ideas regarding vascular caliber stating that the persistence of a drag stimulus would be expected to lead to vasodilation and, subsequently in the long term, to anatomical reorganization of a blood vessel around its enlarged lumen. There is considerable circumstantial evidence for this conclusion, but little direct evidence that shear stress is actually the effective stimulus. Sodium entry, particularly that associated with Na^+ H^+ exchange has been implicated in cell division and growth.[49,50] The dependence of shear stress-induced changes in tone on sodium raises interesting implications in this regard.

Summary

There is a considerable body of evidence that flow-induced changes in vascular tone occur in the circulation, although there is some divergence of opinion regarding their locus of origin in the vascular wall. In this respect, there are experimental findings consistent with the conclusion that the mechanotransducer and the origin of the flow-induced signal is in the endothelium, and other evidence that these are in the vascular smooth muscle cell and/or its extracellular matrix environment. Recent experiments implicate a crucial role for the sodium ion. The evidence for flow-induced dilation is supported by experiments of considerable variety, both in vivo and in vitro from a number of labs. That for flow-induced constriction is limited, and depends exclusively on in vitro studies. The responses observed in many cases are large, implying that they must be included in any analysis of circulatory hemodynamics and in any description of the action of drugs on the circulation.

Acknowledgment

This study was supported by U. S. Public Health Service Grants no. HL 32985 and 32383.

References

[1] Rodbard, S. (1975) *Cardiology* **60**, 4–49.
[2] Kamiya, A., Bukhari, R., and Togawa, T. (1984) *Bull. Math. Biol.* **46**, 127–137.
[3] Zamir, M. (1976) *J. Gen. Physiol.* **67**, 21–222.
[4] Zweifach, B. W. (1974) *Circ. Res.* **34**, 843–857.
[5] Meininger, G. A., Fehr, K. L, and Yates, M. B. (1987) *Microvasc. Res.* **33**, 81–97.
[6] Segal, S. S. and Duling, B. R. (1987) *Circ. Res.* **61**, II-20–II-25.
[7] Schretzenmayer, A. (1933) *Pflugers Arch. Ges. Physiol.* **232**, 743–748.
[8] Johnson, P. C. and Intaglietta M. (1976) *Am. J. Physiol.* **231**, 1686–1698.
[9] Griffith, T. M., Edwards, D. H., Davies, R. L., Harrison, T. J., and Evans, K. T. (1987) *Nature* **329**, 442–445.
[10] Griffith, T. M., Edwards, D. H., Davies, R. L., and Henderson, A. H. (1989) *Microvasc. Res.* **37**, 162–177.
[11] Toda, N. (1990) *Am. J. Physiol.* **258** (*Heart Circ. Physiol.* 27), H57–H63.

[12] Gaw, A. J. and Bevan, J. A. (1990) *Blood Vessels* **27**, 37 (abstract).

[13] Bevan, J. A. and Joyce, E. H. (1988) *Blood Vessels* **25**, 101–104.

[14] Bevan, J. A., Joyce, E. H., and Wellman, G. C. (1988) *Circ. Res.* **63**, 980–985.

[15] Rubanyi, G. M., Romero, J. C., and Vanhoutte, P. M. (1986) *Am. J. Physiol.* **250** *(Heart Circ. Physiol.),* H1145–H1149.

[16] Kaiser, L., Hull, S. S., and Sparks, H. V. (1986) *Am. J. Physiol.* **250** *(Heart Circ. Physiol.),* H974–H981.

[17] Smiesko, V., Kozik, J., and Dolezel, S. (1985) *Blood Vessels* **22**, 247–251.

[18] Koller, A. and Kaley, G. (1990) *Am. J. Physiol.* **258**, H916–H920.

[19] Garcia-Roldan, J.-L. and Bevan, J. A. (1990) *Circ. Res.* **66**, 1445–1448.

[20] Halpern, W., Osol, G., and Coy, G. S. (1984) *Ann. Biomed. Eng.* **12**, 463–479.

[21] Kuo, L., Davis, M. J., and Chilian, W. M. (1990) *Am. J. Physiol.* **259** *(Heart Circ. Physiol.),* H1063–H1070.

[22] Bognar, J. and Gow, B. S. (1990) *Proc. Austr. Physiol. Pharmacol. Soc.* **21**, 61P (absrtact).

[23] Bevan, J. A. and Joyce, E. H. (1988) *Blood Vessels* **25**, 261–264.

[24] Bevan, J. A. and Joyce E. H. (1990) *Am. J. Physiol.* **259** *(Heart Circ. Physiol.),* H23–H28.

[25] Bevan, J. A., Garcia-Roldan, J.-L., and Joyce, E. H. (1990) *Blood Vessels* **27(Suppl.)**, 202–207.

[26] Sipkema, P., vander Linden, P. J. W., Hoogerwerf, N., and Westerhof, N. (1990) *Blood Vessels* **26**, 368-376.

[27] Young, M. A. and Vatner, S. F. (1986) *Am. J . Physiol.* **250** *(Heart Circ. Physiol. 19),* H892–H897.

[28] Hull, S. S., Kaiser, L., Jaffe, M. D., and Sparks, H. V. (1986) *Blood Vessels* **23**, 183–198.

[29] Pohl, U., Holtz, J., Busse, R., and Bassenge, E. (1986) *Hypertension (Dallas),* **8**, 37–44

[30] Kuo, L., Chilian, W. M., and Davis, M. J. (1990) *Circ. Res.* **60**, 860–866.

[31] Osol, G., Cipolla , M., and Osol, R. (1990) *Blood Vessels* **27**, 51.

[32] Davies, P. F. (1989) *NIPS* **4**, 22–25.

[33] Owen, M. P. and Bevan, J. A. (1985) *Experientia* **41**, 1057, 1058.

[34] Olesen, S.-P., Clapham, D. E., and Davies, P. F. (1988) *Nature* **331**, 168–170.

[35] Garcia-Roldan, J.-L. and Bevan, J. A. (1990) *Circulation* **82**, III-704 (abstract).

[36] Garcia-Roldan, J.-L. and Bevan, J. A. (1991) *Hypertension* (in press).

[37] Bevan, J. A., Wellman, G. C., and Joyce, E. H. (1990) *Blood Vessels* **27**, 369–372.

[38] Siegel, G., Ehehalt, R., Gustavsson, H., and Fransson, L. A. (1977) *Excitation-Contraction Coupling in Smooth Muscle*, (Casteels R., Godfraind T., and Ruegg, J. C., eds.), Elsevier, North-Holland Biomedical, Amsterdam, pp. 279–288.

[39] Siegel, G., Walter, A., Rettig, W., Kampe, C. H., Ebeling, B. J., and Bertsche, O. (1980) *Intracellular Electrolytes and Arterial Hypertension*, (Zumkley H. and Losse, H., eds.), George Thieme Verlag, New York, pp. 31–50.

[40] Aalkjaer, C. and Mulvany, M. (1983) *J. Physiol.* **343**, 105–116.

[41] Junker, J. L., Wasserman, A. J., Berner, P. F., and Somlyo, A. P. (1984) *Circ. Res.* **54**, 254–266.

[42] Gustavsson, H., Siegel, G., Lindman, B., and Fransson, L.-A. (1981) *Biochim. Biophys. Acta* **677**, 23–31.

[43] Bevan, J. A. and Laher, I. (1991) *FASEB J.* (in press).

[44] Hwa, J. J. and Bevan, J. A. (1986) *Am. J. Physiol.* **250**, H87–H95.

[45] Smiesko, V., Lang, D. J., and Johnson, P. C. (1989) *Am. J. Physiol.* **257** (*Heart Circ. Physiol. 26*), H1958–H1965.

[46] Bevan, J. A. and Joyce, E. H. (1990) *Am. J. Physiol.* **258** (*Heart Circ. Physiol. 27*), H663–H668.

[47] Bevan, J. A., Garcia-Roldan, J.-L., Joyce, E. H. (1989) *Neurotransmission and Cerebrovascular Function I*, (Seylaz, J. and MacKenzie, E. T., eds.), Elsevier, Biomedical Division, Amsterdam, pp. 119–122.

[48] Mayrovitz, H. and Roy, J. (1989) *Am. J. Physiol.* **245** (*Heart Circ. Physiol. 14*), H1031–H1038.

[49] Bobik, A., Grooms, A., Grinpukel, S., and Little, P. J. (1988) *J. Hypertension* **6(Suppl. 4)**, 5219–5221.

[50] Panet, R., Amir, I., Snyder, D., Zonenshein, L., Atlan, H., Laskov, R., and Panet, A. (1989) *J. Cell Physiol.* **140**, 161–168.

[51] Zweifach, B. W. and Lipowsky, H. H. (1977) *Circ. Res.* **41**, 380–390.

[52] Fronek, K. and Zweifach B. W. (1975) *Am. J. Physiol.* **228**, 791–796.

[53] Zweifach, B. W., Kovalcheck, S., DeLano, F., and Chen, P. (1981) *Hypertension* **3**, 601–614.

[54] Schmid-Schönbein, G. W. and Zweifach, B. W. (1975) *Microvasc. Res.* **10**, 153–164.

[55] Richardson, D. R., Intaglietta, M., and Zweifach, B. W. (1971) *Microvasc. Res.* **3**, 69.

[56] Palmer, A. A. (1979) *Hemodynamics and Blood Vessel Walls* (Stebbins, W. E., ed.), Thomas, Springfield, IL, pp. 157–237.

[57] Rosenblum, W. I. (1971) *Circ. Res.* **29**, 96–103.

[58] Meininger, G. A. (1987) *Microvasc. Res.* **34**, 29–45.

[59] Tangelder, G. J., Slaaf, D. W., Muijtjens, A. M. M., Arts, T., Egbrink,

M. G. A. O., and Reneman, R. S. (1986) *Circ. Res.* **59,** 505–514.

[60] Meininger, G. A., Fehr, K. L., Yates, M. B., Borders, J. L., and Granger, H. J. (1986) *Hypertension* **8,** 66–75.

[61] Davis, M. J., Ferrer, P. N., and Gore, R. W. (1986) *J. Physiol.* **250,** H291–H303.

[62] Hill, M. A. and Larkins, R. G. (1989) *Am. J. Physiol.* **257,** H571–H580.

[63] Bohlen, H. G. and Hawkins, K. D. (1982) *Diabetologia* **22,** 344–348.

[64] Harper, S. L. and Bohlen, H. G. (1984) *Hypertension* **6,** 408–419.

[65] Stacy, D. L., Joyner, W. L., and Gilmore, J. P. (1987) *Hypertension* **10,** 82–92.

Chapter 12

Conduction
in the Resistance-Vessel Wall

Contributions to Vasomotor Tone
and Vascular Communication

Brian R. Duling,
Takamichi Matsuki, and Steven S. Segal

Introduction

The vasculature functions as an integrated system composed of a series of vessels of different forms and functions. One might logically ask if such a series of longitudinal parts is integrated along its longitudinal axis? Beginning in the early 1920s, occasional observations made it clear that information capable of coordinating vasomotor responses could be transferred along the vascular bed. Krogh observed and reported responses that spread well beyond the site of direct stimulation of an arteriole,[1] and conduit vessels have been seen to dilate at locations far removed from the vascular segments controlling flow.[2-5] In 1970, two papers were published that made it clear that the preceding results reflected the existence of two quite different mechanisms for longitudinal communication in the vasculature. First, flow *per se* was found to be a determinant of vasomotor tone in the conduit vessels.[6] Second, a conducting

From *The Resistance Vasculature*, J. A. Bevan et al., eds. ©1991 Humana Press

pathway intrinsic to the walls of the arterioles was described, which caused the effects of locally applied vasoactive agents to be expressed over distances of hundreds of microns.[7]

The foregoing observations and some of their implications are synthesized in Fig. 1, which shows that multiple processes contribute to the so-called local control of blood flow. These processes include proximate effects of local dilators on the smooth muscle or endothelium of the resistance vessels,[8] conduction along the arteriolar wall from a site of activation to a more proximal part of the vasculature,[1,7,9] and flow-dependent dilation arising secondarily to changes in luminal-wall shear stress following the dilation of distal microvessels.[6,10,11] All of these processes, as well as intrinsic vasomotion, and humoral and neural drives, are ultimately integrated at the final common effector, the vascular smooth-muscle cell, to modulate vasomotor tone.

This chapter summarizes the current state of knowledge of one of the sources of vasomotor tone, conduction, and attempts to place this knowledge in a framework of available information on the cell biology of the vessel wall. The nature of the conducted process is discussed, the possible cell types involved are analyzed, and the electrophysiological basis of conduction is explored.

Longitudinal Conduction in Arteriolar Microvessels

Mechanical Length Constant

The conducted response in the peripheral vasculature is perhaps most easily appreciated by observing the microvessels as in Fig. 2. When micropipets are used to apply agonists to short sections of arterioles, a vasomotor response that spreads over several hundred microns of vessel length in a fraction of a second is often induced.[7,9,12] The spread of response is bidirectional and appears to be conducted along the vessel wall.[7,9,13] The conducted response can be clearly distinguished from both diffusion of agonist[7,9,14] and any flow changes accompanying the vasomotor responses.[12,15] Furthermore, it can be shown to be independent of wall motion of the arteriole.[12]

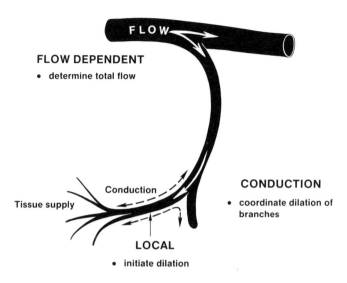

FLOW DEPENDENT
- determine total flow

CONDUCTION
- coordinate dilation of branches

Tissue supply

Conduction

LOCAL
- initiate dilation

Fig. 1. Schematic diagram of the processes contributing to longitudinal integration of vasomotor tone in "local" regulation of blood flow.

The simplest interpretation of the data is that the conducted response is the result of electrotonic spread of current along the microvessel axis, through one or more of the cell types composing the vessel wall. If so, then decremental conduction along the vessel axis would be predicted. Decremental conduction can, in fact, be demonstrated by ejecting drugs from micropipets and observing the subsequent response at distant sites along the length of the vessel. The vasomotor responses decay both with time after application of the drug and with distance from the site of the application.[9,15] Following long-established practice used in electrophysiological analysis of cable theory, the spatial decay of the response has been characterized by a mechanical length constant (λ, the distance over which the magnitude of the response decays to $1/e$ of the initial value). Typically, length constants on the order of 2 mm are observed, both with dilators and constrictors,[9,15] and the length constant is largely independent of the agonist, except in those cases in which no conduction is induced.[15]

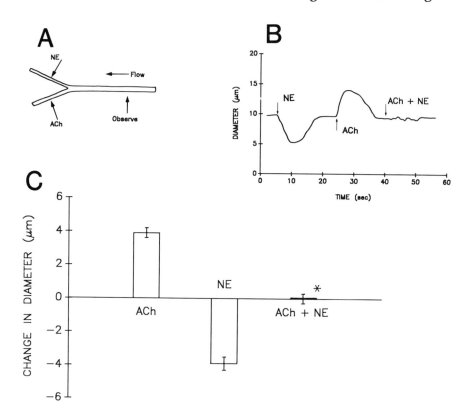

Fig. 2. Summation of the conducted vasomotor response. Micropipets were placed on the small daughter vessels at the points indicated by the arrows labeled NE and ACh. Distances between the sites of drug application and the site of observation were typically on the order of several hundred microns. Placing the pipets downstream from the site of observation assured that convection of the drug did not contribute to the conducted response.[9]

Conduction into Branches and Summation

A second prediction from the hypothesis that the conducted response is based on simple, nonregenerative electrotonic spread is that pairs of responses should sum when they arrive simultaneously at a point. This prediction is most easily tested in the vicinity of a vascular branch, as shown in Fig. 2. The 2-mm length constant noted above is substantially longer than the typical interbranch interval in microvascular networks,[16,17] thus, the conducted responses induced in an arteriole commonly spread into branches, both up- and downstream from the site of application. A typical experimental observation is shown in this figure.

Microvessel segments with branches of approx equal size were chosen for study, a stimulating pipet was applied to each of the daughter branches, and the response of the parent vessel several hundred microns upstream from the pipets was observed. One pipet contained norepinephrine (constrictor), and a pipet on the other branch contained acetylcholine (dilator). The strengths of the individual stimuli were adjusted to produce equivalent changes in diameter with opposite sign when the agonists were ejected from the pipets. When the two pipets were activated simultaneously, the paired signals of opposite sign canceled one another at the site of observation. A comparable experiment with the same agent in the two pipets leads to a doubling of the response in the parent vessel.[9] These observations show that the conducted signals that induce the diameter changes sum linearly, an outcome that is expected if the conducted response represents simple electrotonic spread of current along the vessel wall.[9]

The experiment in Fig. 2 also suggests quite strongly that the conducted response is not propagated by a regenerative process. If this were so, then spreading vasodilation would imply regenerative hyperpolarization, and we are unaware that such a process occurs. If regenerative hyperpolarization did occur, it would be unlikely to generate the same length constant as a regenerative depolarization.[15] Therefore, in the present work, in contrast to earlier comments on the subject, we have eliminated reference to propagation in favor of the word "conduction."[7,9,18]

Velocity of Conduction

By noting the delay in appearance of an acetylcholine-induced dilation at points separated by known distances along the length of arterioles, Duling and Berne[7] estimated the conduction velocity to be about 0.02 cm/s in arterioles of the hamster cheek pouch. More recently, Segal and Duling[12] estimated that the velocity must be in excess of 0.2 cm/s, a value 10 times that earlier reported by Duling and Berne under what would seem to be identical conditions. Measurements of a similar type in the rat mesentery[14] yielded estimates of 0.002 cm/s, even lower than those of Duling and Berne.

The cause for the disparate velocity measurements obtained in the different experiments is unknown, but it is quite possible that two or more processes are involved in the conducted response. This assertion is based on observations of the time-course of the conducted diameter changes following stimulation. It is apparent to an observer of the microcirculation that there is typically an initial, very rapid response that is followed by a secondary, longer-lasting diameter change (unpublished observations). This has been difficult to define quantitatively, but one has the impression that the two phases of the response may represent two different processes with different conduction velocities. Obviously, a clear elucidation of the issue will depend on careful measurement of electrical potential during conduction, a determination yet to be made.

Electrophysiological Correlates of the Conducted Response

There are well-established electrophysiological correlates to the processes of spatial decay and summation seen with the conducted vasomotor response; many of these have been reviewed and summarized by Hirst and Edwards.[19] Electrical length constants of 2 mm, approximately the same value as the mechanical length constants, have been measured in arterioles of the guinea pig submucosa, and bidirectional spread of electrical current has been reported.[20,21] In addition, current spreads into branches in ways entirely consistent with the movement of vasomotor responses across branches. Finally, analysis of the spread of current in the vessel wall of the arteriolar tree shows that the microvessels can be treated as a branching network of one dimensional cables.[20]

Implicit in the concept of electrotonic spread of the conducted vasomotor response is a requirement, often overlooked, that the vessel tone must vary in a reasonably direct way with the level of membrane potential. In other words, a hyperpolarization must be associated with a dilation, and so on. Experimental evidence from large vessels demonstrates that membrane-potential changes can be directly related to vasomotor responses, and the sensitivity can be great.[22-28]

Thus, the conducted vasomotor response can be tentatively attributed to the following sequence of events:

1. Attachment of agonist to cell-surface receptor,
2. Alteration in the membrane potential,
3. Conduction by one or another of the cells in the vessel wall (*see below*),
4. Alteration of intracellular Ca^{2+} with the local change in potential, and
5. Activation or inactivation of the contractile proteins.

Independence of the Local and the Conducted Responses

When very low doses of vasoactive agents are applied by micropipet to arterioles, a local response can be initiated with little evidence of conduction,[7,15,29] (unpublished observations). Commonly with greater stimulus strength, the dilation spreads as described above, suggesting that the local and the conducted vasomotor responses may reflect different processes. Support for this possibility stems from a recent survey of the pharmacology of the conducted response.

A group of 10 agonists with a range of cellular targets, receptors, and second messengers were selected for comparative study, and the conduction process was examined using microapplication of these agonists to the vessel wall. Supramaximal doses of the agonists were used to ensure that, if an agonist were capable of inducing electromechanical coupling (*see below*), it would do so. All of the agonists studied produced local responses, but there was great diversity in the ability of the agonists to induce the conducted response. Some agonists invariably caused conduction (acetylcholine), some totally failed to do so (vasopressin), and unexpectedly some agonists (adenosine) induced conduction on a highly variable subset of the vessels studied. Remarkably, when conduction was induced by any of the agonists, the length constant for the response was independent of the agonist used.[15] In any one arteriole, there could be a high degree of variability in the particular combinations of agonists that did and did not induce conduction. There appeared to be no consistent pattern of failure or success that could be related to either the cell type on which the agonists are known to act or the intracellular mechanism of action, nor was there any definable characteristic of the arteriole with predictive value.

What is relevant to the present issue is that the local and conducted responses were subject to quite different rules. Since three of the agents invariably induced conduction (acetylcholine, muscarine, and phenylephrine), and since the length constants did not vary with the agonists used, one can assume that differences in the conduction pathway (e.g., changes in gap-junctional transmission) are not the source of the variability in response. It seems most reasonable to suppose that the failure of an agonist to induce conduction simply reflects an absence of a change in the membrane potential at the site of drug application. If this is so, then it can be hypothesized that the variability in response reflects the relative abilities of the different agonists to induce vasomotor responses by electromechanical or pharmacomechanical coupling.[30] Agonists that act locally but fail to induce the conducted response would be presumed to act entirely by pharmacomechanical coupling, and those that induce conduction in a variable fashion would likely be inducing variable electrical responses.

Differences in electromechanical coupling from agonist to agonist and under different circumstances seems a reasonable interpretation on two grounds. First, agonists do appear to utilize both chemomechanical and electromechanical coupling in the activation of smooth muscle.[27,30,31] Second, we have noted that norepinephrine can induce a localized arteriolar response at low concentrations, whereas it invariably induces conducted responses at high doses (unpublished observations). This pattern mimics that seen in studies of pharmacomechanical coupling.[30-32]

The variability in the conducted response from vessel to vessel is explained less easily. Electromechanical coupling operates by opening rather specific membrane channels that allow corresponding ions to flow down their electrochemical gradient. Perhaps the variability in the conducted response is a reflection of changes in the level of the resting potential at the time of stimulation, or of the intracellular concentrations of such ions as potassium and chloride, depending on the membrane channels that are opened. Again, electrophysiological measurements will be critical to explaining these variations in vasomotor tone.

Conduction in Vessels Other than Arterioles

Arteries and Veins

For the most part, our knowledge of the conducted vasomotor response is based on the behavior of small and medium-sized arterioles. However, both the arteries and the veins have well-developed conducting mechanisms inherent in their walls. The media of these vessels behaves as an electrical syncitium,[33-36] *(see ref. 19 for review)* and the conducting systems can support both spontaneous and induced contractions that are coordinated over long vascular segments.[34-39] A mechanical response can spread several millimeters along the length of the arterial vessels, and the mechanical length constant appears to depend on the type of vessel studied.[37]

It is noteworthy that the mechanical spread of vasomotor response and the electrical spread of current need not show the same pattern.[37] As mentioned above, the rabbit aorta manifests cable-like electrical properties, but little evidence of a conducted mechanical response. Furthermore, when norepinephrine is applied to the rabbit-ear artery, a two-phase response is observed; the initial phase of the norepinephrine-induced contraction is propagated, but the second is not. How conduction of the different phases of the norepinephrine contraction may be dissociated and why a mechanical propagation fails in a tissue that is reported to be an electrical syncitium remain to be established.

Although the electrical length constant in the larger vessels is similar to that in the arterioles,[36] the implications for function are quite different in the two classes of vessels. In the conduit vessels, sites of vasomotor response may be centimeters, rather than millimeters, away from the point of initiation of a signal in the tissues. Obviously, length constants of a few millimeters severely limit the role that can be played by the conducted response in activating the larger vessels; thus, it seems most likely that the conducting system of the large vessels is important in coordinating cellular mechanical activity especially oscillatory behavior, but not in conducting responses from the periphery to the conduit vessels.[35,36,38]

It is apparent that the large vessels do not have an absolute requirement for an intact conducting system in order for them to be able to participate in regulatory responses. Longitudinal integration can be shown to persist in the conduit vessels even after conduction is interrupted by transection. When the vessel is sectioned distal to a site of diameter measurement and the segments reconnected by a cannula, both functional and reactive dilations in the conduit vessel remain intact.[6,10]

The foregoing comments provide a teleological explanation for the need for a flow-dependent response in the large vessels;[6,10,11] the flow-dependent mechanism can function as an ancillary process of vascular coordination that is essentially independent of distance, since flow will change in the conduit vessels regardless of the separation between them and the resistance vessels they feed. In contrast, conducted response can operate over a distance of only a few millimeters.

Capillaries

Capillary contraction appears to be a feature of the amphibian vascular bed, but not of mammals.[14,40,41] Thus, in most of the circumstances alluded to above, movement of a conducted vasomotor response along the capillaries could not have occurred. However, the capillaries may act as sensors in the initiation of the conducted response. Dietrich,[14] Gaehtgens et. al.,[42] and Tigno et al.[13] have reported that a conducted response can be initiated in the capillaries, and can subsequently spread retrogradely into the feeding arteriole. Then, modification of arteriolar tone, in turn, alters the capillary flow patterns without visible change in the capillary dimensions.

The interpretation of the preceding experiments depends on being able to distinguish between conduction and diffusion of the agonist as the means of activation of the upstream arterioles. In general, this is a problem in any study of the conducted response, and it is particularly difficult in studies on the mesentery, since both the conduction velocity and the response times are very slow. Thus long latencies between application of agonist and vasomotor response provide time for extensive diffusion.[14] The role played by diffusion is further clouded by the fact that the pipets used for drug application are commonly filled with solutions whose concentrations are several orders of magnitude higher than the physiological concentration.[13–15] The expected delay in the response attributable to diffusion is usually based on estimates of average diffusion time between the point of application and the arteriolar site of observation. Because of the high concentrations in the pipets, it is the leading edge of the diffusion front, not the average value, that determines the latency required for the drug concentration to rise to a level that is physiologically significant.[14] Thus, further work will be necessary to unequivocally establish the participation of the capillary in the conducted response.

Venules

There have been no systematic studies of the possibility of longitudinal communication in the venules. However, venular morphology is consistent with a conducted response (*see discussion below*),[43,44] and the larger venous vessels are capable of electrical activity and conduction.[35]

Recently, a process quite different from conduction has been suggested as an additional potentially important link between the venous and the arterial sides of the capillary bed. Tigno et al.[13] microinjected norepinephrine into small precapillary microvessels. No direct effect of the drug was observed at the site of injection, either on the capillary diameters or on the postcapillary venules. However, where the venule carrying the outflow from the injected capillary bed crossed the arteriolar feed vessel, a bidirectional conducted response was induced in the arteriole.

This observation suggests that the venule is sufficiently permeable to allow the drug to exit in adequate quantities to activate the arteriolar smooth muscle. This, in turn, prompts the idea of a rather unique means of communication along the vascular axis, in which material is assumed to be picked up by the capillaries, brought upstream by convection, and delivered to the arteriole by the postcapillary venule. The foregoing proposition is particularly interesting in view of the fact that the arteriolar endothelium is remarkably tight to small solute molecules,[45] and movement directly from blood in the arteriolar lumen to arteriolar smooth muscle is restricted.

Cellular and Intercellular Pathways for Conduction

Based on their physiology, morphology, and electrical properties, four cell types are likely participants in the process of conduction along the vessel wall: nerve, vascular smooth muscle, pericyte, and endothelium. It is also possible that the conducted response is triggered in one cell, for example, the endothelium, and is conducted through other cells, e.g., the smooth muscle. Some support for this idea is provided by the observation that microinjection of acetylcholine into the lumen of an arteriole induces a conducted dilation, apparently via activation of the endothelial cells.[46] Furthermore, agonists that act primarily on smooth muscle as well as those that act on the endothelial cell can induce the conducted vasomotor response.[15]

The distances over which the conducted responses spread are large compared to the cellular dimensions; thus, conduction must rely on a multicellular pathway. This fact, and the absence of polarity in the direction of conduction, strongly suggests that the intercellular pathway by which electrical current passes from cell to cell is the gap junction, rather than via the release of some secondary chemical intermediate. This idea is supported by the fact that CO_2, octanol, and hypertonic solutions, all of which block or attenuate gap-junctional transmission, are effective in attenuating the conducted responses in the arterioles.[47]

The following sections briefly describe the facts salient to the cellular basis for conduction along nerves, vascular smooth muscles, pericytes, and endothelial cells, and review the anatomical, chemical, and electrical evidence supporting or refuting the existence of gap-junctional communication between the cells of the vessel wall. Though anatomical documentation is often taken as *prima facie* evidence for the presence or absence of cell–cell communication, both anatomical and functional data are presented for several reasons. First, small but nevertheless critical areas of gap-junctional contact may be easily missed in an electron-micrographic analysis.[44,48] Furthermore, points of close apposition of membranes are not necessarily gap-junctions, and, for the most part, careful anatomical studies, including appropriate stains and freeze-fracture, have yet to be applied to many of the homocellular and heterocellular junctions of the vessel wall. Finally, one must consider the observation made by Gabella[49] that "the evidence at hand . . . suggests that in some tissues electrical coupling is achieved without the presence of gap-junctions."

Nerves

Neural conduction has not been shown to be a primary pathway for conduction of vasomotor responses. In the resistance vessels, neither local anesthetics[42] nor tetrodotoxin[47] have been effective in blocking the conducted response. Furthermore, the response is conducted bidirectionally, implying the absence of synaptic relays. Additionally, the fact that so many agonists induce conduction suggests that nerves are, at least, not the detector.

However, nerves apparently can play a role in some vessels, since tetrodotoxin does block conduction in rabbit-ear artery.[50] Parenthetically, it should be noted that lidocaine was reported to block the conducted response in a previous study of the cheek pouch.[7] The disparate reports on the effects of lidocaine remain unexplained, but clearly tetrodotoxin is the more specific agent. Also, it should be noted that the conduction velocities reported in the two experiments on the cheek pouch were so different as to suggest that different mechanisms may have been involved.[7,47]

Vascular Smooth Muscles and Pericytes

Gap-junctional connections between smooth-muscle cells are common[49,51,52] and are presumed to provide the pathway for the electrical continuity described above in the large vessels. The gap-junctional coupling extends along the entire media in a variety of species.[44,52,53] In addition, medial cells are coupled in such a manner as to allow the diffusional movement of small molecules.[54]

In the capillary bed, the smooth muscle is replaced by the pericyte as the surrounding cell, and gap-junctions exist between pericytes and endothelial cells.[55] There is little known of the longitudinal connection between pericytes in vivo, although it has been noted that "... specialized intercellular junctions between pericytes have not been observed."[56] However, Larson et al.[48] report that there is molecular transfer through pericytic gap-junctions in culture. It is of some interest that the presence or absence of gap-junctions between pericytes be established in view of the paucity of gap-junctions reported between capillary endothelial cells *in situ*.[44] Since the pericytes invest essentially all the capillaries, gap-junctional communication between these cells would provide an additional potential conducting pathway from the capillaries to the arterioles.[14,57]

Endothelium

The endothelial cells of the arterial intima, the arterioles, and the larger postcapillary venules and veins are extensively connected by gap-junctions,[44,58–60] and electrical and chemical transfer between the endothelial cells has been clearly shown both in culture[48,61,63] and *in situ*.[54,64,65] The endothelium also in some cases possesses the electrophysiological properties necessary to participate in the conduction process. It has a significant membrane potential that varies both with physical stimuli[66,67] and with application of agonists.[68–72]

To date, no anatomically definable junctions have been found in the capillary and the immediate postcapillary venular endothelium.[44,59] However, molecular transfer between adjacent endothelial cells in the capillaries has been shown.[73] It has been surmised that this dichotomy reflects the presence of gap-junctions that are missed in the usual electron-microscopic examination.[44,48]

Myoendothelial Junctions

Heterocellular connections between smooth-muscle cells or pericytes and the endothelium are also common in the vessel walls.[74] In his classical studies of the morphology of the vessel wall, Rhodin[43,53] noted the high incidence of junctional contacts between the endothelium and the smooth muscle of the arterioles. He speculated that these myoendothelial junctions might form ". . . part of a receptor mechanism for humoral transmitter substances,"[43] a speculation now known to be fact.

Although myoendothelial junctions have been reported in larger vessels,[60,74,75] they are more common and more elaborate in the arterioles.[43,44,53,60,76] Endothelial-cell pericyte junctions are also common, especially in the central nervous system.[52,56,77,78] In fact, based on the frequency of perforations in the internal elastic lamina of the arteriole, it has been estimated that each smooth-muscle cell makes myoendothelial contacts with numerous endothelial cells, and vice versa.[79] Whether or not these points of contact actually contain gap-junctions remains to be seen.

There is unequivocal evidence for functional myoendothelial junctions in culture. The two cell types readily establish communication, and both electrical and molecular coupling have been repeatedly shown.[48,74,80] The situation in vivo is less clear, however. There is a report of dye transfer between endothelial cells and "pericytes" of microvessels in the rat omentum,[73] and this appears to be the only functional demonstration *in situ*. In larger vessels there is a report of uridine transfer from cultured endothelium to aortic smooth muscle in organ culture, indicating that endothelial cells and the smooth muscle can make functional contacts.[74] On the other hand, there is a reported failure of dye transfer and in electrical continuity between the two cell types in the coronary artery of the pig.[54]

Can Cell–Cell Communication Be Regulated?

The process of conduction along the vessel wall might well be immutable. However, in view of the fact that gap-junctional communication in other tissues is regulated, it seems timely to consider the ways in which such regulation might play a role in the determination of vascular function. The gap-junctional conductance is potential-sensitive; a transjunctional potential difference of 14 mV is reported to reduce the electrical coupling to neighboring cells by half.[81] Gap-junctional conductance is also controlled by a variety of allosteric regulators. Ca^{2+} appears to be of key importance, and

H$^+$ and cAMP also participate in regulation, probably by modulating the Ca^{2+} sensitivity of the gap-junctional protein.[82–85] Covalent control of gap-junctional conductance is exerted by modulation of phosphorylation of the protein.[82]

Synthesis and/or insertion of the proteins into the membrane or modification of their dwell time in the membrane provide yet other forms of modulation of junctional transmission and, potentially, of conduction along the vessel wall.[80,84–87] The quantity of gap-junctional protein present can be remarkably variable and is under physiological control in some systems. In uterine smooth muscle, the cell area occupied by gap-junctions rises by a factor of 10 during gestation.[88,89] This change in anatomical and chemical presence of gap-junctions is associated with a similarly large change in the cell–cell coupling in the muscle.[90]

To date, there has been little concern for the possibility that gap-junctional function may be modulated as a normal part of the vasomotor control. Obviously, gap-junctions form during development, and it appears that their localization is under the control of, or determined in concert with, other perivascular cells.[78] There is also evidence that, in myocardial tissue, electrical coupling can be altered by vasoactive materials, but, as yet, no such demonstration has been made in vascular tissue.[91,92] Unfortunately, experimental tests of gap-junctional transmission based on electrical measurements are difficult to carry out, and thus there has been little exploration of the possibility of regulation of transmission.

Physiological Significance
of the Conducted Response

Coordination of Cells of the Vessel Wall

The cells of the vascular wall are small, and they can easily carry out many of their functions in isolation when in the culture dish, and perhaps function as single cells in the terminal arteriole.[93] However, if the smooth-muscle cells are to function appropriately in the control of the resistance vessels, adjacent cells must contract in concert to maintain a smooth microvessel lumen and to permit precisely graded changes in flow. This realization led Gabella[49] to note in relation to a similar concern that ". . . a marked ability to cooperate is an essential property of these diminutive cells, and intercellular links must be at work to summate and transmit their mechanical activities."

The communication among cells of the vessel wall may be graphically expressed in the coordinated, rhythmic contraction of arterioles that seems to be a central aspect of their behavior.[93–96] This cyclic vasomotion has been reported to originate in fixed pacemaker sites, perhaps at branches along the length of the vasculature.[94,95] If so, then cell–cell coupling becomes an essential element in coordinating the process, and perhaps in the determination of vascular resistance.[96,97]

Neural Coupling in the Autonomic Nervous System

Conducted responses in the arterioles are also likely to play a role in the modulation of the effects of sympathetic-nerve discharge as described by Hirst in this volume. Excitatory junction potentials can be clearly seen in the microvessels, and they spread along the vessel length.[19–21] The distance from sympathetic-nerve vesicle to smooth-muscle cell is exceedingly short, and the cleft may well restrict the direct action of the nerve transmitter to those few smooth muscle cells lying directly under the vesicle.[19,20,98,99] If so, then the extent of the effect of the nerve discharge is likely to be determined by the magnitude of the conducted vasomotor response. As noted by Bevan,[98] ". . . the smaller the vessel, the more intimate the neuro-effector relationship, the more localized the action of the released transmitter, and the more important myogenic conduction . . . ."

Coordination of Local Stimuli and Total Flow

A simple idea has dominated much of the literature on vascular control in the past century; the concept might be termed "action at a point." It was assumed that a dilator metabolite was released by the tissue and that the concentration of the metabolite at a particular locus on a resistance vessel determined vascular tone and thereby established the necessary balance between flow and metabolism.[8,100] However, as Hilton[3] noted many years ago, ". . . the concept of a vasodilator metabolite diffusing through the interstitial fluid to act directly on the vessel walls has never adequately accounted for the vasodilatation observed in a contracting muscle; for this dilatation is not confined to the capillaries and arterioles, but extends to the small intramuscular arteries and even to larger arteries outside the muscle." The idea that control of the vasculature is an integrated, multilevel process was introduced with Fig. 1, which emphasized that flow-dependent, conducted, and local vasomotor processes work in concert to regulate blood flow. The participation of the flow-dependent process in regulation is discussed elsewhere in this volume, and its relation to the conducted response has been previously mentioned.[101] Here it will be noted only that the flow-dependent process can involve vessels in a physiological response at any distance from the site of flow control, so long as the flowing blood passes through the vessel in question. Neither diffusion of a dilator metabolite nor conduction of the vasomotor response is needed.

The conducted vasomotor response obviously provides an ideal means of involving more proximal microvessels in what may originate as a purely local control process. Resistance to flow is distributed widely along the length of the striated muscle vascular bed,[102,103] and thus a purely local dilation of a short arteriolar segment can have little effect on the magnitude of local microflow.[8] Conduction greatly extends the portion of the vasculature that operates as part of a functional unit composed of the parenchymal cell and its associated supply vessels. In addition, it has been noted that the conduction process may play a significant role in the minimization of microvascular flow heterogeneity.[9,101]

Once it is recognized that the local regulatory process is distributed in space, the possibility arises for sensors and effectors to be placed at quite different locations along the vessel axis. This might involve placing sites of special reactivity on an arteriole, as Jackson[104] has suggested for the O_2 sensor. Alternatively, different segments of the vasculature could have diverse but integrated functions in control. For example, Silver[57] and Honig et. al.[105] have emphasized the need for a very rapid means of communication between the immediate environment of the capillaries and the more proximal vessels that control their perfusion. If capillaries can act as sensors that communicate by conduction with their parent arterioles,[42,106] this process could serve to explain some long-standing issues in relation to establishing a match between tissue metabolic demand and tissue O_2 supply.

Summary and Prospectus

Though recognized for 70 years, the process of longitudinal integration of vascular function has received rather little experimental attention. This is particularly true in relation to the terminal vessels of the vasculature, in which the small size, thin intima and media, and short interbranch intervals make the conducted process particularly important. Almost nothing is known of the interactions that might occur between adjacent segments along the length of microvessels or between parallel branches during normal physiological stimuli, such as functional hyperemia, baroreceptor reflex, or elevation of central venous pressure.

Few of the tools of the modern cell biologist or electrophysiologist have been brought to bear on this issue. Current technology permits accurate measurement of the conduction velocity of the spreading vasomotor response, relatively specific interruption of gap-junctional transmission, and identification of the location of the gap junctions using molecular biological approaches. It is anticipated that the next few years will see a burgeoning of interest in the subject and a deepening in our understanding of physiological integration of vascular function.

Acknowledgments

This work was supported by U.S. Public Health Service grant no. HL12792 and no. 19242.

References

[1] Krogh, A., Harrop, G., and Rehberg, P. (1922) *J. Physiol. (Lond.)* **56**, 179–189.
[2] Fleisch, A. (1935) *Arch. Int. Physiol.* **41**, 141–167.
[3] Hilton, S. (1959) *J. Physiol. (Lond.)* **149**, 93–111.
[4] Rosenthal, S. and Guyton, A. C. (1968) *Circ. Res.* **23**, 239–248.

[5] Schretzenmayer, A. (1933) *Arch. Ges. Physiol.* **232**, 743–748.

[6] Lie, M., Sejersted, O., and Kiil, F. (1970) *Circ. Res.* **27**, 727–737.

[7] Duling, B. and Berne R. (1970) *Circ. Res.* **26**, 163–170.

[8] Gorczynski, R., Klitzman, B., and Duling, B. (1978) *Am. J. Physiol.* **235**, H494–H504.

[9] Segal, S., Damon, D., and Duling, B. (1989) *Am. J. Physiol.* **256**, H832–H837.

[10] Hintze, T. and Vatner, S. (1984) *Circ. Res.* **54**, 50–57.

[11] Pohl, U., Holtz, J., Busse, R., and Bassenge, E. (1986) *Hypertension* **8**, 37–44.

[12] Segal, S. and Duling, B. (1986) *Science* **234**, 868–870.

[13] Tigno, X., Ley, K., Pries, A., and Gaehtgens, P. (1989) *Pflugers Arch.* **414**, 450–456.

[14] Dietrich, H. (1989) *Microvasc. Res.* **38**, 125–135.

[15] Delashaw, J. and Duling, B. (1991) *Am. J. Physiol.* (in press).

[16] Engelson, E., Schmid, S. G., and Zweifach, B. (1986) *Microvasc Res.* **31**, 356–374.

[17] Koller, A., Dawant, B., Liu, A., Popel, A. S., and Johnson, P. (1987) *Am. J. Physiol.* **253**, H154–H164.

[18] Segal, S. and Duling, B. (1987) *Circ. Res.* **61**, 20–25.

[19] Hirst, G. and Edwards, F. (1989) *Physiol. Rev.* **69**, 546–604.

[20] Hirst, G. and Neild, T. (1978) *J. Physiol. (Lond.)* **280**, 87–104.

[21] Hirst, G. D. and Neild, T. O. (1980) *J. Physiol. (Lond.)* **303**, 43–60.

[22] Casteels, R., Kitamura, K., Kuriyama, H., and Suzuki, H. (1977) *J. Physiol. (Lond.)* **271**, 63–79.

[23] Harder, D. (1982) *Vascular Smooth Muscle, Metabolic, Ionic, and Contractile Mechanisms* (Crass, M. F., and Barnes, C. D., eds.), Academic, New York, pp. 71–97.

[24] Hermsmeyer, K., Trapani, A., and Abel, P. (1981) *Vasodilation* (Vanhoutte, P. M. and Leusen, I., eds.), Raven, New York, pp. 273–284.

[25] Heusler, G. (1978) *Blood Vessels* **15**, 46–54.

[26] Mekata, F. J. (1986) *Physiol. (Lond.)* **371**, 257–265.

[27] Neild, T. and Keef, K. (1985) *Microvasc. Res.* **30**, 19–28.

[28] Neild, T. and Kotecha, N. (1987) *Circ. Res.* **60**, 791–795.

[29] Duling, B., Berne, R., and Born, G. (1968) *Microvasc. Res.* **1**, 158–173.

[30] Somlyo, A. and Somlyo, A. (1968) *J. Pharm. Exp. Ther.* **159**, 129–145.

[31] Trapani, A, Matsuki, N., Abel, P., and Hermsmeyer, K. (1981) *Eur. J. Pharmacol.* **72**, 87–91.

[32] Neild, T. and Zelcer, E. (1982) *Prog. Neurobiol.* **19**, 141–158.

[33] Abe, Y. and Tomita, T. (1968) *J. Physiol. (Lond.)* **196**, 87–100.

[34] Araki, H., Sakaino, N., Furusho, N., and Nishi, K. (1989) *Circ. Res.* **64**, 73441.

[35] Johansson, B. and Ljung, B. (1967) *Acta Physiol. Scand.* **70**, 312–322.

[36] Mekata, F. (1974) *J. Physiol. (Lond.)* **242,** 143–155.

[37] Bevan, J. A. and Ljung, B. (1974) *Acta. Physiol. Scand.* **90,** P 703–715.

[38] Biamino, G. and Kruckenburg, P. (1969) *Am. J. Physiol.* **217,** 376–382.

[39] Jackson, W., Mulsch, A., and Busse, R. (1991) *Am. J. Physiol.* **260,** H248–H253.

[40] Weigelt, H. and Lubbers, D. (1984) *Adv. Exp. Med. Biol.* **169,** 731–737.

[41] Wolff, E. and Dietrich, H. (1985) *Microcirc. End. Lymphatics* **2,** 607–615.

[42] Gaehtgens, P., Tigno, X., Thies, K., and Ley, K. (1989) *Funkt. Biol. Syst.* **19,** 87–95.

[43] Rhodin, J. (1968) *J. Ultrastruct. Res.* **25,** 452–500.

[44] Simionescu, M. and Simionescu, N. (1984) *Handbook of Physiology, Section 2: The Cardiovascular System, vol. 4: Microcirculation,* (Renkin, E. M. and Michel, C. C., eds.), American Physiological Society, Bethesda, MD, pp. 41–101.

[45] Lew, M., Rivers, R., and Duling, B. (1989) *Am. J. Physiol.* **257,** H10–H16.

[46] Rivers, R. (1990) *Evidence for a Functional Role for Endothelium–Derived Relaxing Factor in the Microcirculation.* PhD Thesis. University Microfilms International, Ann Arbor, MI, p. 142.

[47] Segal, S. and Duling, B. (1989) *Am. J. Physiol.* **256,** H838–H845.

[48] Larson, D., Carson, M., and Haudenschild, C. (1987) *Microvasc. Res.* **34,** 184–199.

[49] Gabella, G. (1979) *Br. Med. Bull.* **35,** 213–218.

[50] Vonderlage, M. (1981) *Circ. Res.* **49,** 600–608.

[51] Barr, L., Berger,W., and Dewey, M. (1968) *J. Gen. Physiol.* **51,** 347–368.

[52] Forbes, M. (1984) *Physiology and Pathophysiology of the Heart,* (Sperelakis, N., ed.), Martinus Nijhoff, Boston, MA, pp. 659–685.

[53] Rhodin, J. J. (1967) *Ultrastruct. Res.* **18,** 181–223.

[54] Beny, J. (1990) *Endothelium-Derived Relaxing Factors,* (Rubanyi, G. M. and Vanhoutte, P. M., eds.), S. Karger, Basel, pp. 117–123.

[55] Cuevas, P., Gutierrez, D., Reimers, D., Dujovny, M., and Diaz, F. (1984) *Anat. Embryol. (Berl.)* **170,** 155–159.

[56] Forbes, M., Rennels, M., and Nelson, E. (1977) *Am. J. Anat.* **149,** 47–69.

[57] Silver, I. (1978) *Cerebral Vascular Smooth Muscle, Ciba Foundation Symposium 56,* Elsevier Excerpta Medica, Amsterdam, pp. 61–68.

[58] Simionescu, M., Simionescu, N., and Palade, G. (1975) *J. Cell Biol.* **67,** 863–885.

[59] Simionescu, M., Simionescu, N., and Palade, G. (1976) *Thromb. Res.* **8,** 247–256.

[60] Simionescu, M., Simionescu, N., and Palade, G. (1976) *J. Cell Biol.* **68,** 705–723.

[61] Larson, D., Kam, E. Y., and Sheridan, J.(1983) *J. Membr. Biol.* **74,** 103–113.

[62] Larson, D. and Sheridan, J. (1982) *J. Cell Biol.* **92,** 183–191.

[63] Larson, D., Sheridan, J. (1985) *J. Membr. Biol.* **83,** 157–167.

[64] Beny, J. L. and Gribi, F. (1989) *Tissue Cell* **21**, 797–802.

[65] Sheridan, J. and Larson, D. (1982) *Functional Integration of Cells in Animal Tissues, British Society for Cell Biology Symposium 5* (Pitts, J. D. and Finbow, M. E., eds.), Cambridge University Press, pp. 263–283.

[66] Lansman, J., Hallam, T., and Rink, T. (1987) *Nature* **325**, 811–813.

[67] Olesen, S–P., Clapham, D., and Davies, P. (1988) *Nature* **331**, 168–170.

[68] Bregestovski, P. D. and Ryan, U. S. (1989) *J. Mol. Cell Cardiol.* **21**, 103–108.

[69] Johns, A., Lategan, T. W., Lodge, N. J., Ryan, U. S., Van Breemen, C., and Adams, D. J. (1987) *Tissue Cell* **19**, 733–745.

[70] Northover, B. J. (1980) *Adv. Microcirc.* **9**, 135–160.

[71] Northover, A. and Northover, B. (1987) *Int. J. Microcirc. Clin. Exp.* **6**, 137–148.

[72] Olesen, S., Davies, P., and Clapham, D. (1988) *Circ. Res.* **62**, 1059–1064.

[73] Sheridan, J. (1980) *J. Cell Biol.* **87**, 61a.

[74] Davies, P. F., Olesen, S. P., Clapham, D. E., Morrel, E. M., and Schoen, F. J. (1988) *Hypertension* **11**, 563–572.

[75] Spagnoli, L. G., Villaschi, S., Neri, L., and Palmieri, G. (1982) *Experientia* **38**, 124,125.

[76] Taugner, R., Kirchheim, H., and Forssmann, W. (1984) *Cell Tissue Res.* **235**, 319–325.

[77] Forbes, M., Rennels, M., and Nelson, E. (1977) *Am. J. Anat.* **149**, 47–70.

[78] Shivers, R., Arur, F., and Bowman, P. J. (1988) *Submicrosc. Cytol. Pathol.* **20**, 1–14.

[79] Duling, B. and Sleek, G. (1988) *Vascular Neuroeffector Mechanisms*, (Bevan, J. A., Majewski, H., Maxwell, R. A., and Story D. F., eds.), IRL Press, Washington, DC, pp. 23–30.

[80] Larson, D., Haudenschild, C., and Beyer, E. (1990) *Circ. Res.* **66**, 1074–1080.

[81] Spray, D., White, R., De, C. A., Harris, A., and Bennett, M. (1984) *Biophys. J.* **45**, 219–230.

[82] Arellano, R. O., Rivera, A., and Ramon, F. (1990) *Biophys. J.* **57**, 363–367.

[83] Burt, J. M. (1987) *Am. J. Physiol.* **253**, 607–612.

[84] Ramon, F. and Rivera, A. (1986) *Prog. Biophys. Mol. Biol.* **48**, 127–153.

[85] Spray, D., White, R., Mazetm F., and Bennett, M. (1985) *Am. J. Physiol.* **248**, H753–H764.

[86] Saez, J. C., Gregory, W. A., Watanabe, T., Dermietzel, R., Hertzberg, E. L., Reid, L., Bennett, M. V., and Spray, D. C. (1989) *Am. J. Physiol.* **257**, P C1–11.

[87] Spray, D. Bennett, M. (1985) *Ann. Rev. Physiol.* **47**, 281–303.

[88] Garfield, R., Sims, S., and Daniel, E. (1977) *Science* **198**, 958–960.

[89] Puri, C. and Garfield, R. (1982) *Biol. Reprod.* **27**, 967–975.

[90] Sims, S., Daniel, E., and Garfield, R. (1982) *J. Gen. Physiol.* **80**, 353–375.

[91] Burt, J. M. and Spray, D. C. (1988) *Am. J. Physiol.* **254**, 1206–1210.

[92] Hermsmeyer, K. (1980) *Circ. Res.* **47**, 524–529.

[93] Nicoll, P. and Webb, R. (1955) *Angiology* **6**, 291–308.

[94] Colantuoni, A., Bertuglia, S., and Intaglietta, M. (1985) *Pflugers Arch.* **403**, 289–295.

[95] Meyer, J., Lindbom, L., and Intaglietta, M. (1987) *Am. J. Physiol.* **253**, 568–573.

[96] Slaaf, D., Vrielink, H., Tangelder, G., and Reneman, R. (1988) *Am. J. Physiol.* **255**, H1240–H1243.

[97] Boarders, J. (1980) *Vasomotion Patterns in Skeletal Muscle in Normal and Hypertensive Rats.* PhD Thesis. University Microfilms International, Ann Arbor, MI, p. 132.

[98] Bevan, J. (1979) *Circ. Res.* **45**, 161–171.

[99] Luff, S., McLachlan, E., and Hirst, G. (1987) *J. Comp. Neurol.* **257**, 578–594.

[100] Mellander, S. (1970) *Ann. Rev. Physiol.* **32**, 313–344.

[101] Duling, B. (1990) *The Lung, Scientific Foundations*, (Crystal, R. G., West, J. B., Barnes, P. J., Cherniack, N. S., and Weibel, E. R., eds.), Raven, New York, pp. 1497–1506.

[102] Smaje, L., Zweifach, B., and Intaglietta, M. (1970) *Microvasc. Res.* **2**, 96–110.

[103] Chilian, W. M., Eastham, C. L., and Marcus, M. L. (1986) *Am. J. Physiol.* **251**, H779–H788.

[104] Jackson, W. F. (1987) *Am. J. Physiol.* **253**, 1120–1126.

[105] Honig, C., Odoroff, C., and Frierson, J. (1980) *Am. J. Physiol.* **238**, H31–H42.

[106] Dietrich, H., Weigelt, H., and Lubbers, D. (1984) *Adv. Exp. Med. Biol.* **180**, 701–709.

Chapter 13

Neural Control
of Resistance Arteries

Tim O. Neild and Joseph E. Brayden

I. Neural Control of Vasoconstriction

Introduction

Information on the effects of vasoconstrictor nerves on resistance vessels has come from studies on intact vascular beds and isolated resistance arteries. It was from studies on the resistance of vascular beds that it first became clear that most resistance arteries must receive a vasoconstrictor innervation from the sympathetic nervous system. Following the discovery that norepinephrine was released by sympathetic nerves, many workers showed that antagonists of the action of norepinephrine on vascular smooth muscle also greatly reduced the effect of sympathetic nerves on the resistance of vascular beds. There seemed to be no reason to doubt that norepinephrine was the sympathetic neurotransmitter in this situation. However, studies on the vas deferens, which has many properties in common with arteries, led some workers to question whether norepinephrine was the neurotransmitter. This doubt spread to the field of resistance arteries, and the controversy is still

From *The Resistance Vasculature*, J. A. Bevan et al., eds. ©1991 Humana Press

not settled. Indeed, it has been complicated by the realization that the sympathetic nerves are capable of releasing neuropeptide Y (NPY) and perhaps ATP, in addition to norepinephrine.

The Vasoconstrictor Effects
of Norepinephrine on Resistance Vessels
α_1 and α_2 Adrenoreceptors

Resistance arteries generally constrict in response to norepinephrine. In some vascular beds, the response of the resistance arteries seems to be very similar to that of larger systemic arteries (rabbit ear resistance arteries;[1] rat cremaster muscle arterioles in vivo[2]). In others, the resistance arteries are less sensitive to the vasoconstrictor effects of norepinephrine than larger arteries, and the concentration–response curves are also steeper, with a just-maximal concentration of norepinephrine being of the order of 10 times the threshold concentration (rat mesenteric small artery in vitro,[3,4] guinea-pig intestinal arteriole in vitro[5]). However, in the rat mesenteric resistance arteries and arterioles in vivo, Furness and Marshall[6] reported a higher sensitivity to norepinephrine than was found in the in vitro experiments referred to above. They also found that the smallest arterioles (12 μm diameter) were more sensitive than the larger resistance arteries (330 μm diameter), confirming an earlier observation by Altura.[7]

In contrast to larger arteries, which generally have only α_1 adrenoceptors mediating vasoconstriction, both α_1 or α_2 receptors are involved in the constrictor response of resistance arteries to norepinephrine. This has been shown in experiments on the effect of α_1 and α_2 agonists or antagonists on the peripheral vascular resistance in the whole rat,[8] the monkey brain,[9] and the dog heart.[10] When resistance vessels in the exposed rat cremaster muscle were observed, it was found that the proportion of α_2 receptors contributing to the response was greatest for the smallest arterioles, which had mostly α_2 receptors.[2,11]

Low-Affinity Norepinephrine Receptors-
"Extraceptors" and γ Receptors

Work by two different groups has led to the proposal that resistance arteries (and larger arteries) have a type of norepinephrine receptor that is activated only by high concentrations of norepinephrine, i.e., it has a low affinity for norepinephrine. Low affinity receptors in general are characteristic of neurotransmitter receptors at synapses and neuromuscular junctions where the effect of the neurotransmitter must not be prolonged. At these junctions, such as the skeletal neuromuscular junction, the neurotransmitter briefly attains a high concentration in the vicinity of the postjunctional receptors and binds to them. The low affinity of the receptors then assures the rapid unbinding of the neurotransmitter, so that the system does not remain activated or refractory for a long period. For this reason, low affinity receptors are the most likely to be used for nerve-muscle transmission, whereas high affinity receptors are most appropriate for producing a prolonged response to a humoral factor that is present in low concentrations.

In most arteries, stimulation of the sympathetic nerves gives rise to an excitatory junction potential (e.j.p.) in the arterial smooth muscle. With one exception,[12] which we shall discuss below, the e.j.p. has been found to be resistant to α-adrenoceptor antagonists. This observation led Hirst and Neild[13] to search for a low-affinity norepinephrine receptor that might be responsible for the e.j.p. They applied norepinephrine to the adventitial surface of intestinal arterioles by iontophoresis from a micropipet and identified certain areas on the arterioles at which a depolarization could be produced. Other areas were found at which the only response that could be obtained was a small local α-adrenoceptor mediated constriction. (α-adrenoceptor activation can cause depolarization in these arterioles, but it can only be observed when norepinephrine is applied to a greater area of muscle than is possible using iontophoresis.) When the preparation was stained to show the sympathetic nerves it was found that the areas from which depolarization was obtained were always very close to sympathetic nerves. Hirst and Neild speculated that the depolarizing responses were caused by the activation of a junctionally-located norepinephrine receptor (γ receptor) that was normally activated by nerve-released norepinephrine to produce the e.j.p. It was suggested that the receptor had a low affnity and therefore its effects would not easily be observed in experiments in which norepinephrine was applied to the whole arteriole, as this would activate α-adrenoceptors.

The possibility that there might be low-affinity norepine-phrine receptors on resistance arteries emerged independently from the study of Owen, Quinn, and Bevan[1] on small arteries from the rabbit ear tip. Although most of the contractile responses of these arteries to nerve stimulation was abolished by phentolamine (an antagonist of both α_1 and α_2 receptors), a small part remained. They suggested that this might be a result of norepinephrine acting on some other kind of receptor. This idea was investigated further by Laher, Nishamura, and Bevan,[14] who confirmed the existence of non-α-norepinephrine receptors on the rabbit aorta. They named the receptors "extraceptors," and showed that prazosin was an antagonist of these receptors. They felt that these receptors did not play a major role in neurotransmission, as α-adrenoceptor antago-nists blocked most of the contraction caused by nerve stimulation.

The Role of Norepinephrine Receptors in Neuromuscular Transmission

Given the pharmacological characterization of norepine-phrine receptors outlined above, it might appear that relatively simple experiments could establish which of these receptors is involved in nerve-muscle transmission in resistance arteries. In practice, there is evidence for the involvement of all the receptors, depending on the protocol of the experiment considered. The majority of studies have used long trains of nerve stimulation and shown that contractile responses of resistance vessels or the resis-tance increase in a vascular bed can be almost completely blocked by α-adrenoceptor antagonists. This has led to the general feeling that norepinephrine acting on α-adrenoceptors is the physiologi-cal mechanism. On the other hand, smooth muscle electrophysi-ologists, who for practical reasons usually stimulate nerves with single pulses or short trains, agree that the e.j.p. is not mediated by receptors sensitive to α-adrenoceptor antagonists, and tend to look for another process. Rather than try to review all the work that has given rise to these points of view, we shall consider two studies designed specifically to determine the role of α-adrenoceptors in neuromuscular transmission in resistance vessels, but which have come to different conclusions.

Hirst and Lew[15] measured the hind limb vascular conductance in anesthetized, ganglion-blocked, propranolol-treated rabbits. They observed, as expected, that hind limb conductance was reduced by intraarterial injection of norepinephrine or trains of lumbar sympathetic nerve stimulation (trains of 1–100 pulses at 5 Hz). After the rabbit was treated with the irreversible α_1 and α_2 antagonist benextramine, the responses to norepinephrine were greatly reduced, but the responses to sympathetic nerve stimulation were slightly enhanced. Tests on arteries taken from benextramine-treated rabbits showed that the benextramine had blocked the response to exogenous norepinephrine of mesenteric resistance arteries completely, and reduced the sensitivity of the aorta considerably. There was therefore no doubt that the majority of α-adrenoceptors had been successfully blocked in these experiments, but this had not blocked neurally-induced vasoconstriction in the hind limb. The conclusion from these experiments and similar ones using phenoxybenzamine instead of benextramine was that "at least in the rabbit, sympathetic neural vasoconstriction is largely independent of α-adrenoceptors."

In contrast, Angus, Broughton and Mulvany[16] studied isolated mesenteric resistance arteries from rats, guinea-pigs, and rabbits. They tested the response of the arteries to a supramaximal concentration of norepinephrine (10 μM) and trains of nerve stimulation (mostly 25 Hz for 3 s), and examined the effects of α-adrenoceptor antagonists including benextramine and phenoxybenzamine. α_2-adrenoceptors were blocked to eliminate the effects of the prejunctional α_2-adrenoceptors whose activation reduces norepinephrine release. (Without this precaution, the α-adrenoceptor antagonists increase neurotransmitter release, and it is possible that this might compensate for any blockade of postjunctional receptors.) It appeared in their experiments that a variety of α-adrenoceptor antagonists were equally effective in blocking the responses to both nerve stimulation and exogenous norepinephrine, and there was no important role for another receptor type.

In discussing the discrepancy between their conclusion and that of Hirst and Lew,[15] they pointed out that Hirst and Lew had not eliminated effects on prejunctional α-adrenoceptors. However, Hirst and Lew had considered this and pointed out that the responses to a single pulse of nerve stimulation in their study were resistant to α-adrenoceptor antagonists, and with a single pulse the auto-inhibition mediated by prejunctional α_2-adrenoceptors could not occur.

Another difference between the two studies was the frequency of nerve stimulation used. Most of the data shown by Angus et al. was from experiments in which the nerves were stimulated at 25 Hz, whereas Hirst and Lew stimulated at 5 Hz. Higher frequencies of stimulation would tend to increase the probability of spillover of norepinephrine from the junctional region to α-adrenoceptors on nonjunctional areas of the muscle. The maximum discharge rates of vasoconstrictor neurones supplying the viscera during strong reflex activation do not exceed 10 Hz,[17] suggesting that Hirst and Lews' experiments may have come closer to approx physiological conditions.

Unexpected Effects of α–Adrenoceptor Antagonists

The interpretation of much of the work on noradrenergic neuromuscular transmission relies on the assumption that proven α-adrenoceptor antagonists act mainly, if not exclusively, on α-adrenoceptors. There are, however, some reports that α-adrenoceptor antagonists can have other effects that reduce the contractile responses of arterial smooth muscle. Hirst, DeGlaria, and Van Helden[18] presented evidence that prazosin (1 μM) changes the relationship between membrane depolarization and contraction in arteriolar smooth muscle, so that the arteriole becomes less responsive to depolarizing stimuli. Prazosin appeared to change the voltage dependence of membrane conductances activated by depolarization, and thus prevent action potential generation in submucosal arterioles. This action of prazosin was probably responsible for the report by Morgan[12] that 3 μM prazosin reduced the amplitude of the e.j.p. in arterioles of the cat small intestine. The response shown in this paper was probably an e.j.p. with a superimposed small action potential. Prazosin appears to block the action potential, leaving the underlying e.j.p.

In a larger artery (rabbit ear artery), Holman and Surprenant[19] showed that prolonged exposure to 3 μM phentolamine prevented contraction of the artery in response to nerve stimulation without reducing the amplitude of the e.j.p. or smooth muscle action potential. Both prazosin and phentolamine can therefore block neuromuscular transmission in arteries by mechanisms that do not involve competition with the neurotransmitter for its receptors.

The Vasoconstrictor Effects of ATP and NPY, and Their Possible Role in Neurotransmission

ATP

The failure of α-adrenoceptor antagonists to block sympathetic neuromuscular transmission led some groups to search for a non-α-adrenoceptor as outlined above, while others favored the hypothesis that there was another neurotransmitter in addition to norepinephrine. Most of the early work was done on the vas deferens.[20,21,22] When antagonists of P2 purine receptors became available, they were found to reduce both the component of the contractile response that was resistant to α-adrenoceptor antagonists and the excitatory junction potential in the vas deferens.[23,24] Similar observations on large arteries[25,26] led to the suggestion that the neurotransmitter responsible for the e.j.p. and rapid phase of contraction in the vas deferens and blood vessels was ATP, acting on P2 purine receptors. This work has not been extended to many resistance vessels, but work on small rat cerebral arteries has brought into question the specificity of one of the commonly-used P2 receptor antagonists, αβ-methylene-ATP.[27,23] This antagonist also failed to abolish e.j.ps in the afferent arteriole of the kidney.[29]

There is some doubt that the nerves release sufficient ATP for it to serve as a neurotransmitter.[30] However, it remains a possibility that ATP is the neurotransmitter responsible for the e.j.p. in resistance arteries.

NPY

NPY is found in the sympathetic nerves around all large arteries (except in marsupials[31]), and also in most resistance arteries. There is some evidence that in the guinea-pig uterine artery, NPY might act as an excitatory neurotransmitter,[32] but no evidence that it does so in other vessels. The most prominent action of NPY is to potentiate the response to a wide variety of vasoconstrictor substances (*see* Potter[33] for a review), and it is probable that in larger arteries it is a neuromodulator of the action of other substances rather than an excitatory neurotransmitter in its own right.

In resistance arteries, NPY is not universally present in the sympathetic nerves as it is in larger arteries. In the guinea-pig ear, it is absent from the sympathetic nerves supplying arteriovenous anastomoses and precapillary arterioles, but present in the nerves around other small arteries.[34] Such detailed studies have not been carried out in other tissues, but is likely that similar differences will emerge elsewhere.

The responses of resistance arteries to NPY are similarly diverse. In the rabbit kidney,[35] the smallest resistance arteries constricted in response to NPY, whereas larger arteries in the same vascular bed did not. The potentiating effect of NPY was not observed in the small resistance arteries, but could be demonstrated in the larger arteries. This raises the possibility that NPY might be an excitatory neurotransmitter in the resistance arteries of the kidney, but not in the larger arteries. NPY is also a potent vasoconstrictor of cerebral resistance arteries,[36] and may be a neurotransmitter there also, although experiments to test this point in arteries from the rat suggest that it is not.[37]

In contrast, NPY causes no constriction of arterioles from the guinea-pig small intestine, although it potentiated responses to vasoconstrictors as it does in larger arteries.[38] Clearly here, NPY could only function as a neuromodulator. There is too little data on the exact localization and actions of NPY on resistance arteries to generalize, but enough to warn that there may be considerable diversity of function in different tissues or species.

The Role of Smooth Muscle
Depolarization in Vasoconstriction

It is now clear that most vasoconstrictor substances cause depolarization of the smooth muscle membrane. It is equally clear that at low concentrations, they can cause a small constriction without depolarization, showing that the depolarization is not a prerequisite for the constriction. However, recent work on the membrane properties of the smooth muscle of resistance vessels indicates that moderate depolarization, whatever its cause, will increase calcium influx into the cell. This suggests that depolarization probably contributes to the vasoconstrictor action of many substances, even though it may not be the only process they activate.

In resistance arteries, there are voltage-dependent calcium channels that are opened by depolarization (*see* Hirst and Edwards[39] for a detailed review) and that can give rise to a readily recognizable action potential.[16,40,41] In some cases, however, activation of voltage-dependent conductances gives rise to smaller regenerative responses that do not resemble the large action potentials in other tissues,[41–43] but that still lead to constriction. The failure of these tissues to produce an action potential is probably a result of the simultaneous activation of a potassium conductance,[41,44] but this will not prevent the calcium entry from contributing to tension generation by the muscle.

Reviewing the data on activation of calcium channels from several arteries and arterioles, Hirst and Edwards[39] concluded that depolarization to –50 mV or less would lead to calcium influx, although not necessarily to action potential generation. This could easily be achieved by summation of e.j.ps during sympathetic nerve activity. It is most likely to be important in resistance vessels because of their relatively thin layer of muscle compared to larger arteries.[45] In large arteries, the depolarizing current flowing through channels opened by the neurotransmitter is dissipated through many layers of smooth muscle cells, and thus causes less depolarization than it does in thin-walled resistance arteries.

Depolarization also plays a role in the response of arteries to norepinephrine acting on α-adrenoceptors. As illustrated by the work of Mulvany, Nilsson, and Flatman,[4] α-adrenoceptor activation causes depolarization as well as force generation. When the artery was exposed to a high-potassium solution to depolarize the smooth muscle, the threshold concentration of norepinephrine for force generation was lower and the concentration–response relationship was not as steep as when the muscle was not depolarized. The concentration–response relationship in this situation is probably an indication of the ability of norepinephrine to release calcium ions from intracellular stores. The threshold concentration was lower because intracellular calcium was already raised above the threshold for contraction by the depolarization. The steep concentration–response curve seen in normal conditions represents the combination of at least two processes, calcium release from internal stores and transmembrane influx following the opening of depolarization-sensitive channels. In all resistance arteries that have a steep norepinephrine concentration–response curve (*see* refs. 3–5), it has been shown that norepinephrine causes a large depolarization over the concentration range in which the curve is steepest. It may be that in the other resistance arteries for which the curve is not as steep,[1,2] norepinephrine does not cause the same large change in membrane potential.

In Vivo Recordings
of Smooth Muscle Membrane Potential in Resistance Arteries

In a few instances, recordings of smooth muscle membrane potential have been made from anesthetized animals in which the nerve supply to the arteries was intact and active. Speden[46] recorded from guinea-pig mesenteric arteries and observed oscillations of membrane potential and smooth muscle action potentials. This membrane electrical activity was least in deeply anesthetized animals and increased as the depth of anesthesia was reduced, suggesting that it might be associated with nerve activity. In a more detailed study on rat mesenteric artery, Steedman[47] showed that ongoing nerve activity in moderately anesthetized rats was depolarizing the smooth muscle by 8 mV compared to deeply anesthetized rats. In small arteries of the ear tip of anesthetized rabbits, Neild and Keef[43] showed that blocking nerve activity hyperpolarized the smooth muscle by 6 mV and caused the artery to dilate. Block of α_1-adrenoceptors with prazosin caused less dilatation and no hyperpolarization. From these observations, they concluded that vasoconstrictor nerves were acting partly by release of norepinephrine onto α-adrenoceptors, but there was insufficient α-adrenoceptor activation to cause depolarization or explain all the nerve-mediated tone.

Conclusions

The mechanisms exist for vasoconstrictor nerves to act in two ways: by depolarizing the smooth muscle by the summation of excitatory junction potentials, and by releasing norepinephrine onto α-adrenoceptors (which may also cause depolarization if sufficient receptors are activated). It is not yet clear which of the two processes is the more important in the physiological situation.

In vivo studies indicate that sympathetic nerve activity at the rates occurring in anesthetized animals does depolarize the smooth muscle of resistance arteries by 6–10 mV, bringing the resting potential into the range at which some voltage-sensitive calcium channels might be opened. This suggests that the depolarizing effect of the nerves, usually seen in in vitro experiments as e.j.ps, is important. Further elucidation of the role of nerve-induced depolarization will be easier when the neurotransmitter transmitter mediating the e.j.p. is identified and blockers for its receptor on the muscle are found.

In vivo experiments also show that α-adrenoceptor activation can occur at natural rates of nerve activity, and that the level of activation is too low to cause depolarization. In vitro studies tend to show that the α-adrenoceptor-mediated component of neurogenic vasoconstriction is clearest when long stimulus trains or high frequencies are used. Many studies have used stimulation regimes outside the physiological range, and may therefore have overestimated the importance of α-adrenoceptors in the physiological nervous control of resistance artery diameter.

II. Neural Control of Vasodilation

Introduction

Neurally-mediated vasodilation can occur by direct action of neurotransmitters at the postsynaptic site (active neurogenic vasodilation) or via mechanisms that involve inhibition of tonic sympathetic vasoconstrictor activity (passive vasodilation). Passive vasodilation can occur, for instance, when the release of an excitatory neurotransmitter, such as norepinephrine is inhibited by the presynaptic action of a second transmitter, such as acetylcholine.[48] However, passive vasodilator mechanisms can be difficult to demonstrate, particularly at the level of the resistance vasculature using in vitro techniques, which often are the most effective way of establishing the identity of the transmitters involved in a neurally mediated event. In contrast, active neurogenic vasodilation, involving potent endogenous substances, such as acetylcholine or vasoactive intestinal polypeptide (VIP), has been observed in many tissues using both in vivo and in vitro techniques. Consequently, this discussion will focus on active neurogenic vasodilation and will review the evidence in support of a vasodilator neurotransmitter role for a variety of substances at the level of the resistance vasculature.

Specific Transmitters
Involved In Neurodilation of Resistance Arteries

Evidence for functional, nonadrenergic dilator innervation of the vasculature has been reviewed recently.[49] Among the dozen or so putative dilator transmitters considered at that time, evidence of a functional neurodilator role was most convincing for only a few substances, namely acetylcholine, VIP, dopamine, and substance P. Data supporting a role for these dilator neurotransmitters in small arteries are extremely limited, but do suggest that the above-mentioned substances as well as calcitonin gene-related peptide (CGRP) are dilator neurotransmitters in the resistance vasculature.

Acetylcholine

Several indices of cholinergic innervation of blood vessels have been measured during the past two decades. These include biochemical indicators of innervation, such as choline acetyltransferase (ChAT) activity, uptake of labeled choline, and release of acetylcholine.[50–52] Most of such measurements have been made on large arteries. However, these biochemical markers of cholinergic innervation are also expressed in small pial arteries[53,54] and in intraparenchymal vessels (mixed population of arterioles, capillaries, and veinules[55,56]) in the cerebral circulation. In addition, ChAT-immunoreactivity has been identified by Saito, Wu, and Lee[57] in periarterial nerves that distribute to small pial arteries and by Eckenstein and Baughman[58] and Armstrong[59] in intraparenchymal nerves closely associated with arteries. Comparable measures for extracerebral resistance arteries have not been reported.

Numerous in vivo and perfused tissue studies suggest the presence of functional cholinergic neurodilator systems in the resis-tance vasculature. For instance, in skeletal muscle of cats and dogs, an atropine-sensitive vasodilation in response to stimulation of the sympathetic nerves was first noted by Bulbring and Burn[60] over 50 years ago. Canine coronary resistance is decreased substantially following vagal stimulation, and this response is blocked by atropine.[61] More recent studies in humans revealed an atropine-sensitive decrease in vascular resistance in the skeletal muscle of the forearm during mental stress.[62] This response is presumably mediated by neural outflow from the central nervous system. Stimulation of the fastigial nucleus results in an increased cerebral blood flow in the rat that is inhibited by muscarinic antagonists and by selective ablation of neuronal pathways originating in the fastigial nucleus.[55,63] Dilator responses to nerve-stimulation in the feline gastrointestinal tract, which involve the resistance vasculature and which are partially sensitive to atropine, have been described.[64] Similar responses, almost certainly involving some degree of participation by resistance arteries, have also been observed in the feline lung[65] and salivary glands.[66] The hepatic microcirculation of the rat also is subject to parasympathetic vasodilator control mechanisms.[67]

Although these studies imply the presence of a neurally-mediated cholinergic dilator system in the resistance vasculature of several tissues, confirmation of this possibility, through direct observation of such a response in isolated resistance arteries, has only been reported very recently. Neild and coworkers[68] found that electrical stimulation of single nerves or ganglia in the guinea-pig submucosal plexus caused dilation of arterioles within the plexus. The vasodilation was blocked by tetrodotoxin and by muscarinic receptor antagonists. Pharmacological characterization of the receptor involved suggested that it was of the M_3 subtype. Exogenously applied muscarinic agonists also caused arteriolar dilation and neither this response nor the neurally evoked dilation was reduced by gossypol, an inhibitor of the production of endothelium-derived relaxing factor. These studies clearly demonstrate a cholinergic neurodilator system in resistance arteries of the guinea-pig in which acetylcholine acts via postsynaptic receptors located directly on the vascular smooth muscle cells. This system is comparable to that reported for two larger arteries, namely the feline auricular artery[69] and the rabbit lingual artery.[70] Neurally mediated dilation of these cephalic arteries is inhibited by atropine, potentiated by physostigmine, and is independent of the endothelium. Small branches (lumen diameter: 100–200 μm) of the rabbit lingual artery show similar behavior when studied in vitro (Brayden, unpublished observation). Considered together, these observations strongly support the suggestion that the resistance vasculature of some parts of the head, gut, and perhaps other tissues, is subject to cholinergic neurovasodilator influence.

VIP

Since its identification by Said and Mutt,[71] VIP has been implicated as a neurally derived vasodilator in many vascular beds. With respect to the resistance vasculature, immunohistochemical studies have demonstrated VIP within perivascular nerves that supply small arteries (internal diameter <200 μm) in the cerebral,[72,73] coronary, gastrointestinal, skeletal, renal,[74] genitourinary, upper respiratory, ocular,[75] and salivary gland circulations.[66,73]

Nonadrenergic, noncholinergic vasodilation following stimulation of the autonomic nervous supply has been observed in several vascular beds, including those supplying the brain,[76,77] the salivary glands,[66] the tongue,[78] the nasal mucosa,[79] the uterus,[80] and the gastrointestinal tract.[81] This response is mimicked by exogenous VIP and release of VIP from some of these tissues during activation of nerves has been measured.[66,81] The comprehensive studies of Lundberg[66] provide convincing evidence of a neurodilator function for VIP in the vasculature of the feline submandibular salivary gland. VIP is synthesized in cell bodies within the submandibular ganglia and is released from the parasympathetic nerve terminals on neural activation, resulting in inhibition of vascular tone. It seems likely that much or all of this response occurs at the level of the resistance vasculature, because the small arteries and arterioles in this tissue are primary recipients of VIP-immuno-reactive perivascular nerves. A neurotransmitter role for VIP in small arteries has also been demonstrated in studies of isolated pial vessels of the cat. VIP has been identified at the ultrastructural level within neuronal vesicles in pial perivascular nerves.[77] VIP is released from cerebral perivascular nerves[82] and is a potent dilator of these arteries. In addition, the magnitude of neurally evoked dilation correlates with the degree of innervation of the pial arteries, and an antibody raised against VIP inhibits the response to nerve-stimulation and to exogenous VIP.[83] The neurotransmitter role of VIP in resistance arteries from other vascular beds remains to be determined. However, based on the distribution of VIP-immunoreactive perivascular nerve fibers and potent dilator action of this peptide, a neurovasodilator contribution to control of vascular resistance in many regional vascular beds seems likely.

Substance P

Substance P is a potent vasodilator peptide that has been identified in periarterial nerves. Such nerve fibers distribute to small arteries and arterioles in central and peripheral vascular beds,[84,85] although in general, the densest innervation seems to occur on larger, muscular arteries.

Substance P is found primarily in sensory C fibers and can be depleted by treatment with capsaicin.[86] This peptide has been implicated in a nocisensor reflex response in cutaneous tissues[87] and in the phenomenon of antidromic vasodilation.[88] Vasodilation induced by exogenous substance P or by stimulation of sensory nerves is blocked by substance P-antagonists such as [D-Pro^2,D-Phe^7,D-Trp^9]substance P and [D-Pro^2,D-Trp^{7,9}]substance P.[88,89] Similar action of substance P antagonists has been observed in the dental pulp.[89,91] These responses probably have a significant component at the level of the resistance vasculature but definitive studies identifying a specific role for substance P as a dilator transmitter in resistance arteries are only beginning to emerge. A recent report by Galligan and coworkers[92] demonstrates that under conditions when the extrinsic innervation of the submucosal circulation of the guinea-pig is eliminated, a significant substance P-mediated neural dilator system is revealed. In this preparation, dilator responses of submucosal arterioles to exogenous substance P and to activation of intrinsic substance P immunoreactive nerve fibers, were inhibited by the substance P antagonists spantide and [D-Arg^1, D-Phe^5,D-Trp^{79},Leu^{11}]Substance P. Although the functional significance of this system under physiological conditions is not clear, this study provides clear evidence of a substance P neurodilator system in the resistance vasculature.

CGRP

CGRP is a 37 amino acid peptide produced in neural tissue as a result of tissue-specffic, alternative processing of mRNA transcribed from the calcitonin gene.[93] CGRP-immunoreactive nerves are distributed throughout the cardiovascular system and have been identified in perivascular nerves that supply small diameter arteries in the respiratory tract, gastrointestinal tract, and genitourinary tract,[94] and in the coronary, cutaneous, skeletal muscle,[95,96] and cerebral circulations.[97] In some species CGRP is found in sensory nerves and, at least in the guinea-pig, coexists with substance P in many of these nerve fibers.[95,98,99] CGRP has potent vasodilator activity on most arteries, including small cerebral and systemic blood vessels.[96,97] In these arteries, half-maximal dilation occurs in response to 10–30 nM CGRP.

A recent study of the rat mesenteric vascular bed provides the strongest evidence for a neurodilator role of CGRP in resistance arteries.[100] In these experiments, the mesenteric vascular bed was isolated and perfused at constant flow. Activation of periarterial nerves evoked a frequency-dependent decrease in vascular resistance, which was abolished by tetrodotoxin. The vasodilation was mimicked by exogenous CGRP but not by substance P. Arterioles in this vascular bed receive a dense supply of CGRP-immunoreactive nerves, but only a few substance P-immunoreactive nerves. CGRP-immunoreactive nerves and the nerve-evoked dilation were eliminated by capsaicin treatment. In addition, frequency-dependent release of CGRP from this vascular bed during electrical stimulation has been demonstrated.[101] The nurodilator role of CGRP in the mesenteric vascular bed of the rat has recently been confirmed in studies describing the specific antagonist action of a CGRP analogue (CGRP 8–37) on neurally evoked dilations in isolated, perfused rat mesenteric arteries.[102] Taken together, these data strongly suggest a specific neurodilator role for CGRP in this vascular bed. Availability of a specific CGRP-antagonist should allow extension of this observation to other vascular beds and other species.

Dopamine

Evidence has been presented that indicates a possible functional dopaminergic dilator innervation of small arteries in the renal cortex and in cutaneous arteriovenous anastamoses (for review, *see* Bell[103]). For technical reasons, primarily because all adrenergic nerves contain appreciable amounts of dopamine, it is difficult to determine the presence of distinct dopaminergic nerve fibers using direct techniques such as immunohistochemistry. Nevertheless, indirect measures support the existence of such nerves, primarily in the dog. For instance, ratios of dopamine to norepinephrine in the kidney and skin are much higher than those observed in other tissues receiving only adrenergic nerves.[104,105] Also, a distinct population of nerves within ganglia that distribute nerves to the kidney and skin does not contain the enzyme dopamine beta hydroxylase, which is necessary for synthesis of norepinephrine, but does contain DOPA decarboxylase, required for synthesis of dopamine from DOPA.[106] A micro spectrofluorimetric technique indicated the presence of renal perivascular nerves that contain dopamine, but not norepinephrine, in renal cortical and juxtaglomerular arterioles.[107]

Dopamine decreases arterial tone in many tissues, but this response is particularly well-developed in the mesentery,[108] kidney,[109] and in cutaneous tissues;[110] the dilator action of dopamine is blocked by dopamine antagonists, such as haloperidol, ergometrine and sulpride. Activation of sympathetic outflow to the canine kidney and skin can cause vasodilation that also is inhibited by dopamine antagonists.[111,112] These data suggest a specific dopaminergic neurodilator influence on renal and cutaneous resistance arteries. In vitro studies of isolated renal afferent and efferent arterioles have demonstrated a potent dilator effect of exogenous dopamine that is inhibited by the dopamine receptor antagonist metoclopramide.[113] Similarly, exogenous dopamine dilates human subcutaneous and omental resistance arteries in vitro and this response is inhibited by dopamine receptor antagonists.[114] However, evidence from in vitro studies that would indicate the existence of a neurogenic dopaminergic dilator system in resistance arteries is not yet available.

Other Putative Dilator Transmitters

Several other substances should be considered as possible dilator neurotransmitters in the resistance vasculature. For instance, activation of beta adrenoceptors results in vasodilation in several vascular beds, in particular those in skeletal muscle[115] and the heart[116] and in veins.[117] Stimulation of perivascular nerves induces a adrenoceptor-mediated relaxation of large coronary arteries,[118] and of the facial vein of the rabbit.[117]

However, there is no clear evidence of beta-mediated neurogenic vasodilation in any resistance artery. Histamine, serotonin, and purines (adenosine, ADP, ATP) relax arteries, but few other specific criteria that would help establish these agents as dilator neurotransmitters have been satisfied.[49] Information pertaining to the role of these substances in the resistance vasculature is virtually nonexistent.

Mechanisms of Action

Recent evidence indicates that membrane hyperpolarization may be a common mechanism of action of many vasodilators,[119] including several of the neurotransmitters discussed above. Acetylcholine, VIP, and CGRP can induce large hyperpolarizations (10–20 mV) in vascular smooth muscle. The ATP-sensitive potassium channel (K_{ATP}), a specific potassium conductance pathway that is inhibited by ATP and by sulfonyl urea compounds such as glibenclamide,[120] is centrally involved in this vasodilator mechanism. This channel has been identified and characterized in vascular smooth muscle using the patch clamp technique and is activated by vasodilator neurotransmitters.[119,121] Glibenclamide abolishes the hyperpolarizations induced by dilator transmitters and inhibits the vasodilator response. Membrane hyperpolarization will close voltage-dependent calcium channels, and thereby decrease calcium influx. In light of the large dependence of small arteries on calcium influx through voltage-dependent calcium channels for maintenance of force,[122,123] alteration of calcium entry pathways could have substantial effects on vascular tone in resistance arteries. In this regard, the effects of vasodilator neurotransmitters on calcium entry in resistance arteries will be an important area of future investigation.

Although the vasodilator action of neurotransmitter substances is reduced following blockade of membrane hyperpolarization, a residual component of dilation has been reported.[121] This finding suggests the presence of additional mechanisms of action such as second messenger-mediated increases in calcium extrusion or sequestration or perhaps direct effects on calcium channels resulting in their closure. Cyclic AMP (cAMP) is a potential second messenger involved in such a mechanism—CGRP and VIP both cause elevation of cAMP in vascular smooth muscle.[124,125] However, these measurements have been made only on larger arteries and the involvement of such changes and the precise mechanism of action in resistance arteries remain to be determined.

Conclusions

Neurally mediated vasodilation of resistance arteries has been observed in many different vascular beds in a number of species. The most likely neurotransmitters involved in this response are acetylcholine, VIP, dopamine, CGRP, and substance P. Membrane hyperpolarization and resultant decrease in calcium influx may be an important mechanism of action of these substances, but other pathways of dilation are probably involved as well. Although the physiological conditions under which these neural systems are activated remain to be elucidated, it is likely that dilator neurotransmitters play an important role in regulation of vascular resistance, and perhaps in the etiology of vascular pathologies, such as migraine headache.[126]

References

[1] Owen, M. P., Quinn, C., and Bevan, J. A. (1985) *Am. J. Physiol.* **249,** H404–H414.

[2] Faber, J. E. (1988) *Circ. Res.* **62,** 37–50.

[3] Nilsson, H. (1984) *Acta Physiol. Scand.* **121**, 353–361.

[4] Mulvany, M. J., Nilsson, H., and Flatman, J. A. (1982) *J. Physiol.* **332**, 363–373.

[5] Neild, T. O. and Kotecha, N. (1989) *Microvasc. Res.* **38**, 186–199.

[6] Furness, J. B. and Marshall, J. M. (1974) *J. Physiol.* **239**, 75–88.

[7] Altura, B. M. (1984) *Microvasc. Res.* **3**, 361–384.

[8] Docherty, J. R. and McGrath, J. C. (1980) *Naunyn–Schmiedeberg's Arch. Pharmacol.* **312**, 107–116.

[9] Goadsby, P. J., Lambert, G. A., and Lance, J. W. (1985) *Brain Res.* **326**, 213–217.

[10] Heusch, G., Deussen, A., Schipke, J., and Thamer, V. (1984) *J. Cardiovasc. Pharmacol.* **69**, 61–68.

[11] Faber, J. E. and McGillivray, K. M. (1988) *Resistance Arteries* (Halpern, W., Pegram, B. L., Brayden, J. E., Mackey, K., McLaughlin, M. K., and Osol, G., eds.) Perinatology, Ithaca, NY, pp. 19–128.

[12] Morgan, K. G. (1983) *Am. J. Physiol.* **244**, H540–H545.

[13] Hirst, G. D. S. and Neild, T. O. (1981) *J. Physiol.* **313**, 343–350.

[14] Laher, L., Nishimura, S., Bevan, J. A. (1986) *J. Pharmacol. Exp. Ther.* **239**, 846–852.

[15] Hirst, G. D. S. and Lew, M. J. (1987) *Brit. J. Pharmacol.* **90**, 51–60.

[16] Angus, J. A., Broughton, A., and Mulvany, M. J. (1988) *J. Physiol.* **403**, 495–510.

[17] Janig, W. (1988) *Ann. Rev. Physiol.* **50**, 525–539.

[18] Hirst, G. D. S., De Glaria, S., and van Helden, D. F (1985) *Experientia* **41**, 874–879.

[19] Holman, M. E. and Surprenant, A. (1980) *Br. J. Pharmacol.* **71**, 651–661.

[20] Ambache, N. and Zar, M. A. (1971) *J. Physiol.* **216**, 359–389.

[21] vonEuler, U. S. and Hedqvist, P. (1975) *Acta Physiol. Scand.* **93**, 572–573.

[22] McGrath, J. C. (1980) *Nature* **288**, 301–302.

[23] Sneddon, P. and Westfall, D. P. (1984) *J. Physiol.* **347**, 561–580.

[24] Sneddon, P., Westfall, D. P., and Fedan, J. S. (1982) *Science* **218**, 693–694.

[25] Sneddon, P. and Burnstock, G. (1984) *Eur. J. Pharmacol.* **106**, 149–152.

[26] Kugelgen, I. V. and Starke, K. (1985) *Physiol.* **367**, 435–455.

[27] Byrne, N. G. and Large, W. A. (1986) *Br. J. Pharmacol.* **88**, 6–8.

[28] Edwards, F. R., Hards, D., Hirst, G. D. S., and Silverberg, G. D. (1989) *Br. J. Pharmacol.* **96**, 785–788.

[29] Buhrle, C. P., Scholz, H., Nobiling, R., and Taugner, R. (1986) *Pflugers Arch.* **406**, 578–586.

[30] Freid, G., Langercrantz, H., Klein, R., and Thureson-Klein, A. (1984) *Catecholamines, Basic and Peripheral Mechanisms,* (Usdin, B., Carlsson, A., Dahlstrom, S., and Eagel, M., eds.), Liss, New York, pp. 45–53.

[31] Morris, J. L., Gibbins, I. L., and Murphy, R. (1986) *Neuroscience Letters* **71**, 264–270.

[32] Morris, J. L. and Murphy, R. (1988) *J. Auton. Nerv. Syst.* **24,** 241–249.

[33] Potter, E. K. (1988) *Phamac. Ther.* **37,** 251–273.

[34] Gibbins, I. L. and Morris, J. L. (1990) *J. Auton. Nerv. Syst.* **29,** 137–150.

[35] Owen, M. P. and Taphorn, M. C. (1988) Resistance Arteries, Halpern, W., Pegram, B. L., Brayden, J. E., Mackey, K., McLaughlin, M. K., and Osol, G., eds.) Perinatology, Ithaca, NY, pp. 74–79.

[36] Edvinsson, L., Emson, P., McCulloch, J., Tatemoto, K., and Uddman, R. (1984) *Acta Physiol. Scand.* **122,** 155–163.

[37] Brayden, J. E. and Conway, M. A. (1988) *Regul. Pept.* **22,** 253–265.

[38] Neild, T. O. and Kotecha, N. (1990) *J. Auton. Nerv. Syst.* **30,** 29–36.

[39] Hirst, G. D. S. and Edwards, F. R (1988) *Physiol. Rev.* **69,** 546–604.

[40] Hirst, G. D. S. (1977) *J. Physiol.* **273,** 263–275.

[41] Hirst, G. D. S., Silverberg, G. D., and VanHelden, D. F. (1986) *J. Physiol.* **371,** 289–304.

[42] Bolton, T. B., Lang, R. J., and Takewaki, T. (1984) *J. Physiol.* **351,** 549–572.

[43] Neild, T. O. and Keef, K. (1985) *Microvasc. Res.* **30,** 19–28.

[44] Bolton, T. B. and Large, W. A. (1986) *J. Exp. Physiol.* **71,** 1–28.

[45] Neild, T. O. (1984) *Proc. Roy. Soc. B.* **220,** 237–249.

[46] Speden, R. N. (1964) *Nature* **202,** 193,194.

[47] Steedman, W. M. (1966) *J. Physiol.* **186,** 382–400.

[48] Cohen, R. A., Shepherd, J. T., and Vanhoutte, P. M. (1984) *J. Pharmacol. Exp. Ther.* **229,** 417–421.

[49] Bevan, J. A. and Brayden, J. E. (1987) *Circ. Res.* **60,** 309–326.

[50] Florence, V. M. and Bevan, J. A. (1979) *Circ. Res.* **45,** 212–218.

[51] Duckles, S. P. (1981) *J. Pharmacol. Exp. Ther.* **217,** 544–548.

[52] Bevan, J. A., Buga, G. M., Jope, C. A., Jope, R. S., and Moritoki, H., (1982) *Circ. Res.* **51,** 421–429.

[53] Hamel, E., Assumel-Lurdin, C., Edvinsson, L., and MacKenzie, E. T. (1986) *Acta Physiol. Scand.* **127 (Suppl. 552),** 13–16.

[54] Hamel, E., Assumel-Lurdin, C., Edvinsson, L., Fage, D., and MacKenzie, E. T. (1987) *Brain Res.* **420,** 391–396.

[55] Arneric, S. P., Iadecola, C., Underwood, M. D., and Reis, D. J. (1987) *Brain Res.* **411,** 212–225.

[56] Shimon, M., Egozi, Y., Kloog, Y., Sokolovsky, M., and Cohen, S. J. (1988) *Neurochem.* **50,** 1719–1724.

[57] Saito, A., Wu, J.-Y., and Lee, T. J.-F. (1985) *J. Cereb. Blood Flow Metab.* **5,** 327–334.

[58] Eckenstein, F. and Baughman, R. W. (1984) *Nature* **309,** 153–155.

[59] Armstrong, D. M. (1986) *J. Comp. Neurol.* **250,** 81–92.

[60] Bulbring, E. and Burn, J. H. (1936) *J. Physiol.* **87,** 254–274.

[61] Feigl, E. O. (1969) *Circ. Res.* **25,** 509–519.

[62] Sanders, J. S., Mark, A. L., and Fergason, D. W. (1989) *Circulation* **79,** 815–824.

[63] Iadecola, C., Underwood, M. D., and Reis, D. J. (1986) *Brain Res.* **368**, 375–379.

[64] Andersson, P. O., Bloom, S. R., and Jarhult, J. (1983) *J. Physiol.* **334**, 293–307.

[65] Nandiwada, P. A., Hyman, A. L., and Kadowitz, P. J. (1983)*Circ. Res.* **53**, 86–95.

[66] Lundberg, J. M. (1981) *Acta Physiol. Scand.* **112 (Suppl. 496)**, 1–57.

[67] Koo, A. and Liang, I. Y. S. (1979) *J. Exp. Physiol.* **64**, 149–159.

[68] Neild, T. O., Shen, K. Z., and Surprenant, A. (1990) *J. Physiol.* **420**, 247–265.

[69] Brayden, J. E. and Bevan, J. A. (1985) *Circ. Res.* **56**, 205–211.

[70] Brayden, J. E. and Large W. (1986) *Br. J. Pharmacol.* **89**, 163–171.

[71] Said, S. I. and Mutt, V. (1970) *Science* **169**, 1217,1218.

[72] Hokfelt, T., Schultzburg, M., Lundberg, J., Fuxe, K., Mutt, V., Fahrendrug, J., and Said, S. I. (1982) Vasoactive Intestinal Polypeptide (Said, S. I., ed.), Raven, New York, pp. 65–90.

[73] Gibbins, I. L., Brayden, J. E., and Bevan, J. A. (1984) *Neuroscience* **13**, 1327–1346.

[74] Della, N. G., Papka, R. E., Furness, J. B., and Costa M. (1983) *Neuroscience* **9**, 605–619.

[75] Uddman, R., Alumets, J., Edvinsson, L, Hakanson, R., and Sundler, F. (1981) *Acta Physiol. Scand.* **112**, 65–70.

[76] Lee, T. J.-F. and Bevan, J. A. (1975) *Experientia* **31**, 1424–1425.

[77] Lee, T. J.-F., Saito, A., and Berezin I. (1984) *Sciencr* **224**, 898–901.

[78] Lundberg, J. M., Anggard, A., and Fahrenkrug, J. (1982) *Acta Physiol. Scand.* **116**, 387–392.

[79] Lundberg, J. M., Anggard, A., Emson, P., Fahrenkrug, J., and Hokfelt, T. (1981) *Proc. Natl. Acad. Sci. USA* **78**, 5255–5259.

[80] Fahrenkrug, J. and Ottesen, B. (1982) *J. Physiol.* **331**, 331–460.

[81] Bloom, S. R. and Edwards, A. V. (1980) *J. Physiol.* **229**, 437–452.

[82] Bevan, J. A., Moskowitz, M., Said, S. I., and Buga, G. M. (1984) *Peptides* **5**, 385–388.

[83] Brayden, J. E. and Bevan, J. A. (1986) *Stroke* **17**, 1189–1192.

[84] Edvinsson, L., McCulloch, Uddman, R. (1981) *J. Physiol.* **318**, 251–258.

[85] Furness, J. B., Papka, R. E., Della, N. G., Costa, M., and Eskay, R. L. (1982) *Neuroscience* **7**, 447–459.

[86] Jessell, T. M., Iversen, L. L., and Cuello, A. C. (1978) *Brain Res.* **152**, 183–188.

[87] Pernow, B. (1983) *Pharmacol. Rev.* **35**, 85–141.

[88] Lembeck, F. and Holzer, P. (1979) *Naunyn-Schmiederberg's Arch. Pharmacol.* **310**, 175–183.

[89] Rosell, S., Olgart, L., Gazelius, B., Panopoulos, P., Folker, K., and Horig, J. (1981) *Acta Physiol. Scand.* **111**, 381,382.

[90] Lembeck, F., Donnerer, J., and Bartho, L. (1982) *Eur. J. Pharmacol.* **85,** 171–176.

[91] Gazelius, B. and Olgart, L. (1980) *Acta Physiol. Scand.* **108,** 181–186.

[92] Galligan, J. J., Ming-ming, Jiang, Sjen, K. Z., and Surprenant, A. (1990) *J. Physiol.* **420,** 267–280.

[93] Rosenfeld, M., Mermod, J., Amara, S., Swanson, L., Sawchenko, P. E., Rivier, J., Vale, W. W., and Evans, R. M. (1983) *Nature* **304,** 129–135.

[94] Uddman, R., Edvinsson, L., Ekblad, E., Hakanson, R., and Sundler, F. (1986) *Reg. Peptides* **15,** 1–23.

[95] Gibbins, I. L., Furness, J. B., Costa, M., MacIntyre, I., Hillyard, C. J., and Girgis, S. (1985) *Neurosci. Lett.* **57,** 125–130.

[96] Hanko, J., Hardebo, J. E., Kahrstrom, J., Owman, C., and Sundler, F. (1985) *Neurosci. Lett.* **57,** 91–95.

[97] Edvinsson, L., Ekman, R., Jansen, I., McCulloch, J., and Uddman, R. (1987) *J. Cer. Blood Flow Metab.* **7,** 720–728.

[98] Gulbenkian, S., Merighi, A., Wharton, J, Varndell, I. M., and Polak, J. M. (1986) *J. Neurocytology* **15,** 535–542.

[99] Wharton, J., Gulbenkian, S., Mulderry, P., Ghatei, M., McGregor, G, Bloom, S. P., and Polak, J. (1986) *J. Autonom. Nervous Syst.* **16,** 289–309.

[100] Kawasaki, H., Takasaki, K., Saito, A., and Goto K. (1988) *Nature* **335,** 164–167.

[101] Fujimori, A., Saito, A., Kimura, S., Watanabe, T., Uchiyama, Y., Kawasaki, H., and Goto, K. (1989) *Biochem. Biophys. Res. Comm.* **165,** 1391–1398.

[102] Han, S.-P., Naes, L., and Westfall, T. C. (1990) *Biochem. Biophys. Res. Comm.* **168,** 786–791.

[103] Bell, C., Burnstock, G., and Griffith, S. G., eds. (1988) *Nonadrenergic Innervation of Blood Vessels,* vol. 1. CRC Press, Boca Raton, FL, pp. 41–64.

[104] Bell, C. and Gillespie, J. S. (1981) *Neurochem.* **36,** 703–706.

[105] Bell, C., Lang, W., and Laska, F. (1978) *J. Neurochem.* **31,** 77–83.

[106] Bell, C. and Muller, B. D. (1982) *Neurosci. Lett.* **31,** 31–35.

[107] Dinerstein, R. J., Vannice, J., Henderson, R. C., Roth, L., Goldberg, L. I., and Hoffman, P. C. (1979) *Science* **205,** 497–499.

[108] Nichols, A. J. and Hiley, C. R. (1985) *J. Pharm. Pharmacol.* **37,** 110–115.

[109] Toda, N. and Goldberg, L. I. (1973) *J. Pharm. Pharmacol.* **25,** 587–589.

[110] Bell, C. and Lang, W.J. (1979) *Br. J. Pharm.* **67,** 337–343.

[111] Lang, W.J, Bell, C., and Conway, F. L (1976) *Circ. Res.* **38,** 560–566.

[112] Bell, C. and Lang, W.J. (1973) *Nature New Biol.* **246,** 27–29.

[113] Edwards, R. M. (1985) *Am. J. Physiol.* **248,** F183–F189.

[114] Hughes, A. D. and Sever, P. S. (1989) *Br. J. Pharm.* **97,** 950–956.

[115] Mellander, S. and Johansson, B. (1968) *Pharmacol. Rev.* **20,** 117–196.

[116] Ross, G. (1976) *Circ. Res.* **39,** 461–465.

[117] Pegram, B. L., Bevan, R. D., and Bevan, J. A. (1976) *Circ. Res.* **39,** 854–860.

[118] Cohen, R. A., Shepherd, J. T., and Vanhoutte, P. M. (1983) *Circ. Res.* **52,** 16–25.

[119] Standen, N. B., Quayle, M., Davies, N. W., Brayden, J. E., Huang, Y., and Nelson, M. T. (1989) *Science* **245,** 177–180.

[120] Ashcroft, F. M. (1988) *Annu. Rev. Neurosci.* **11,** 97–118.

[121] Nelson, M. T., Huang, Y., Brayden, J. E., Hescheler, J., and Standen, N. B. (1990) *Nature* **344,** 770–773.

[122] Towart, R. (1981) *Circ. Res.* **8,** 650–657.

[123] Cauvin, C., Saida, K., and van Breeman, C. (1984) *Blood Vessels* **21,** 23–31.

[124] Edvinsson, L., Fredholm, B. B., Hamel, E., Jansen, I., and Verrecchia, C. (1985) *Neurosci. Lett.* **58,** 213–217.

[125] Itoh, T., Sasafuri, T., Makita, Y., Kanmura, Y., and Kuriyama, H. (1985) *Am. J. Physiol.* **249,** H231–H240.

[126] Moskowitz, M. A. (1984) *Ann. Neurol.* **16,** 157–168.

Chapter 14

Local Metabolic Influences on Resistance Vessels

David E. Mohrman

Introduction

In many important organs such as the brain, heart, and skeletal muscle, it is clear that the control of resistance-vessel tone is dominated by local metabolic mechanisms that automatically adjust organ blood flow to match changing metabolic needs of the tissue. This chapter will highlight results from microcirculatory approaches to identifying the mechanisms important in local vascular control. Also considered are the interactions that must exist in vivo between local metabolic mechanisms and the other known influences on resistance vessels, described in other chapters of this work.

Mechanisms of Metabolic Vasodilation

General Aspects

The search for the mechanism of metabolic vasodilation is well over 100 years old[1] and is yet unfulfilled. The current consensus is that several of the host of internal chemical changes that occur within a tissue as a result of increased metabolic activity act in con-

From *The Resistance Vasculature*, J. A. Bevan et al., eds. ©1991 Humana Press

cert upon resistance vessels to produce metabolic vasodilation.[2,3] Among the local tissue factors thought to be important in local vascular control are P_{O_2}, P_{CO_2} and/or pH, K^+, osmolarity, and adenosine.[4] It has proven difficult to assign relative importance to potential mediators of metabolic vasodilation for two reasons. First, since the overall process is one of negative feedback, the elimination of any one factor may be immediately compensated for by automatic adjustments in others. Second, there may be synergistic interactions between factors (*see*, e.g., Mohrman and Regal[15]) that make it difficult to determine experimentally the vasodilatory potency of any single factor.

Graded arteriolar dilation responses to increasing metabolic rate have been demonstrated in a number of microcirculatory preparations of different tissues, including skeletal muscle[6–10] heart,[11] brain,[12] and intestine.[13] In general, arterioles of all sizes seem to participate in the response. Some report that the smallest terminal arterioles show the largest percentage increase in diameter in response to a given metabolic stimulus,[6,11,14] but this is not a universal finding.[9] At least in skeletal muscle, the time and rate of onset of vasodilation in large and small arterioles are similar, whereas the smaller arterioles seem to recover much more rapidly than larger vessels.[6,8,15]

In an especially important study, Gorczynski et al.[8] observed arteriolar responses to stimulation of single or small groups of skeletal muscle fibers in the hamster cremaster preparation. They found that vasodilator responses were confined to those segments of arterioles in direct apposition to the contracting muscle cells. This seems clear proof that arteriolar segments do respond to the conditions in their immediate local environment and that these responses are not automatically propagated to adjacent arteriolar segments. Gorczynski et al. also found distinctly biphasic dilator responses to muscle stimulation, which supports the idea of multifactorial mechanisms for metabolic vasodilation. One important advantage of activating single skeletal-muscle fibers as Gorczynski et al. did is that the resulting vascular response is confined to short vessel segments and therefore does not automatically cause large increases in tissue flow, which complicate the interpretation of the results by secondarily modifying the conditions in the tissue.

Oxygen

For obvious reasons, oxygen has long been considered a primary candidate for involvement in local vascular control in a number of tissues.[3,12,13,16,17] Without doubt, arteriolar diameter measured in microcirculatory preparations changes in response to changes in the P_{O_2} of the suffusing solution.[18-23] The difficult question is whether this is a result of a direct effect of decreased P_{O_2} on arteriole vessels or a secondary result of some effect of reduced P_{O_2} on parenchymal cells or other elements of the tissue.[22] To evaluate the direct role of P_{O_2} in metabolic vasodilation, it is necessary to establish (1) how sensitive resistance vessels are to changes in P_{O_2} and (2) what changes in periarteriolar P_{O_2} actually occur during periods of increased metabolic activity.

There are conflicting reports on the sensitivity of arterioles to oxygen. Early work by Duling[24] indicated that arterioles are insensitive to oxygen. Later work from the same laboratory, [21] which involved in vitro experiments on 50–μm arterioles as well as novel *in situ* experiments on arteriolar segments from which the surrounding parenchyma had been removed, indicated that arterioles are indeed directly sensitive to changes in oxygen tension. Recently, on the basis of further in vivo studies, which included measurements of periarteriolar and luminal P_{O_2}, Jackson[22] concluded that arterioles larger than 15 m in hamster cheek pouch are not intrinsically sensitive to oxygen.

It is well established that oxygen passes through the walls of precapillary vessels so that blood oxygen content and perivascular P_{O_2} decrease significantly from arterial values with progression toward the capillaries.[9,25,26] Because metabolic vasodilation implies increased flow of oxygen-rich blood through arterioles, it has long been questioned on theoretical grounds whether arterioles would necessarily be exposed to decreased P_{O_2} during periods of increased metabolic activity.[3,27] Gorczynski and Duling[28] first measured periarteriolar P_{O_2} during muscle contraction and found that it remained essentially constant despite significant decreases in tissue P_{O_2} in general. This finding has since been confirmed by others.[9,29] Consequently, it seems unlikely that a direct influence of oxygen on arteriolar vessels plays a significant role in the production of functional hyperemia in skeletal muscle. Gorczynski and Duling[28] did find that exercise-induced vasodilation was reduced when tissue P_{O_2} was elevated by suffusing the hamster cremaster muscle preparation with an oxygen-rich solution. This suggests that general tissue P_{O_2} may be an important variable in exercise hyperemia. Klitzman et al.[30] found a similar result in the same preparation. Lash and Bohlen,[9] however, report that elevation of tissue and periarteriolar P_{O_2} by oxygen-rich superfusion does not reduce exercise-induced arteriolar dilation in rat spino-trapezius muscle. Kontos et al.[12] found that local suffusion of cat pial arterioles with oxygen-rich solutions can completely eliminate their response to systemic hypoxia or hypotension, as well as attenuate their response to drug-induced seizure.

Many mechanisms have been proposed by which oxygen might directly or indirectly affect resistance-vessel tone (see Pittman[31] for review). One strong candidate for an indirect mechanism involves hypoxia-induced release of adenosine from parenchymal cells or other tissue components (see below). Recent work by Jackson[32] suggests that a leukotriene may mediate the arteriolar oxygen reactivity in the hamster-cheek pouch preparation. In addition, recent findings of Pohl and Busse[33] on isolated rabbit large artery segments indicates that hypoxia may act on endothelial cells to cause the release of endothelial-derived relaxing factor (EDRF).

Adenosine

Adenosine is a potent endogenous vasodilator substance released in many tissues during periods of inadequate oxygen supply, which has received much attention as a potential mediator of local flow control.[34] The results from microcirculatory tests of adenosine's role in controlling flow in skeletal muscle are mixed. Proctor and Duling[35] found that treating the hamster-cremaster muscle with adenosine deaminase (which presumably lowers interstitial adenosine levels) had no effect on arteriolar constrictor responses to hyperoxic suffusion, but did reduce by 20–25% the arteriolar dilation response to cremaster muscle exercise. In a later study, Proctor[36] observed similar results after treating the hamster-cremaster preparation with either adenosine deaminase or the competitive adenosine-receptor antagonist theophylline. Proctor investigated exercise rates from 2–10 Hz (all of which produced nearly maximal vasodilation) and found evidence for relatively greater adenosine involvement in the vascular responses to higher exercise rates. Mohrman and Heller,[7] however, found no effect of aminophylline, a competitive adenosine-receptor blocker, on arteriolar responses to 0.5–2 Hz contractions in rat cremaster muscle. The implication seems to be that adenosine is certainly not the primary mechanism of metabolic vasodilation in skeletal muscle, but may assume greater relative importance with severe imbalances between metabolism and nutrient supply.

Morff and Granger[37] found that theophylline treatment had no effect on arteriolar autoregulation (dilation responses to reduced arterial pressure) in rat-cremaster-muscle preparations bathed with oxygen-rich solutions. Theophylline did, however, blunt an exaggeration of the arteriolar autoregulatory response to arterial hypotension, which these investigators observed under oxygen-deficient bathing conditions. Again, the implication is that adenosine may be involved in local vascular control in skeletal muscle only during severe disturbances to homeostasis.

The case for adenosine involvement in the local regulation of cerebral arterial tone is generally stronger than that for skeletal muscle.[38,39] Treatment with the adenosine receptor blocker theophylline produces pial arteriolar vasoconstriction in normoxic animals and attenuates hypoxic pial arteriolar dilation,[40] indicating that adenosine is involved in the maintenance of resting cerebral-resistance vessel tone as well as in responses to hypoxia. Moreover, the pial arteriolar dilator response to arterial hypotension is inhibited by topical application of adenosine deaminase,[41] which suggests that adenosine is importantly involved in cerebral autoregulation. In addition, adenosine appears to be involved in counteracting cerebral vasoconstrictor responses to hypocarbia, since such responses are potentiated by theophylline and attenuated by dipyridamole treatment.[42] Although negative findings have been reported,[43,44] the bulk of the evidence supports an important role for adenosine in the local regulation of cerebral resistance vessels.

Carbon Dioxide

It is well accepted that carbon dioxide plays a major role in the local regulation of the cerebral circulation,[45,46] but its role in the local vascular control of many other organs is less obvious.[3] Carbon dioxide has a direct vascular influence that is thought to be mediated through changes in pH.[46,47] Studies on isolated cerebral arteries indicate that reduced intracellular pH caused by P_{CO_2} elevation is accompanied by an increased K$^+$ conductance and hyperpolarization of vascular smooth-muscle cells,[48] whereas reduced P_{CO_2} causes the opposite effects.[49] In contrast to the pial arteriolar dilation response to hypoxia, the dilator response to hypercarbia is not attenuated by the adenosine-receptor blocker theophylline.[40]

The response to CO_2 of resistance vessels in tissue preparations is complex, because vascular responses to direct influences cause flow changes that themselves alter the conditions within the tissue and possibly alter vascular tone. Regarding oxygen, the situation is especially complicated because P_{CO_2} can influence oxygen levels both by changing flow and by its influence on the oxyhemoglobin dissociation curve.[50] It is clear that P_{CO_2} has important indirect influences on cerebral resistance vessels, because the response to P_{CO_2} can be altered by manipulations of adenosine and arachadonic acid metabolism systems.[51] In addition, there are indications from organ-level studies on heart and skeletal muscle that there could be an important synergistic interaction between P_{CO_2} and P_{O_2} in local vascular regulation.[5,52] This issue has not yet been directly pursued at the resistance-vessel level.

Interaction with Other Mechanisms

As indicated in other chapters of this work, resistance vessels are continuously subjected to several other influences in addition to local metabolic factors. The ultimate challenge will be to understand how these various mechanisms interact in vivo to produce a system that operates to satisfy the overall homeostatic needs of tissues and the organism.

Metabolic and Neurogenic Mechanisms

Sympathetic-nerve stimulation causes dramatic constriction of arterioles in resting skeletal muscle.[53-56] Especially in the smallest terminal arterioles, there is a modest time-dependent escape from the constrictor response to sympathetic-nerve stimulation, which potentially represents competition from local metabolic vasodilator mechanisms elicited by the initial reduction in blood flow.[54-56] This contention is supported by the fact that the degree of escape is attenuated during oxygen-rich superfusion of the preparation. These microvascular findings generally support organ-level work[57] and indicate that in skeletal muscle, local metabolic influences are only marginally effective in counteracting influences of sympathetic nerves on resistance-vessel tone.

Metabolic and Myogenic Mechanisms

Both metabolic and myogenic mechanisms[58] are thought to be important contributors to the regulation of resistance-vessel tone and thereby organ blood flow. In many physiological responses, such as cerebral autoregulation to systemic arterial hypotension, metabolic and myogenic mechanisms presumably would act synergistically to reduce resistance-vessel tone. The relative importance of each mechanism in such a response is difficult to assess directly, since no means are presently known to block either mechanism effectively. Under normal conditions in the cerebral circulation, venous pressure elevation elicits arteriolar dilation.[59] Since one would have expected vasoconstriction from the myogenic mechanism alone in this situation, the observed dilation suggests that metabolic influences normally dominate over myogenic influences on cerebral arterioles. The opposite may be true, however, in resting rat cremaster muscle. Here, Meininger et al.[60] demonstrated strong myogenic vasoconstrictor responses in small arterioles in response to simultaneous arterial and venous pressure elevation, even though they are being accompanied by significantly decreased blood flow and perivascular P_{O_2}.

References

[1] Gaskell, W. H. (1877) *J. Anat.* **11**, 360–402.

[2] Sparks, H. V. Jr. and Belloni, F. L. (1978) *Annu. Rev. Physiol.* **40**, 67–92.

[3] Sparks, H. V. Jr. (1980) *Handbook of Physiology. Section 2: The Cardiovascular System, vol. 2: Vascular Smooth Muscle.* American Physiological Society, Bethesda, MD, pp. 475–513.

[4] Bassenge, E. and Munzel, T. (1988) *Am. J. Cardiol.* **62**, 40E–44E.

[5] Mohrman, D. E. and Regal, R. R. (1988) *Am. J. Physiol.* **255**, H1004–H1010.

[6] Marshall, J. M. and Tandon, H. C. (1984) *J. Physiol. (Lond.)* **350**, 447–459.

[7] Mohrman, D. E. and Heller, L. J. (1984) *Am. J. Physiol.* **246**, H592–H600.

[8] Gorczynski, R. J., Klitzman, B., and Duling, B. R. (1978) *Am. J. Physiol.* **235**, H494–H504.

[9] Lash, J. M. and Bohlen, H. G. (1987) *Am. J. Physiol.* **252**, H1192–H1202.

[10] Bjornberg, J., Maspers, M., and Mellander, S. (1989) *Acta Physiol. Scand.* **135**, 83–94.

[11] Kanatsuka, H., Lamping, K. G., Eastham, C. L., Dellsperger, K. C., and Marcus, M. L. (1989) *Circ. Res.* **65**, 1296–1305.

[12] Kontos, H. A., Wei, E. P., Raper, A. J., Rosenblum, W. I., Navari, R. M., and Paterson, J. L. Jr. (1978) *Am. J. Physiol.* **234**, H582–H591.

[13] Bohlen, H. G. (1980) *Am. J. Physiol.* **238**, H164–H171.

[14] Lindbom, L. (1986) *Microvasc. Res.* **31**, 143–156.

[15] Mohrman, D. E. and Heller., L. J. (1988) *Physiologist* **31**, A26.

[16] Rubin, M. J. and Bohlen, H. G. (1985) *Am. J. Physiol.* **249**, H540–H546.

[17] Lombard, J. H. and Stekiel, W. J. (1985) *Microvasc. Res.* **30**, 346–349.

[18] Duling, B. R. (1972) *Circ. Res.* **31**, 481–489.

[19] Prewitt, R. L. and Johnson, P. C. (1976) *Microvasc. Res.* **12**, 59–70.

[20] Sullivan, S. M. and Johnson, P. C. (1981) *Am. J. Physiol.* **241**, H547–H556.

[21] Jackson, W. F. and Duling, B. R. (1983) *Circ. Res.* **53**, 515–525.

[22] Jackson, W. F. (1987) *Am. J. Physiol.* **253**, H1120–H1126.

[23] Boegehold, M. A. and Bohlen, H. G. (1988) *Hypertension* **12**, 184–191.

[24] Duling, B. R. (1974) *Am. J. Physiol.* **227**, 42–49.

[25] Duling, B. R. and Berne, R. M. (1970) *Circ. Res.* **23**, 669–678.

[26] Ivanov, K. P., Derry, A. N., Vovenko, E. P., Samoilov, M. O., and Semionov, D. G. (1982) *Plfuger's Arch.* **393**, 118–120.

[27] Roth, A. C. and Wade, K. (1986) *Microvasc. Res.* **32**, 64–83.

[28] Gorczynski, R. J. and Duling, B. R. (1978) *Am. J. Physiol.* **235**, H505–H515.

[29] Proctor, K. G. and Bohlen, H. G. (1981) *Blood Vessels* **18**, 58–66.

[30] Klitzman, B., Damon, D., Gorczynski, R. J., and Duling, B. R. (1982) *Circ. Res.* **52**, 711–721.

[31] Pittman, R. N. (1986) *Can. J. Cardiol.* **2**, 124–131.

[32] Jackson, W. F. (1989) *Am. J. Physiol.* **257**, H1565–H1572.

33 Pohl, U. and Busse, R. (1989) *Am. J. Physiol.* **256,** H1595–H1600.

34 Berne, R. M., Winn, H. R., Knabb, R. M., Ely, S. W., and Rubio, R. (1983) *Regulatory Function of Adenosine* (Berne, R. M., Rall, T. W., and Rubio, R., eds.) Nijhoff, The Hague, pp. 293–317.

35 Proctor, K. G. and Duling, B. R. (1982) *Am. J. Physiol.* **242,** H688–H697.

36 Proctor, K. G. (1984) *Am. J. Physiol.* **247,** H195–H205.

37 Morff, R. J. and Granger, H. J. (1983) *Am. J. Physiol.* **244,** H567–H576.

38 Wei, E. P. and Kontos, H. A. (1981) *J. Cereb. Blood Flow Metab.* **1**(Suppl. 1), S395–S396.

39 Winn, H. R., Welsh, J. E., Rubio, R., and Berne, R. M. (1981) *J. Cereb. Blood Flow Metab.* **1,** 239–244.

40 Morii, S., Ngai, A. C., Ko, K. R., and Winn, H. R. (1987) *Am. J. Physiol.* **253,** H165–H175.

41 Kontos, H. A. and Wei, E. P. (1985) *Ann. Biomed. Eng.* **13,** 329–334.

42 Ibayashi, S., Ngai, A. C., Meno, J. R., and Winn, H. R. (1988) *J. Cereb. Blood Flow Metab.* **8,** 829–833.

43 Dora, E., Koller, A., and Kovach, A. G. (1984) *J. Cereb. Blood Flow Metab.* **4,** 447–457.

44 Dora, E. (1986) *Acta Physiol. Hung.* **68,** 183–197.

45 Strandgaard, S. and Paulson, O. B. (1984) *Stroke* **15,** 413–416.

46 Heistad, D. D. and Kontos, H. A. (1983) *Handbook of Physiology, Section 2: The Cardiovascular System, vol. 3: Peripheral Circulation and Organ Blood Flow.* American Physiological Society, Bethesda, MD, pp. 137–182.

47 Kontos, H. A., Raper, A. J., and Patterson, J. L. Jr. (1977) *Stroke* **8,** 358–360.

48 Harder, D. R. (1982) *Pflugers Arch.* **394,** 182–185.

49 Harder, D. R. and Madden, J. A. (1985) *Pflugers Arch.* **405,** 402–404.

50 Pittman, R. N. and Duling, B. R. (1977) *Microvasc. Res.* **13,** 211–224.

51 Wagerle, L. C. and Mishra, O. P. (1988) *Circ. Res.* **62,** 1019–1026.

52 Case, R. B., Felix, A., Wachter, M., Kyriakidis, G., and Castellana, F. (1978) *Circ. Res.* **42,** 410–418.

53 Eriksson, E. and Lisander, B. (1972) *Acta Physiol. Scand.* **84,** 295–305.

54 Marshall, J. M. (1982) *J. Physiol. (Lond.),* **332,** 169–186.

55 Boegehold, M. A. and Johnson, P. C. (1988) *Am. J. Physiol.* **254,** H919–H928.

56 Boegehold, M. A. and Johnson, P. C. (1988) *Am. J. Physiol.* **254,** H929–H936.

57 Thompson, L. P. and Mohrman, D. E. (1983) *Am. J. Physiol.* **245,** H66–H71.

58 Johnson, P. C. (1980) *Handbook of Physiology. Section 2: The Cardiovascular System, vol. 2: Vascular Smooth Muscle.* American Physiological Society, Bethesda, MD, pp. 409–422.

59 Wei, E. P. and Kontos, H. A. (1984) *Circ. Res.* **55,** 249–252.

60 Meininger, G. A., Mack, C. A., Fehr K. L., and Bolin, H. G. (1987) *Circ. Res.* **60,** 861–870.

Chapter 15

Monovalent-Ion Pumps and Carriers in Resistance Arteries

Christian Aalkjaer

Introduction

The function of the resistance arteries is influenced by the extracellular and intracellular ion milieu. This is evident not only for calcium, but also for the monovalent ions. The importance of these ions lies in their contribution to the membrane potential and to various co- and countertransport systems. The membrane potential of the vascular smooth muscle cells (VSMC) reflects the distribution and permeability of the monovalent-ions over the cell membrane, and the permeability is determined in the main by the characteristics of the ion channels. Recently a lot of new information about the characteristics of ion channels has become available from patch-clamp experiments. This aspect is dealt with elsewhere in this book (chapter 16, this volume).

The transmembranal distribution of the ions "exploited" by the ion channels to control the membrane potential is set up through the activity of various ion-transport systems, which transport the ions against their electrochemical gradient. The ion gradients are

From *The Resistance Vasculature*, J. A. Bevan et al., eds. ©1991 Humana Press

also used to transport substances involved in cell metabolism, including protons and bicarbonate and possibly lactate, glucose, and amino acids, and is also involved in the regulation of cell volume. This aspect of ion homoeostasis is assessed using radioisotopes of the various ions, by measurements of force production, by conventional electrophysiology, and more recently, by microspectrofluorometry with ion-specific fluorescent dyes. There is therefore a methodological and, to a certain extent, a functional basis for making a distinction between, on the one hand, ion channels and, on the other hand, ion pumps and ion carriers. Since any ion takes part in several transport processes, a change in one transport mechanism is likely to affect the others. Thus an integrated understanding of the physiological importance of these transport systems is difficult to obtain.

The purpose of the present report is to describe briefly the methods employed in studying ion pumps and ion carriers in resistance arteries and to discuss the pathways known to be present in these vessels and their possible physiological significance.

Methodology

Force Development

On the basis of the effect of changes in ion concentrations or the effect of ion-transport inhibitors on force development, predictions about ion transport and its relation to vascular smooth muscle cell (VSMC) contraction are sometimes made. Since one of the purposes of understanding ion transport is to understand muscle contraction, it is obvious that it is important to correlate the characteristics of ion transport to contraction. However, the excitation–contraction coupling of VSMC is complex, and, as already pointed out, changes in ion concentrations affect different transport pathways, giving a complex pattern. Furthermore, since ion-transport inhibitors are only rarely specific for one transport system, it is often difficult to deduce changes in ion transport from measurements of force.

Radioisotopes

Measurements of fluxes with radioisotopes is the classical way of assessing ion transport in smooth muscles.[1] In resistance arteries, radioisotopes of sodium, chloride, potassium, and calcium have been used to assess ion transport.[2,3] In nonvascular smooth-muscle preparations in which an independent measure of ion transport has been obtained with ion-selective electrodes (the cells are frequently larger), the results suggest that flux measurements with radioisotopes give results that are compatible with the data obtained with electrophysiological techniques.[4,5] On the other hand based on measurements with electron probes,[6,7] it has been suggested that fluxes measured with radioisotopes may be difficult to interpret in view of the time lag caused by the inevitable washout of extracellular label before transmembranal transport can be assessed. A further drawback in flux measurements with radioisotopes is that they are not made under mechanically well-defined conditions. Since some of the characteristics of force development and also the electrophysiological characteristics are dependent on the degree of passive extension of the VSMC,[8–10] it might be expected that the fluxes are affected by the amount of stretch. It is, therefore, always important to obtain measurements of radioisotope transport in resistance arteries under mechanically well-defined conditions.

Electrophysiology

As already pointed out, electrophysiological techniques are useful to define and characterize ion channels. Measurements of membrane potential are also important, to determine the stoichiometry of the ion-transport pathways. Furthermore, with double-barreled microelectrodes, it has been possible in nonvascular smooth muscles to obtain information about intracellular ion activity; recently, this technique has also been used in measurements of intracellular pH in strips of femoral artery of the guinea pig.[11] It will be of interest to see whether this elegant technique can be applied to resistance arteries.

Microspectrofluorometry

Over the last five to six years, a number of ion-specific fluorescent dyes for calcium, protons, sodium, potassium, chloride, and magnesium have become available. With these dyes it is possible to assess simultaneously ion activity and force development in resistance arteries.[12] It is potentially possible to combine these measurements with electrophysiological techniques[13] and obtain simultaneous measurements of membrane potential, ion turnover, and force. Another prospect is to obtain video images of the intracellular distribution of the ion activity. Although these latter applications must be developed further before they can be applied to resistance arteries, it must be expected that microspectrofluorometry will provide a new source of information with respect to ion-transport pathways in these vessels.

Na–K Pump

Not unexpectedly, the resistance arteries from all beds studied have a Na–K pump,[14,15] which, by pumping sodium out of the cells and potassium into the cells, sets up the gradient for sodium and potassium. This fuels a number of other transport mechanisms, in addition to being important for the establishment of the membrane potential. The presence of a Na–K pump has been documented by ouabain-sensitive ^{22}Na and ^{86}Rb fluxes and by demonstration of ouabain-sensitive ATP hydrolysis of vesicles from plasmalemma.[16] The Na–K pump in the resistance arteries does not appear to be different from that of other smooth muscles. From binding studies with ^{3}H-ouabain, the concentration of binding sites is of the same order as in taenia coli and aorta[15]. The rate of sodium transport mediated by the pump is also similar to that of other smooth muscles, and, as in other cells, the Na–K pump is likely to be electrogenic, as judged from a depolarization of up to 11 mV that is associated with its inhibition.[15,17]

Although the Na–K pump mediates an efflux of sodium from the vessels, a large proportion (varying between 50 and 80%) of the sodium efflux still occurs after complete inhibition of the Na–K pump with either ouabain or omission of extracellular potassium.[15,18] This ouabain-insensitive efflux, which is also seen in large arteries, has a steep temperature dependence, indicating that it is not a passive leak of sodium.[18] This interpretation is supported by thermodynamic considerations which suggest that this flux is not a passive leak of sodium, but is mediated by either a carrier or a pump.[18] A more precise definition of this flux is lacking, although the fact that it mediates a large part of the sodium efflux under normal conditions makes it a potentially important phenomenon.

A point of substantial interest is the mechanism responsible for the effect of ouabain on the level of tone in resistance arteries. Despite an increase in intracellular sodium associated with incubation with ouabain, no maintained tone develops in resting resistance arteries over several hours.[19] This is in contrast to conduit arteries, in which tone develops.[20] On the other hand, the tone of isolated resistance arteries is at least transiently enhanced by ouabain, and in vivo the vascular resistance increases during the infusion of ouabain. The question has been raised whether this potentiating effect of ouabain is attributable to a rise of intracellular sodium and a consequent rise in intracellular calcium through Na–Ca exchange, or to the depolarizing effect of ouabain and consequent influx of calcium through voltage dependent calcium channels. Evidence in favor of the latter mechanism comes from observations that the potentiation of tone with ouabain is inhibited by calcium antagonists,[17,21] which exert their action through inhibition of the potential-operated calcium channels, rather than by interference with Na–Ca-exchange.

Na–Ca Exchange

Evidence for the presence of Na–Ca exchange in resistance arteries comes from the observation that substitution of extracellular sodium with sucrose causes a reduction of the efflux of ^{45}Ca from rat mesenteric resistance vessels[2] and that substitution of extracellular calcium with manganese causes a reduction in the efflux of ^{22}Na from sodium loaded vessels.[3] The functional role of the exchange system in these vessels, however, is debated.[22,23]

In aorta and other large arteries, evidence has been provided for a potentially significant role of Na–Ca exchange in the control of tone.[20,24] In rat mesenteric resistance arteries, it has not been possible to find a correlation between changes in the sodium electrochemical gradient and noradrenaline or potassium-induced vessel tone.[17] If, however, extracellular sodium is reduced to 25 mM or less while intracellular sodium is increased, this nonphysiological shift of the sodium gradient induces tone that is not inhibited with calcium antagonists.[17] Furthermore, the low extracellular sodium reduces the rate of relaxation after washout of an agonist or may even abolish the relaxation.[2,25] These mechanical effects are probably mediated by the Na–Ca exchange. However, the large changes in the sodium gradient necessary to provoke them makes it unlikely that they play an important role in vivo.

Na–Na Exchange

Under circumstances in which the smooth-muscle cells of resistance arteries are loaded with sodium (the intracellular sodium concentration rising to about 60 mmol/L of cells), substitution of extracellular sodium with magnesium causes a significant reduction of the ouabain-insensitive sodium efflux.[3] This reduction of the ouabain-insensitive sodium efflux is not seen in smooth muscles that are not loaded with sodium.[18] These findings suggest that the smooth muscles of resistance arteries have a mechanism for Na–Na exchange that is expressed under suitable conditions. The characteristics of the Na–Na exchange have been studied in great detail in *taenia coli*,[26] and it has been suggested that it is an expression of Na–Li exchange and Na–H exchange.[27] In resistance arteries little is known about the role of this system.

Na–K–Cl Cotransport

The Na–K–Cl cotransport, which is important for the regulation of cell volume,[28] has been demonstrated in rat and rabbit aorta[29,30] and in vascular smooth muscles in culture.[31] In rat mesenteric resistance arteries, the ouabain-insensitive ^{86}Rb uptake is inhibited by furosemide (Aalkjaer and Mulvany, unpublished observation). This observation is consistent with the presence of this transport system in these vessels, but its physiological importance is unknown. Interestingly, in vitro experiments have shown that furosemide has vasodilator effects in conduit arteries of the rabbit,[30,32] although the relation of this effect of furosemide to inhibition of the Na–K–Cl cotransport and the extent to which it occurs in resistance arteries is unclear.

Na–H Exchange

An amiloride-sensitive Na–H exchange has been demonstrated in most cell types, including vascular smooth muscle cells.[33] In resistance arteries, an amiloride-sensitive and sodium-dependent recovery of intracellular pH_i after an acid load can be demonstrated under HCO_3– free conditions.[12] Furthermore, an acid load induces a substantial sodium uptake, which can be inhibited with amiloride and 5-(N-ethyl-N-isopropyl)-amiloride.[12] These observations unambiguously demonstrate the presence of Na–H exchange in resistance arteries. The importance of this mechanism in the regulation of pH_i is far from understood. There is no doubt that bicarbonate transport plays an important role in the regulation of pH_i *(see below)* and, in the presence of bicarbonate, it is difficult to demonstrate Na–H exchange in rat mesenteric resistance arteries of 13- to 18-wk-old WKY rats. However, there is some evidence that, in resistance arteries from younger rats, Na–H exchange may play a more dominant role in the regulation of pH_i.[34] It is an interesting possibility that the relative importance of Na–H exchange and Na–HCO_3–cotransport *(see below)* changes with development. In the absence of HCO_3–, the acid produced during force development is pumped out by Na–H exchange, although steady-state PH_i is reduced; in resting vessels, the activity of the Na–H exchange is rather low. The latter is in contrast to observations made on strips of aorta[35] or VSMC in culture,[36,37] in which a substantial activity in the Na–H exchange is reported under resting conditions.

Na–HCO$_3$⁻ Cotransport

The importance of bicarbonate transport in the control of pH_i of smooth muscles has been pointed out by Aickin.[38] In mesenteric resistance arteries from WKY rats, a diisothiocyanato-stilbene-disulfonic acid (DIDS) sensitive sodium-coupled uptake of bicarbonate has been shown to be of primary importance for the recovery from an intracellular acid load.[12] This sodium-coupled bicarbonate uptake is not associated with an efflux of chloride, suggesting that the mechanism is a Na–HCO$_3$⁻ cotransport.[39] Of particular interest is the observation that the Na–HCO$_3$⁻ cotransport appears electroneutral,[39] which is in contrast to what has been reported for nonsmooth-muscle cells.[40,41] However, the suggestion that the Na–HCO$_3$⁻ cotransport of vascular smooth-muscles is electroneutral is supported by recent reports that, in the smooth muscle of the guinea pig ureter[38] and in a smooth-muscle cell line (BC3H-1[42]) the Na,HCO$_3$⁻ cotransport appears electroneutral. Although a Na–HCO$_3$⁻ cotransport has been demonstrated in several smooth muscle types, it is not ubiquitous, since in two cell lines derived from smooth muscles (A10 and A7r5), no sodium-dependent bicarbonate transport could be detected.[43,44]

In the last six to seven years, the importance of hormonal activation of transmembranal proton transport mechanisms has been discussed. It was originally proposed that an increase in pH_i consequent upon hormonal activation of Na–H exchange served a second-messenger function, mediating the response to activation with growth factors.[45] More recently it has been proposed that the activation of not only Na–H exchange, but also bicarbonate transport, is, rather, a response in anticipation of a metabolic load.[46,47] We found that, in resistance arteries, activation with vasoconstrictor hormones causes no change in pH_i although both Na–HCO$_3$⁻ cotransport and Na–H exchange are activated by the vasoconstrictors.[48] This suggests that, although the force development caused by these vasoconstrictors is associated with a net acid production, the activation of acid-extrusion mechanisms exactly matches the acid production in a perfect feedback control. In contrast, during depolarization with high potassium, pH_i falls, indicating that the homoeostatic control operates only during stimulation with "physiological" activators.[48] From in vivo and in vitro experiments there is evidence that an increase in pCO_2 reduces the tone of resistance arteries (and conduit arteries) and peripheral resistance.[33] Although the precise mechanism responsible for this effect may vary from bed to bed,[33] it is conceivable that a reduction in pH_i plays a role in the reduction of the tone. On this basis, it seems appropriate that the smooth muscles of resistance arteries (which, in contrast to, e.g., skeletal muscles, are constantly in a tonic state that changes in response to exogenous stimuli) can very efficiently extrude the metabolically produced acid, so it does not interfere with the control of tone.

Cl–HCO$_3^-$ Exchange

Omission of chloride from the external medium causes a substantial DIDS-sensitive increase in pH_i of rat mesenteric resistance arteries, but only when bicarbonate is present.[39] Furthermore, omission of chloride reduces the ^{36}Cl efflux by about 75%, but only when bicarbonate is absent. These findings are consistent with the presence in these vessels of a Cl–HCO$_3^-$ exchange that can operate in a Cl–Cl exchange mode.[39] The findings further suggest that the major part of the ^{36}Cl efflux is mediated by the exchange mechanism and only a smaller fraction by a conductive chloride transport. This is consistent with previous observations on the vas deferens,[4] on the basis of which it was concluded that, in resting smooth-muscle cells, chloride permeability (and thus the contribution of chloride distribution to the membrane potential) is smaller than hitherto believed. Under resting conditions, as well as during an alkaline load, the transmembranal gradients for chloride and bicarbonate will result in net efflux of bicarbonate by Cl–HCO$_3^-$ exchange. In some other cell types, the function of the Cl–HCO$_3^-$ exchange is probably related to extrusion of an alkaline load. In rat mesenteric resistance vessels, the recovery of pH_i after an alkaline load is not significantly affected by DIDS or by omission of bicarbonate. This indicates that the Cl–HCO$_3^-$ exchange may not be quantitatively important for the recovery of pH_i from an alkaline load, which must therefore be mediated by a bicarbonate-independent mechanism.[39] This conclusion is supported by the rapid recovery of pH_i seen after an alkaline load induced by omission of CO_2 and bicarbonate.[12] Whether this reduction in pH_i represents a net efflux of acid equivalents or simply reflects a metabolic production of acid is unknown. Another question of interest is why the cells possess a mechanism that apparently transports a lot of chloride across the cell membrane in a self-exchange mode (Cl–Cl exchange). Such a feature seems unnecessary.

Summary

Information about ion carriers and ion pumps in resistance arteries is difficult to obtain and, so far, is derived almost exclusively from studies of rat mesenteric resistance arteries. This obviously limits generalization to different types of resistance arteries. Furthermore, little is known about the processes in endothelial cells and fibroblasts, which are invariably present in the preparations studied. With these reservations in mind, it appears that several of the transport mechanisms defined from the study of other cell types are present in the smooth-muscle of resistance arteries (Fig. 1). In fact, it appears that it is possible, if the right conditions are provided, to demonstrate the presence of most, if not all, known systems. The relevant question, therefore, becomes not so much whether any particular transport mechanism can be expressed, but whether it plays a role in normal or pathological cell function. With respect to this question, there seems little doubt that the Na–K pump, in the resistance arteries as in other tissues, is essential in establishing the sodium gradient on which a number of other transport processes thrive. It may also be important for the membrane potential under some conditions. There is also little doubt that the $Na–HCO_3^-$ cotransport is important in mediating a very tight regulation of pH_i, particularly during contraction, when a net production of acid is pumped out primarily by this mechanism. On the other hand, the role of Na–H exchange in regulation of pH_i is more uncertain. In mesenteric resistance arteries from adult WKY rats, there is evidence against an important role as long as bicarbonate is present and pH_i is in the physiological range, although in similar vessels from young rats it may be more active in pH_i regulation. With regard to $Cl–HCO_3^-$ exchange, it has been diffi-

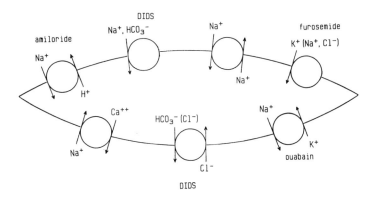

Fig. 1. Schematic representation of a smooth-muscle cell from a resistance artery. The monovalent-ion carriers and pumps and some of their inhibitors are shown.

cult to define a quantitatively important role in pH_i regulation. The observation, however, that this transport system mediates 70–80% of the chloride flux under resting conditions indicates that it must play some role in ion homoeostasis. With respect to the Na–Ca exchange, it has been difficult to ascribe an important role to this exchanger. It is likely that Ca-ATPase is of prime importance in the extrusion of calcium *(see* chapter 17, this volume). The Na–K–Cl cotransport is not characterized to any degree in resistance arteries, and its possible functional importance is unknown. Information is being gathered about these ionic mechanisms, and our understanding of the part they play in controlling the function of resistance arteries is increasing. Much more, however, needs to be done. In this connection, it is disturbing that virtually nothing is known about the mechanism responsible for the influx of sodium as well as a large part of the sodium efflux from resistance arteries.

References

[1] Jones, A. W. (1980) *Handbook of Physiology, Section 2: The Cardiovascular System*, vol. 2: *Vascular Smooth Muscle* (Bohr, D. F., Somlyo, A. P., and Sparks, H. V., eds.), American Physiological Society, Bethesda, MD, pp. 253–299.

[2] Cauvin, C. (1988) *Resistance Arteries* (Halpern, W., Pegram B, Brayden J., and Mackey, K., eds.), Perinatology, Ithaca, NY, pp. 195–211.

[3] Aalkjaer, C. (1988) A review. Thesis, Aarhus University, Aarhus, Denmark.

[4] Aickin, C. C. and Brading, A. F. (1982) *J. Physiol.* **326,**139–154.

[5] Aickin, C. C. and Brading, A. F. (1983) *J. Physiol.* **336,** 179–197.

[6] Junker, J. L., Wasserman, A. J., Berner, P. F., and Somlyo, A. P. (1984) *Circ. Res.* **54,** 254–266.

[7] Wasserman, A. J., McClellan, G., and Somlyo, A. P. (1986) *Circ. Res.* **58,** 790–802.

[8] Nilsson, H. and Sjoblom, N. (1985) *Acta Physiol. Scand.* **125,** 429–435.

[9] Laher, I., VanBreemen, C., and Bevan, J. A. (1988) *Circ. Res.* **63,** 669–772.

[10] Johansson, B. and Mellander, S. (1975) *Circ. Res.* **36,** 76–83.

[11] Aickin, C. (1990) *J. Physiol.* **423,** 49P.

[12] Aalkjaer, C. and Cragoe, E. J. (1988) *J. Physiol.* **402,** 391–410.

[13] Benham, C. D. (1989) *J. Physiol.* **419,** 689–701.

[14] Aalkjaer, C. and Mulvany, M. J. (1983) *J. Physiol.* **343,** 105–116.

[15] Aalkjaer, C. and Mulvany, M. J. (1985) *J. Physiol.* **362,** 215–231.

[16] Wei, J.-W., Janis, R. A., and Daniel, E. E. (1976) *Blood Vessels* **279,** 279–292.

[17] Mulvany, M. J., Aalkjaer, C., and Petersen, T. T. (1984) *Circ. Res.* **54,** 740–749.

[18] Aalkjaer, C. and Mulvany, M. J. (1988) *ICSU Symposium Series. Vascular Neuroeffector Mechanisms* (Bevan, J. A., Majewski, H., Maxwell, R. A., and Story, D. F., eds.), IRL, Washington, DC, pp. 201–208.

[19] Mulvany, M. J., Aalkjaer, C., Nilsson, H. Korsgaard, N., and Petersen, T. T. (1982) *Clin. Sci.* **63,** 45s–48s.

[20] Karaki, H., Ozaki, H., and Urakawa, N. (1978) *Eur. J. Pharmacol.* **48,** 439–443.

[21] Nelson, M. T., Standen, N. B., Brayden, G. E., and Worley, J. F. III. (1988) *Nature* **336,** 382–385.

[22] Mulvany, M. J. (1985) *J. Hypertension* **3,** 429–436.

[23] Blaustein, M. P. (1988) *J. Cardiovasc. Pharmacol.* **12(Suppl. 5),** 556–568.

[24] Ozaki, H. and Urakawa, N. (1981) *Pflugers Arch.* **390,** 107–112.

[25] Petersen, T. T. and Mulvany, M. J. (1984) *Blood Vessels* **21,** 279–289.

[26] Brading, A. F. (1975) *J. Physiol.* **251,** 79–105.

[27] Funder, J. and Wieth, J. O. (1978) *Cell Membrane Receptors for Drugs and Hormones* (Straub, R. W. and Bolis, L., eds.), Raven, New York, pp. 271–279.

[28] Hoffmann, E. K. and Simonsen, L. O. (1989) *Physiol. Rev.* **69,** 315–382.

[29] Kreye, V. A. W., Bauer, P. K., and Villhauer, I. (1981) *Eur. J. Pharmacol.* **73,** 91–95.

[30] Deth, R. C., Payne, R. A., and Peecher, D. M. (1987) *Blood Vessels* **24,** 321–333.

[31] Owen, N. E. (1984) *Biochem. Biophys. Res. Comm.* **125**, 500–508.

[32] Andreasen, F. and Christensen, J. H. (1988) *Pharmacol. Toxicol.* **63**, 324–326.

[33] Aalkjaer, C. (1990) *J. Hypertension* **8**, 197–206.

[34] Izzard, A. S. and Heagerty, A. M. (1989) *J. Hypertension* **7(Suppl. 6)**, s128, s129.

[35] Ek, T. P. and Deth, R. C. (1988) *Hypertension* **12**, 331.

[36] Little, P. J., Cragoe, E. J., and Bobik, A. (1986) *Am. J. Physiol.* **251**, C707–C712.

[37] Hatori, N., Fine, B. P., Nakamura, A., Cragoe, E., and Aviv, A. (1987) *J. Biol. Chem.* **262**, 5073–5078.

[38] Aickin, C. C. (1989) *Verh. Dtsch. Zool. Ges.* **82**, 121–129.

[39] Aalkjaer, C. and Hughes A. (1991) *J. Physiol.* **436**, 57–73.

[40] Boron, W. F. and Boulpaep, E. L. (1983) *J. Gen. Physiol.* **81**, 29–52.

[41] Jentsch, T. J., Korbmacher, C., Janicke, I., Fischer, D. G., Stahl, F., Helbig, H., Hollwede, H., et al. (1988) *J. Membrane Biol.* **103**, 29–40.

[42] Putnam, R. W. (1990) *Am. J. Physiol.* **27**, C470–C479.

[43] Korbmacher, C., Helbig, H., Stahl, F., and Wiederholt, M. (1988) *Pflugers Arch.* **412**, 29–36.

[44] Vigne, P., Breittmayer, J.-P., Frelin, C., and Lazdunski, M. (1988) *J. Biol. Chem.* **263**, 18023–18029.

[45] Moolenaar, W. H., Yarden, Y., DeLaat, S. W., and Schlessinger, J. (1982) *J. Biol. Chem.* **257**, 8502–8506.

[46] Ganz, M. B., Boyarsky, G., Sterzel, R. B., and Boron, W. F. (1989) *Nature* **337**, 648–651.

[47] Thomas, R. C. (1989) *Nature* **337**, 601.

[48] Aalkjaer, C. and Mulvany, M. J. (1991) Submitted for publication.

Chapter 16

Ion Channels
in Resistance Arteries

Mark T. Nelson, John G. McCarron, and John M. Quayle

Introduction

Intravascular pressure causes many resistance arteries to develop (myogenic) tone, which is further modulated by constricting and dilating substances. Myogenic tone is a significant contributor to basal arterial tone as well as to blood-flow autoregulation, which is responsible for maintaining constant blood flow in a number of vascular beds.[1] The development of tone in response to increases in transmural pressure is associated with a depolarization of smooth-muscle cells from about –60 mV to about –40 mV.[2–5] Myogenic tone depends on extracellular calcium and is abolished by calcium channel antagonists and hyperpolarizing vasodilators, suggesting that voltage-dependent calcium channels provide the route for calcium entry on depolarization.[6]

In this review we will concentrate on the role of ion channels in regulating tone in resistance arteries. We will include information from larger muscular arteries, since very few studies on ion channels in resistance arteries have been reported (however, *see*

From *The Resistance Vasculature*, J. A. Bevan et al., eds. ©1991 Humana Press

refs. 7–10). We propose that voltage-dependent calcium channels play an important role in the regulation of tone in resistance arteries and that membrane potential regulates tone, in large part through the voltage dependence of calcium channels. In addition, we discuss the evidence that some of the membrane-potential-independent effects of vasoactive substances are caused by direct modulation of calcium channels.

Voltage-Dependent Calcium Channels

Calcium entry into smooth muscle activates calcium- and calmodulin-dependent myosin light chain kinase, which phosphorylates myosin light chain, allowing myosin–actin interaction and force generation. Since resistance arteries exist in a state of maintained contraction, the balance between calcium entry and extrusion will be of primary importance.

The predominant voltage-dependent calcium channel in arterial smooth muscle is inhibited by dihydropyridine Ca-channel blockers and inactivates incompletely during prolonged depolarizations.[10–17] This Ca channel has been referred to as "L type" or "high threshold." Calcium currents (measured in physiological external Ca and temperature) can be detected in cerebral arterioles at membrane potentials that would occur in vivo (e.g., –50 mV).[7,9] This maintained Ca current increased steeply with membrane depolarization from –60 to –30 mV.[7,16] Myogenic tone is maintained by a steady influx of Ca^{2+} since it is reduced by calcium-channel blockers and membrane hyperpolarization. It seems reasonable to assume that the dihydropyridine-sensitive, voltage-dependent Ca channel is an important entry pathway for Ca^{2+} that is required for tone.

Calcium influx through calcium channels in the cell membrane is determined by the number of functional channels (N), ion flux through an individual channel (measured as single-channel current, i), and the fraction of time a channel is open, or open-state probability (p_{open}) (Ca influx $\propto N \cdot i \cdot p_{open}$). Membrane potential affects *both* i and p_{open}. Membrane depolarization decreases the single channel current since it reduces the calcium electrochemical gradient (e.g., Fig. 2 in ref. 14; *see* Fig. 1). It should be stressed that

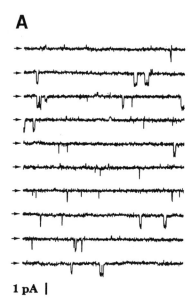

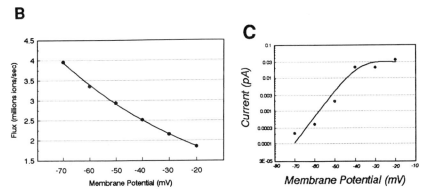

Fig. 1. (**A**) Original continuous recordings of single Ca channels in an on-cell patch of single-smooth muscle cells from a resistance-sized posterior cerebral artery of rat. Barium (10 m*M*) was the charge carrier. Bay R 5417 (500 n*M*) was present. A total of continuous recordings are shown in 1-s traces; holding potential –50 mV. (**B**) Relationship between ion flux through a single open Ca channel and membrane potential from the same experiment shown in panel **A**. (i.e., single-channel current converted to ion flux). (**C**) Relationship between steady-state current and membrane potential. Steady-state current is proportional to mean divalent cation flux and equal to the mean p_{open} multiplied by the single channel current. Data from the experiment shown in **A**. This work was done in collaboration with W. Halpern.

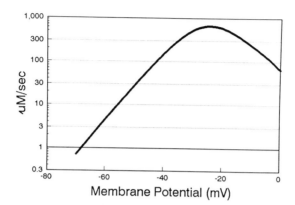

Fig. 2. Simulated relationship between membrane potential and the increase in intracellular Ca^{2+} (in μM) per second in a single smooth muscle of intracellular volume of 1 pl in the absence of Ca^{2+} binding, extrusion, and sequestration. The midpoints of the activation and inactivation curves ($V_{0.5}$) were assumed to be –25 mV, and the slope factors (k) were assumed to be 5 and 7 mV, respectively. The maximum p_{open} in the absence of inactivation was assumed to be 0.3 (p_{funct}). The conductance of the single Ca channel was assumed to be 4 pS at 37°C. The following equation was used to generate the curve:

$$\text{Steady-state } Ca^{2+} \text{ influx} \propto (N)(p_{act})(1-p_{inact})(p_{funct})(i)$$

$$\text{where } p_{act} = [(1 + \exp((V0.5 - Vm)/k))]^{-1}.$$

membrane depolarization would decrease Ca influx through any type of calcium-permeable channel. However, membrane depolarization can greatly increase overall Ca^{2+} influx by causing a substantial increase in p_{open}. Thus, steady state-calcium influx through calcium channels reflects voltage dependence of steady-state p_{open} and the single-channel current. Single-channel recordings provide the most sensitive method for investigating the effect of membrane potential on calcium channels. Thus far, measurements of the voltage dependence of p_{open} Of Ca channels have been made in mesenteric arteries,[14,6] coronary arteries,[17] and cerebral arteries.[10,18] Below, we will discuss three important mechanisms (membrane potential dependence, pharmacological regulation, and vasoconstrictor activation) for the regulation of Ca entry through voltage-dependent Ca channels.

Membrane-Potential Dependence

The membrane potential of smooth muscle cells in resistance arteries appears to be an important determinant of tone[5,6] and to be in the range of –50 to –40 mV.[6,9] Many endogenous and synthetic vasodilators appear to act through membrane hyperpolarization, whereas vasoconstrictors and pressure appear to work at least in part through membrane depolarization.[2-5] It is reasonable to speculate that the membrane-potential dependence of tone reflects the voltage dependence of Ca channels.

Figure 1A illustrates recordings of single Ca channels in an on-cell patch on a single smooth-muscle cell enzymatically isolated from a resistance-sized (100 μm diameter) branch of the cerebral posterior artery from a WKY rat. This is a steady-state recording at a membrane potential (–50 mV) that would be appropriate for an intact resistance vessel with tone. The rate of ion movement through the open channel decreases with membrane depolarization (Fig. 1B). For example, membrane depolarization from –70 to –20 mV reduced the ion flux through an open channel by 53.5% (Fig. 1B). However, as shown in Fig. 1C, the steady-state influx of divalent cations through a single channel increases dramatically with membrane depolarization caused by the steep voltage dependence of p_{open}. The steady-state current (which is proportional to ion flux) increases exponentially with membrane depolarization and then plateaus at a membrane potential that is determined by the positions of the activation and inactivation curves (*see* ref. 6) . In the experiment shown in Fig. 1, p_{open} increases 2.7-fold/5 mV negative to –40 mV and tends to saturate positive to –40 mV. In this case, 10 mM barium was the charge carrier and the Ca-channel agonist Bay R 5417 (500 nM) was present. Bay R 5417 shifts the relationship between membrane potential and p_{open} by about 15 mV to more-negative potentials without changing the shape of the relation. In other words, in the absence of Bay R 5417, divalent cation influx increases steeply until about –25 mV. There are two important features of this relationship:

1. There is no threshold for the activation of Ca channels. Ca^{2+} influx through voltage dependent Ca channels is always finite.

2. The voltage dependence of Ca channels can be very steep over the physiological range of membrane potentials, with a 5-mV depolarization increasing Ca^{2+} influx as much as 2.7-fold.

The studies reported above (as well as those in other tissues) were performed with unphysiological conditions (high [barium], calcium-channel agonist Bay K 8644 or Bay R 5417) to facilitate measurement of single-channel properties. Barium as charge carrier increases single-channel conductance and alters channel kinetics (removing calcium-dependent inactivation). High divalent-cation concentrations shift activation and inactivation curves of calcium channels to more-positive potentials.[16] Bay K 8644 offsets this somewhat by shifting activation curves 10–15 mV negative.[19] However, the similarity between the voltage dependence of steady-state Ca-channel currents with barium (10 mM) as charge carrier at room temperature and Ca-channel currents with physiological Ca (2 mM) at 37°C (compare Fig. 1 with ref. 7; *see* refs. 16,17), suggests that single channel measurements with 10 mM barium may be a reasonable approximation to physiological conditions.

The analysis of single-channel data has been complicated by the observation that dihydropyridine-sensitive, voltage-dependent Ca channels in arterial smooth muscle can exhibit multiple ion-per meation rates (mesenteric artery,[11,15] rabbit basilar artery,[20] rat posterior cerebral artery,[10] coronary arteries,[17] and portal vein[21]). Two prominent levels are often observed, one being half the conductance of the other (in 80 mM barium, 24 pS and 12 pS) . Multiple conductance levels of Ca channels appear to be a widespread phenomenon; for example, the purified Ca channel from skeletal muscle[22] exhibits multiple conductance levels. It is conceivable that Ca channels of different conductance levels represent important, functionally different "isoforms" of dihydropyridine-sensitive Ca channels. In fact, a recent report suggests that endothelin modulates the two conductance levels differentially in isolated cells from portal vein.[21]

Although Ca currents maintained between −60 and −40 mV are a small fraction of the maximal Ca current that can be evoked by depolarizing steps to positive membrane potentials, these steady Ca currents represent substantial Ca^{2+} influx into smooth-muscle cells that have small intracellular volumes (1 pL) .[6,9,15,16,23] Figure 2 simulates the relationship between membrane potential and steady-state calcium entry through voltage-dependent calcium channels for a single smooth-muscle cell with 1000 channels. Even at relatively low p_{open} Ca entry into the cell is sufficient to cause significant increases in intracellular free calcium.[6,9,16] This simulation suggests that calcium-extrusion systems have the capacity to handle substantial Ca influxes.

Calcium-Channel-Blocker Inhibition of Myogenic Tone and Calcium Channels

Myogenic tone and reactivity appear to be inhibited by calcium-channel antagonists.[24] Calcium-channel inhibitors diltiazem and nimodipine abolish myogenic tone of resistance-sized cerebral arteries from normotensive and hypertensive animals.[10,25]

Dihydropyridine (DHP) calcium channel antagonists such as nimodipine,[26] are selective, and usually potent, inhibitors of voltage-dependent Ca channels in arterial smooth-muscle and some other tissues. These drugs bind to the α_1 subunit of the calcium channel and have been used to purify the Ca channel.[22,27] The action of these drugs is dependent on membrane potential, with membrane depolarization increasing inhibition.[15] The affinity of these drugs may also be modulated by neurotransmitters that regulate calcium channels through second-messenger systems, e.g., adrenergic stimulation of voltage-dependent calcium channels in cardiac muscle reduces the inhibitory action of D600.[28,29]

Vasoconstrictor Activation of Arterial Smooth-Muscle Voltage-Dependent Calcium Channels

Recently, norepinephrine (NE),[14,30] endothelin,[21,31] angiotensin II,[32] and serotonin[18] have been shown to increase currents through voltage-dependent calcium channels in arterial smooth muscle. Whereas Benham and Tsien[30] showed that norepinephrine increased Ca currents in rabbit-ear artery, Droogmans et al.[33] reported a reduction of Ca current with NE in the same preparation. The reason for the discrepancy is not clear; however, a reduction in Ca current by NE is at odds with the well-known ability of NE to increase Ca entry and constriction (cf Bülbring and Tomita[34]).

The vasoconstrictors angiotensin II, α1-adrenergic agents, and endothelin stimulate phospholipase C (perhaps through a GTP-binding protein), which results in the production of diacylglycerol (DAG) and inositol trisphosphate (IP_3) from phosphatidylinositol 4,5-bisphosphate. IP_3 stimulates Ca release from the SR, which causes a transient contraction, whereas DAG stimulation of protein kinase C (PKC) may be involved in the maintained contraction through elevating calcium influx and by increasing the effect of calcium on force production. Activation of PKC by phorbol ester has been reported to activate Ca currents in aortic A7r5 cells.[35] In support of the idea that activation of the phosphoinositide pathway in smooth muscle leads to activation of Ca channels, membrane-permeable analogs of DAG (diCg) increased Ca currents in single smooth-muscle cells from toad stomach[36] in a manner similar to acetylcholine stimulation of Ca currents in the same preparation.[37]

Potassium Channels

The membrane potential of an arterial smooth-muscle cell is very dependent on the potassium permeability of the plasma membrane and can also be influenced by the sodium and chloride permeabilities. The membrane potential of smooth-muscle cells in myogenic arteries under physiological pressure appears to be between –50 and –40 mV. The potassium, chloride, and sodium equilibrium potentials have been estimated to be –90 mV, –20 mV, and +50 mV.[9] Presently there is little information on the nature and role of sodium and chloride-permeable channels. There is evidence for at least four types of potassium channels in arterial smooth muscle: (1) inward-rectifying potassium channels, (2) Ca^{2+}-activated potassium channels, (3) ATP-sensitive K^+ channels, and (4) voltage-dependent ("delayed-rectifier") K^+ channels.

Although the inward-rectifier K^+ channel is important in determining the resting potential in some arteries,[9] it may not be the dominant resting potassium permeability in other arteries,[8] and other potassium channels recorded in arterial smooth muscle may also have a role in determining the resting potential (*see* Table 1).

Table 1
Potassium Channels in Arterial Smooth-Muscle[a]

	ATP-Sensitive K$^+$ Channels	Ca^{2+}-Activated K$^+$ Channels
External TEA$^+$	Low-affinity block, $K_d > 7$mM	Relatively high affinity block, $K_d = 150–200$ µM
External TPA+	K_d about 15 µM	K_d, 1.5 mM
Charybdotoxin	?; No effect	High affinity blocker, $K_d = 5$-10 nM
Sulfonylurea, e.g., glibenclamide	Inhibition, $K_d < 10$ µM	No effect
External barium $K_d < 100$µM.	High-affinity block, $K_d > 10$ mM	Very low affinity,
External 4-aminopyridine	?	No effect at < 8 mM
External phencyclidine	?	Little effect at 0.1 mM
External quinidine	?	$K_d = 0.3$mM

[a]See ref. 6 for additional references.

Progress has recently been made in determining the roles of various types of K$^+$ channel through the use of specific blockers[6,38] (*see* Table 1). For example, charybdotoxin appears to be specific for the large-conductance Ca^{2+}-activated potassium channel, whereas the sulfonylurea glibenclamide appears to be relatively selective for ATP-sensitive K$^+$ channels.

Inward-Rectifier Potassium Channels

Elevation of external K$^+$ from 5 to 10 mM causes barium (0.1 mM)-sensitive hyperpolarization[8] and dilation[39] of cerebral arterioles and small arteries. In other preparations, external K$^+$ increases K$^+$ efflux through these channels by elevating its p_{open} and by relieving internal Mg^{2+} block.[40] Another characteristic of inward-rectifier K$^+$ channels is that the single-channel conductance increases steeply negative to the potassium equilibrium potential. Although this channel may be important in the regulation of the membrane potential of certain types of small arteries, it has not been directly identified at the single-channel level in arteries.

Delayed-Rectifier Potassium Channel

Depolarization of vascular smooth muscle activates an outward potassium current that decays on prolonged depolarization.[3,41–43] Voltage-dependent activation followed by slow inactivation is typical of delayed-rectifier potassium channels first described in nerve and skeletal muscle, where the channel participates in the repolarization phase of the action potential.[44] Since smooth muscle cells have very high input resistances, this channel, which would have low p_{open} at physiological membrane potentials in resistance arteries, still could make a significant contribution to the membrane potential.

Figure 3 illustrates a recording of delayed-rectifier potassium channels from coronary arteries. Single channels recorded from cells of rabbit coronary artery in a physiological potassium gradient (6 mM extracellular, 120 mM intracellular potassium) have a single-channel conductance of around 5 pS, with a single-channel current of 0.4 pA at a membrane potential of 0 mV (cf rabbit portal vein[43]).

Calcium-Activated Potassium Channels (K_{Ca})

K_{Ca} channels are activated by depolarization and intracellular calcium.[45] The channels are steeply voltage-dependent, with channel p_{open} increasing e-fold for a depolarization of approx 8-mV, and are activated by physiological concentrations of intracellular calcium. These channels have high single channel conductances in symmetric K^+ and are blocked by charydotoxin and by low concentrations of external tetraethylammonium ions (*see* Table 1).

Modulation by vasoactive substances of K_{Ca} channels from arterial smooth-muscle has also been reported. The vasocontrictor angiotensin II reduces potassium currents in aortic cells and single K_{Ca} channels from coronary arteries reincorporated into lipid bilayers.[32,46] Nevertheless, the role of K_{Ca} channels in resistance arteries is unclear. They may contribute to the membrane potential of smooth-muscle cells in resistance arteries, since, in these arteries, the cells are somewhat depolarized (–40 mV) and intracellular Ca^{2+} is elevated.

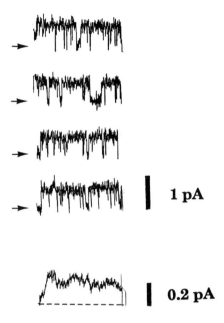

Fig. 3. Original recordings of delayed-rectifier K$^+$ channels in an on-cell patch on a single smooth-muscle cell from rabbit left descending anterior coronary artery. Holding potential was –80 mV and test potential, +20 mV. 6 mM K$^+$ in the pipet. The pulse duration was 0.5 s.

ATP-Sensitive (K_{ATP}) Channels

Potassium channels inhibited by intracellular ATP (K_{ATP} channels) have been identified in arterial smooth muscle (*see* Fig. 4).[6,47,48] This type of channel was first identified in cardiac muscle[49] and subsequently in pancreatic β cells, skeletal muscle, and neurons.[6,38,50] K_{ATP} channels are activated by a variety of synthetic and physiological vasodilators. The ensuing membrane hyperpolarization presumably leads to vasodilation through the closure of voltage-dependent Ca channels.

The membrane hyperpolarization and vasodilation caused by a number of potassium-channel openers (e.g., cromakalim, pinacidil, diazoxide, and nicorandil) are reversed by agents that block K_{ATP} channels (e.g., the antidiabetic sulfonylurea glibenclamide [10

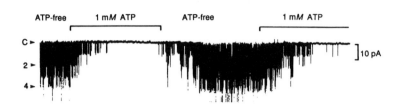

Fig. 4. Original recording of ATP-sensitive K⁺ channels in an inside-out excised patch from a smooth-muscle cell of rabbit mesenteric arteries. The holding potential was –90 mV. In the pipet was 60 mM K⁺, in the bath solution, 120 mM K⁺. Reprinted with permission from Standen, N. B., J. M. Quayle, N. W. Davies, J. E. Brayden, Y. Huang, and M. T. Nelson. Hyperpolarizing vasodilators activate ATP-sensitive K⁺ channels in arterial smooth muscle. *Science* 1989; 245: 177–180. Copyright 1989 by the AAAS.

μM], barium [0.1 mM], and tetrapentyl ammonium ions [5 μM]).[6,47,48] Single K_{ATP} channels are activated by cromakalim and inhibited by glibenclamide in mesenteric-artery smooth-muscle cells.[47] Cromakalim and pinacidil also activate single K_{ATP} channels in cardiac myocytes, and diazoxide activates K_{ATP} channels in pancreatic β cells.[6,38,50] Agents that block other potassium channels, not K_{ATP} channels, are relatively ineffective at reversing the effects of the vasodilators.[6,38]

A number of physiological vasodilators also activate K_{ATP} channels. For example, the hyperpolarization of cerebral arteries by vasoactive intestinal peptide and endothelial-dependent vasodilators (acetylcholine and ADP) is reversed by both glibenclamide and barium.[47,51] In addition, membrane hyperpolarization and part of the dilation in mesenteric arteries caused by the neuropeptide calcitonin-gene-related peptide (CGRP) are blocked by glibenclamide.[52] At the single channel level, CGRP activates ATP-sensitive potassium channels in cell-attached patches. CGRP that was added to the solution bathing the cell did not have direct access to channels in the patch (which are isolated in the patch electrode). Thus, CGRP may work through generation of a second-messenger system or via direct activation of the channel by a G protein.[52]

The increase in coronary flow accompanying hypoxia in isolated, perfused hearts is also blocked by glibenclamide, and the hypoxic dilation is mimicked by cromakalim and metabolic poisons, implicating the K_{ATP} channel in this response.[53] Activation of the channel during hypoxia may occur by a number of mechanisms (e.g., a fall in cytoplasmic P_{O_2} may result in sufficient lowering of cytoplasmic [ATP] to activate channels). However, adenosine dilation of coronary arteries was also largely reduced by gliben-clamide, suggesting that adenosine, which has long been regarded as an important mediator of hypoxic dilation of coronary arteries, may itself activate K_{ATP} channels.[53]

Acknowledgments

This work was supported by grants from the American Heart Association and the National Science Foundation. JMQ was an Agan Fellow of the American Heart Association, Vermont Affiliate. MTN is an Established Investigator of the American Heart Association.

References

[1] Faraci, F. M., Baumbach, G. L., and Heistad, D. D. (1989) *J. Hypertens.* **7** (Suppl. 4), S61–S64.
[2] Harder, D. R. (1984) *Circ. Res.* **55**, 197–202.
[3] Harder, D. R., Gilbert, R., and Lombard, J. H. (1987) *Am. J. Physiol.* **253**, F778–F781.
[4] Harder, D. R., Kauser, K., Roman, R. J., and Lombard, J. H. (1989) *J. Hypertens.* **7 (Suppl. 4),** s11–s1 5.
[5] Brayden, J. E. and Wellman, G. C. (1989) *J. Cereb. Blood Flow Metab.* **9,** 256–263.
[6] Nelson, M. T., Patlak, J. B., Worley, J. F., and Standen, N. B. (1990B) *Am. J. Physiol.* **259,** C3–C18.
[7] Hirst, G. D. S., Silverberg, G. D., and Van Helden, D. F. (1986) *J. Physiol. (Lond.)* **371,** 289–304.
[8] Edwards, F. R., Hirst, G. D. S., and Silverberg, G. D. (1988) *J. Physiol. (Lond.)* **404,** 455–466.
[9] Hirst, G. D. S. and Edwards, F. R. (1989) *Physiol. Rev.* **69,** 546–604.
[10] Quayle, J. M., McCarron, J., Halpern, W., and Nelson, M. T. (1990) *Biophys. J.* **57,** 301a.
[11] Worley, J. F., Deitmer, J. W., and Nelson, M. T. (1986) *Proc. Natl. Acad. Sci. USA* **83,** 5746–5750.

[12] Bean, B. P., Sturek, M., Puga, A., and Hermsmeyer, K. (1986) *Circ. Res.* **59,** 229–235.

[13] Benham, C. D., Hess, P., and Tsien, R. W . (1987) *Circ. Res.* **61 (Suppl. 1),** 110–116.

[14] Nelson, M. T., Standen, N. B., Brayden, J. E., and Worley, J. F. (1988) *Nature* **336,** 382–385.

[15] Nelson, M. T. and Worley, J. F. (1989) *J. Physiol. (Lond.)* **412,** 65–91.

[16] Aaronson, P. L., Bolton, T. B., Lang, R. J., and MacKenzie, I. (1989) *J. Physiol. (Lond.)* **405,** 57–75.

[17] Ganitkevich, V. Ya, and Isenberg, G. (1990) *J. Physiol. (Lond.)* **426,** 19–42.

[18] Worley, J. F., Standen, N. B., and Nelson, M. T. (1989) *Biophys. J.* **55,** 304a.

[19] Sanguinetti, M. C., Krafte, D. S., and Kass, R. S. (1986) *J. Gen. Physiol.* **88,** 369–392.

[20] Quayle, J. M. and Nelson, M. T. (1990) *Biophys. J.* **57,** 301a.

[21] Inoue, Y., Oike, M., Nakao, K., Kitamura, K., and Kuriyama, H. (1990) *J. Physiol. (Lond.)* **423,** 171–191.

[22] Talvenheimo, J., Worley, J. F., and Nelson, M. T. (1987) *Biophys. J.* **52,** 891–899.

[23] Imaizumi, Y., Muraki, K., Takeda, M., and Watanabe, M. (1989) *Am. J. Physiol.* **256,** C880–885.

[24] Nordlander, M. (1989) *J. Hypertension* **7(Suppl. 4),** s141–147.

[25] Osol, G., Osol, R., and Halpern, W. (1986) *Essential Hypertension* (Aoki, K., ed.), Springer-Verlag, Tokyo, pp. 107–114.

[26] Janis, R. A., Silver, S. C., and Triggle, D. J. (1987) *Adv. Drug Res.* **16,** 309–591.

[27] Campbell, K. P., Leung, A. T., and Sharp, A. H. (1988) *Trends Neurosci.* **11,** 425–430.

[28] McDonald, T., Pelzer, D., and Trautwein, W. (1989) *J. Physiol. (Lond.)* **414,** 569–586.

[29] Meisheri, K. D., Sage, G. P., and Cipkus-Dubray, L. (1990) *J. Pharm. Exp. Therap.* **252,** 1167–1174.

[30] Benham, C. D. and Tsien, R. W. (1988) *J. Physiol. (Lond.)* **404,** 767–784.

[31] Goto, K., Kasuya, Y., Matsuki, N., Takuwa, Y., Kurihara, H., Ishikawa, T., Kimura, S., Yanagisawa, M., and Masaki, T. (1989) *Proc. Natl. Acad. Sci. USA* **86,** 3915–3918.

[32] Bkaily, G., Peyrow, M., Sculptoreanu, A., Jacques, D., Chahine, M., Regoli, D., and Sperelakis, N. (1988) *Pflugers Arch.* **412,** 448–450.

[33] Droogmans, G., Declerck, I., and Casteels, R. (1987) *PflugersArch.* **409,** 7–12.

[34] Bulbring, E. and Tomita, T. (1987) *Pharmacol. Rev.* **39,** 49–96.

[35] Fish, D., Sperti, G., Colucci, W., and Clapham, D. (1988) *Circ. Res.* **62,** 1049–1054.

[36] Vivaudou, M. B., Clapp, L. H., Walsh, J. V., and Singer, J. J. (1988) *FASEB J.* **2**, 2497–2504.

[37] Clapp, L. H., Vivaudou, M. B., Walsh, J. V., and Singer, J. J. (1987) *Proc. Natl. Acad. Sci. USA* **84**, 2092–2096.

[38] Quast, U. and Cook, N. S. (1989) *Trends Pharmacol. Sci.* **10**, 431–435.

[39] McCarron, J. G. and Halpern, W. (1990) *Am. J. Physiol.* **259**, H902–H908.

[40] Stanfield, P. R. (1988) *Trends Neurosci.* **11**, 475–477.

[41] Okabe, K., Kitamura, K., and Kuriyama, H. (1987) *Pflugers Arch.* **409**, 561–568.

[42] Beech, D. J. and Bolton, T. B. (1989) *Br. J. Pharmacol.* **98**, 851–864.

[43] Beech, D. J. and Bolton, T. B. (1989) *J. Physiol. (Lond.)* **418**, 293–309.

[44] Hille, B. (1984) *Ionic Channels of Excitable Membranes*, Sinauer, Sunderland, MA.

[45] Benham, C. D., Bolton, T. B., Lang, R. J., and Takewaki, T. (1986) *J. Physiol. (Lond.)* **371**, 45–67.

[46] Toro, L., Amador, M., and Stefani, E. (1990) *Am. J. Physiol.* **258**, H912–H915.

[47] Standen, N. B., Quayle, J. M., Davies, N. W., Brayden, J. E., Huang, Y., and Nelson, M. T. (1989) *Science* **245**, 177–180.

[48] Kovacs, R. J. and Nelson, M. T. (1989) *Circulation* **80**, II–519.

[49] Noma, A. (1983) *Nature* **305**, 147–148.

[50] Ashcroft, F. M . (1988) *Ann. Rev. Neurosci.* **11**, 97–118.

[51] Brayden, J. E. (1990) *Am. J. Physiol.* **259**, H668–H673.

[52] Nelson, M. T., Huang, Y., Brayden, J. E., Hescheler, J. K., and Standen, N. B. (1990) *Nature* **344**, 770–773.

[53] Daut, J., Maier-Rudolph W., von Beckerath, N., Mehrke, G., Gunther, K., and Goedel-Meinen, L. (1990) *Science* **247**, 1341–1344.

Chapter 17

Membrane Biochemistry of Vascular Smooth Muscle of Resistance Blood Vessels and Changes in Hypertension

Chiu-Yin Kwan

Introduction

In vascular smooth-muscle (VSM), the physiological fluctuation of cytoplasmic Ca^{2+} concentration ($[Ca^{2+}]$), which is 1,000- to 10,000-fold lower than the extracellular Ca^{2+} concentration, is governed by finely integrated interactions between short-term regulation at the level of internal membrane compartments and long-term regulation at the level of cell membranes to maintain such a steep electrochemical gradient across the cell membrane.[1] Interpretation of the results from studies using intact muscle preparations to investigate the roles of cell membrane and intracellular membranes in the regulation of cytoplasmic $[Ca^{2+}]$ has often been complicated by the problems arising from the complex geometry of the blood vessels and the heterogeneity of Ca^{2+}-handling mechanisms present in various cellular membrane systems.[1,2] The use of highly purified subcellular-membranes from the VSM

From *The Resistance Vasculature*, J. A. Bevan et al., eds. ©1991 Humana Press

may circumvent some of the difficulties inherent to the intact tissue preparation, thus allowing better characterization of the various modes of Ca^{2+}–membrane interactions.[1,3] The subcellular-membrane approach also permits investigation of ligand–receptor interactions at the receptor recognition sites at which the membrane transduction processes occur and ultimately lead to Ca^{2+}-signaling mechanisms. Better understanding of some membrane-bound enymes not only can facilitate the characterization of fractionated membranes, but also may provide alternative investigatory tools to the study of receptor-activation processes and ion-transport mechanisms. This review represents a summary, based largely on the author's progress during the past decade, concerning the broader aspects of the membrane biochemistry of VSM, particularly those in small and resistance arteries of the mesenteric vasculature. In light of the important role of these vessels in the regulation of systemic blood pressure, the application of this subcellular-membrane technique to the study of the VSM membrane abnormalities in relation to the etiology of hypertension is also described. Although this volume is intended for reviews on the studies using resistance arteries, it is inevitable that some results on larger vessels will also be described, primarily for comparative purposes.

Membrane Fractionation

The use of membrane fractions from VSM, such as intact vesicles of plasma membranes (PM) or endoplasmic reticulum (ER), or intact mitochondria, to study Ca^{2+} binding and transport was introduced in the 1970s. In these earlier studies, crude microsomal membrane fragments isolated from large arteries were preferred for pragmatic reasons, because of their size. The membrane-separation method generally failed to distinguish membrane fragments originating from PM or ER, but an origin from ER was usually assumed by analogy with the sarcoplasmic reticulum (SR) system in striated muscle, and the role of PM vesicles in the active transport of Ca^{2+} has not been given adequate attention. This proved to be erroneous, as shown by the introduction of more markers for various membrane fractions.[1,2,4] Most of the ATP-dependent Ca^{2+} uptake by isolated heterogeneous microsomal fractions from VSM turned out to be associated with the PM.[5-12] This by no means suggests that ER in VSM does not play a significant role in the regulation of cytosolic $[Ca^{2+}]$. However, attempts to study ER have been frustrated by the meager quantities of ER, particularly in the VSM of small vessels; by the lack of an unequivocal marker for ER; by a loss of the Ca^{2+}-handling activities associated with the ER-enriched fraction during isolation;[13,14] and by the difficulty in separating the ER from the PM vesicles.[4,15,16] In spite of the limitations, the subcellular-membrane technique to date has revealed most of the quantitative information about mechanisms of Ca^{2+} binding and transport in VSM.

The fractionation technique involves basically four major steps: (a) trimming of vascular tissues, (b) tissue homogenization (c), membrane separation, and (d) characterization of isolated subcellular fractions. Much of the technical information on subcellular fractionation of VSM has been described and recently reviewed. Among the small vessels studied, membranes from the VSM of mesenteric vasculature have been best characterized in terms of membrane-separation technology and the application of the isolated membranes for the studies of ion-transport, receptor binding, and membrane-bound enzymes. The handling of mesenteric vasculature to remove large amounts of mesenteric connective tissue, paravascular nerve fibers, surrounding fat, and the collateral veins require special trimming procedures and has been described previously in great detail.[7,8,17] Failure to remove effectively the nonarterial muscle tissues will complicate the interpretation of the results obtained.[10,18]

Microsomal membranes, usually quite heterogeneous, containing mainly PM, ER, and perhaps other ill-defined membrane fragments, were frequently obtained by differential centrifugation and used without further purification.[1-3] Better enrichment of the PM and ER fractions can be achieved by further subfractionation of the microsomes on a sucrose density gradient. However, microsomal fraction of VSM with improved purity in PM content has also been obtained by a modified method utilizing only differential centrifugation.[15,16,18a,19] The modification involves a change in the composition of the homogenization medium so that the major contaminating contractile proteins can be largely eliminated by repeated washings with medium containing Mg-ATP and a high salt concentration[18a,20] and the PM fragments can be selectively separated from the contaminating intracellular membranes by cation-induced membrane aggregation.[15,19] Such a cation-induced membrane-aggregation method is, however, quite empirical. The separation technique using free-flow electrophoresis, which separates membranes solely on the basis of the net surface charge difference in membranes of different subcellular origins, has also been attempted in the fractionation of VSM.[16] However, this technique offered no major advantage over the conventional centrifugation technique.

Isolated membrane fractions were usually checked for their subcellular origin and purity by the use of various membrane markers. These markers should be stable and convenient, as well as easy to monitor. The marker enzymes commonly employed include 5'-nucleotidase, phosphodiesterase I, Na^+,K^+-ATPase, K^+-activated p-nitrophenyl phosphatase, Mg^{2+}-ATPase, adenylate cyclase, and alkaline phosphatase for plasma membranes (*see* Membrane-Bound Enzymes section); NADPH cytochrome c reductase for endoplasmic reticulum; and cytochrome c oxidase, succinate dehydrogenase, and succinate cytochrome c reductase for mitochondria (for a comprehensive review on the use of these membrane markers, *see* refs. 1–3). The simplicity in labeling the surface membrane receptors also make it a useful PM marker (*see* Membrane Receptors section).

Ca^{2+}–Membrane Interactions

Ca^{2+}-Handling by Plasma Membranes

Binding and Release of Ca^{2+}

Nakayama et al.[21] reported that, in rabbit cerebral artery, the stretch-induced contraction was resistant to the removal of extracellular Ca^{2+} and was absent in chemically skinned preparations (suggesting the utilization of intracellular Ca^{2+} in this process). They also provided, by the pyroantimonate precipitation technique, further evidence that was validated with electron probe X-ray microanalysis that the intracellular source of the Ca^{2+} was the inner surfaces of the VSM cells. It has long ago been demonstrated that isolated PM fractions from rat and dog mesenteric arteries indeed represent the major Ca-binding site.[7–9,17] More interestingly, a high-capacity, pH-dependent Ca binding, which was first identified in the PM fragments of rat myometrium,[22] has also been reported in PM isolated from rat mesenteric artery.[23] The apparent K_m value for Ca^{2+} ([Ca^{2+}] required for half-maximum binding) was 0.3–0.5 μM and the apparent pK_a value (the inflection point of the pH vs Ca-binding profile) was at pH 6.7. Lowering of pH from 7.2 to 6.4 caused an immediate release of bound Ca^{2+}. These properties suggest that this pH-dependent binding of Ca^{2+} may be a physiologically relevant process. Any physiological stimulus or metabolic perturbation affecting the intracellular pH near the microenvironment of the PM (e.g., via activation of an Na–H exchanger at the surface membrane?) would be expected to cause the release of bound calcium.

Release of Ca^{2+} from preloaded PM vesicles isolated from rat mesenteric arteries has also been studied as a relative measurement of membrane permeability to Ca^{2+}.[7,24] However, interpretation of the resultant multiexponential Ca^{2+} efflux profile is difficult because of the different leak mechanisms that allow the passive Ca^{2+} efflux to occur and because of our meager knowledge about the properties of various Ca channels in isolated PM vesicles.

ATP-Driven Transport of Ca^{2+}

There is substantial evidence to indicate that the ATP-driven Ca^{2+} transport is an intrinsic property of the smooth-muscle plasma membrane vesicles, and such an ATP-dependent Ca accumulation has also been shown to be an active transport process.[7–9,17] The transport of Ca^{2+} against its own electrochemical gradient across the isolated membrane vesicles (presumably the inside-out-oriented PM or right-side-out ER membrane vesicles) has been conveniently and convincingly demonstrated in two manners: first, the ability of oxalate or inorganic phosphate anions that precipitate with Ca^{2+} inside the membrane vesicles to enhance the Ca^{2+} accumulation in the presence, but not in the absence, of ATP; second, the ability of membrane-permeant reagents, such as Ca^{2+}-selective ionophores (e.g., A23187 and ionomycin), membrane-selective pore-forming reagents (e.g., digitonin and saponin), or nonspecific detergents (e.g., Triton X-100), to inhibit the ATP driven Ca^{2+} transport by relaxing the Ca^{2+} gradient across the membrane vesicles. The stimulation of ATP-driven Ca accumulation by inorganic phosphate and oxalate has for some time been widely used as a marker for the presence of ER.[2] This property is now recognized to be nonspecific with respect to PM and ER in VSM membranes.[5,7,9]

The vascular muscle PM represents a physiologically important site for the regulation of cytoplasmic Ca^{2+} concentration, since the apparent K_m values for Ca^{2+} (determined from the apparent initial rate of Ca transport) all fall within 0.1–0.5 μM.[18a,25] The activity of this pump can be modulated by cAMP-dependent protein kinase.[26,27] The activation of this Ca pump in highly purified PM fractions from rat aorta, dog aorta, pig coronary artery, and bovine carotid artery by physiological concentrations of calmodulin further illustrates its physiological modulation via interactions with cytoplasmic effectors.[25] However, not all these vascular muscle plasma membranes showed comparable levels of activation by exogenously added calmodulin. For example, Morel et al.[28] found that rat aortic microsomes not washed with EGTA showed a twofold higher ATP-dependent Ca^{2+} accumulation than to the EGTA-washed microsomes, but did not respond to added calmodulin. After an EGTA wash, the Ca^{2+} accumulation was reduced by 50%, but it was stimulated about twofold by exogenously added calmodulin. On the other hand, Wuytack and Casteels[11] reported that EGTA-washed microsomes isolated from porcine coronary artery showed only 20% stimulation of Ca^{2+} accumulation by calmodulin. In this regard, our unpublished finding, using rat mesenteric artery PM under experimental conditions that ensure the removal of endogenous membrane-bound calmodulin by EGTA washing, also showed only about 20% stimulation of ATP-dependent Ca transport by exogenously added calmodulin. However, under similar conditions, a 250–300% activation by calmodulin was observed using dog aortic muscle membranes. These findings point to the probable presence of heterogeneity of interactions among Ca^{2+}, calmodulin, and PM in vascular smooth muscle, and also caution against the use of calmodulin stimulation of Ca transport as a PM marker.

Na–Ca and Na–H Exchangers

A direct demonstration of the presence of Na–Ca exchange in isolated smooth-muscle PM was first reported in rat myometrium.[29] In VSM, Na–Ca exchange was first demonstrated in rat mesenteric artery PM[4] and was subsequently confirmed by Matlib et al.[30,31] The Na–Ca exchanger was later isolated from the PM of rat mesenteric artery and reconstituted in soybean phospholipid vesicles.[32] Similarly, Na–Ca exchange was also characterized in PM vesicles isolated from dog[32] and bovine[33] mesenteric arteries. In all these studies, such a Na^+-dependent Ca^{2+} transport was shown to be driven by the transient Na^+ gradient, since this Ca^{2+} transport was prohibited by an Na^+-selective ionophore, monensin, or by restoring extravehicular $[Na^+]$ to the same level as the intravesicular $[Na^+]$. Amiloride was also shown to inhibit the VSM Na–Ca exchanger. In fact, Na–H exchange activity was also found to coexist with, and perhaps coupled to, the Na–Ca exchange activity in the PM isolated from bovine mesenteric artery. However, the rate of this Na^+-gradient-dependent Ca^{2+} accumulation was at least an order of magnitude lower than that of the ATP-dependent Ca^{2+} transport. Also, the K_m for Ca^{2+} in Na–Ca exchange was considerably higher than that in ATP-dependent Ca^{2+} transport.[29,34] Although these properties make Na–Ca exchange less likely to be an effective mechanism in the regulation of cytoplasmic Ca^{2+} levels under physiological conditions, this Na–Ca exchange mechanism has received considerable attention because it is an essential component of an attractive hypothesis to relate the defective Na^+ metabolism in VSM to altered vascular reactivity in hypertension.[35] Although the existence of Na–Ca exchange activity can be demonstrated in PM vesicles isolated from VSM under selected experimental conditions, many functional studies using intact arteries provided evidence against it being an important Ca^{2+} control system.[36] Therefore, the issue on the role of Na–Ca exchange in the physiological regulation of cytosolic $[Ca^{2+}]$ is still not settled.

Another Na^+-related, amiloride-sensitive exchanger system, namely the Na–H counter-ion transport system, has recently been identified in PM vesicles isolated from dog[19] and bovine[37] mesenteric arteries. Kahn et al.[33,38] reported that Na–Li exchange and Na–H exchange are mediated by the same transport system in PM vesicles from bovine mesenteric arteries. The Na–H counter-ion transport system is of particular interest because of its importance in intracellular pH and [Na$^+$] regulation and its possible link to cellular growth and proliferation.[39]

Ca Channels

The vasodilatory and antihypertensive effects of various Ca antagonists have attracted considerable interest in the study of the control of Ca^{2+} entry via the voltage-operated Ca channels in vascular smooth muscle. With regard to the properties of voltage operated Ca channels, the presence, but not the function, of voltage-operated Ca channels in isolated VSM microsomes has also been demonstrated with radiolabeled dihydropyridine Ca antagonists. However, the number of channels studied by such a radioligand-binding approach has been considerably less in VSM than in nonvascular smooth-muscle microsomes.[40,41] Morel and Godfraind[42] reported that membrane potential could modulate the binding affinity of some dyhyrdopyridines for the Ca channels in the PM of rat mesenteric arteries. To date, there is little convincing and direct evidence that these dihydropyridine-labeled Ca channels in isolated membranes of VSM functional in such a way that binding of the Ca channel antagonists affects their handling of Ca^{2+}. However, it has been reported[43] that 10 μM of verapamil or nifedipine inhibited up to 40% of ATP-dependent Ca^{2+} accumulation in a poorly defined subcellular fraction from rabbit cerebral arteries. This may represent a nonspecific effect of these Ca antagonists, since the concentrations employed were in excess of those relevant to Ca channel antagonism. Moreover, this property was attributed to the presence of ER fragments although no marker for ER was used; only Na$^+$,K$^+$-ATPase activity was used as the PM marker. Since the preparation of cerebral arteries is likely to be substantially contaminated by membrane fragments of neuronal tissues, which are more enriched in Na$^+$,K$^+$-ATPase activities than the smooth-muscle membranes,[44] the use of this marker alone to monitor the PM of VSM in this preparation could be misleading. Also, since no marker was used to monitor ER, equating the subcellular fraction with relatively lower Na^+,K^+-ATPase activity to vascular-muscle ER is rather naive and unjustified.

Study of the receptor-operated Ca channel has so far been limited to contractility studies and Ca-flux experiments. These aspects are discussed in more detail elsewhere in this volume. Unlike the voltage-operated Ca channel, which can be studied by a wide spectrum of Ca channel blockers or activators as useful pharmacological tools, there are no satisfactory probes or tools available to identify the receptor-operated Ca channel(s). Recently, Guan et al.[45,46] have reported that a group of ginsenosides purified from the *panax notoginseng* possess pharmacological actions on rabbit aortic and dog mesenteric arterial muscle strips that are consistent with its action as a receptor-operated Ca channel blocker. These ginsenosides had no effect on the contractile response of rabbit aorta depolarized by elevated K^+, but they blocked the responses to norepinephrine in normal Ca-containing media without any effect on the contraction induce by norepinephrine in Ca-free media. These results suggest that the ginsenosides purified from *panax notoginseng* do not act through the voltage-operated Ca channel or via the release of intracellular Ca^{2+}. Rather, it may suggest a novel receptor-operated Ca channel blocker.

Ca-Handling by Internal Membranes

It is generally accepted that PM and ER, but not mitochondria, play a major physiological role in regulating cytosolic $[Ca^{2+}]$.[47,48] Numerous attempts have been made to differentiate between the properties of the Ca pumps derived from the PM and ER. This includes the effects of Ca-precipitating anions such as oxalate and inorganic phosphate,[20] calmodulin (discussed in the preceding section), sulfhydryl agents,[47] vanadate ion,[28] agents that bind specifically to such cholesterol molecules as digitonin[9] and saponin,[5,49] radiation inactivation analysis,[50] and the use of antibody raised against the Ca pump ATPase purified from erythrocyte membranes or from skeletal-muscle SR.[48] Interpretation of many of the above findings has been based on the weak assumption that Ca accumulation by ER is selectively enhanced by Ca-precipitating anions (as is the case for skeletal-muscle SR). Chiesi et al.[51] claimed that the microsomal membrane fraction that was isolated from bovine aorta by differential centrifugation after a very vigorous homogenization contained mainly ER. This was based on the observation that the molecular size of the phosphointermediate of the putative Ca-pump protein and its sensitivity to La^{3+} were similar to those of SR isolated from striated muscle. However, this Ca-pump protein from dog aortic microsomes was immunologically different from that isolated from the SR of skeletal or cardiac muscle. It is rather surprising that Chiesi et al.[51] were unable to detect the presence of PM even in the initial crude membrane fraction. It is reasonable to expect that the vigorous homogenization procedure required to disrupt the large elastic artery will cause vesiculation of PM to various sizes. It is possible, but unlikely that all PM fragments formed large sheets or sacs under experimental conditions used by Chiesi et al., and were removed completely during the initial low-speed centrifugation. This hypothesis could have been easily and unequivocally tested by morphological or biochemical means. Nevertheless, some progress has been made using nonvascular smooth muscle, which is easily obtained (in greater quantity) from large animal organs (such as porcine stomach) with the aid of antibodies raised against Ca-pump ATPases purified from erythrocytes and skeletal muscle SR.[48] It was evident that the ER Ca pump, even in this smooth muscle tissue, could be identified only in substantially smaller quantities than to the PM Ca pump. It should also be noted, however, that such a sensitive technique may also identify the ER fragments derived from the contaminating tissues, such as the intrinsic nerve fibers, nerve plexus, endothelium, and fibroblasts, even if they are present in very minute quantities.

It has become clear that inositol-1,4,5-trisphosphate (IP_3), which is produced from PM upon receptor activation, causes release of Ca^{2+} from internal stores in many cell types. It has been demonstrated that IP_3 caused a small release of Ca^{2+} from preloaded VSM membrane fractions obtained from large blood vessels.[52] This Ca^{2+}-releasing fraction is presumably a result of the presence of ER. IP_3-induced release of Ca^{2+} from ER was also demonstrated by Yamamoto and Van Breemen[52a] using a saponin-skinned monolayer of primary aortic culture cells. Such a Ca^{2+}-releasing action or the existence of IP_3 receptors has yet to be demonstrated in membrane fraction derived from VSM of small arteries, in which the ER content is presumed to be much smaller than that in large arteries. Another commonly used Ca^{2+} releasing reagent, ryanodine, has been shown to attenuate the vascular response to norepinephrine or caffeine in the absence of extracellular Ca^{2+}.[53,54] Ryanodine probably acts by promoting the release of Ca^{2+} from the intracellular store. A similar action of ryanodine has also been observed in rabbit ear artery,[55] dog mesenteric artery,[56] and rat vas deferens smooth muscle.[57] However, ^3H–ryanodine binding sites can be readily identified in the membrane fractions coenriched in ER marker enzyme, NADPH-cytochrome c reductase, from vas deferens, but not mesenteric artery smooth muscle.

Alterations in Hypertension

Since the contractile apparatus in vascular muscle is activated by a rise of cytoplasmic $[Ca^{2+}]$, derangement of the regulation of Ca^{2+} leading to elevated intracellular Ca^{2+} activity in the VSM cells[58–60] may provide a logical interpretation of some of the altered contractile properties of VSM seen in spontaneously hypertensive rats (SHR), including elevated muscle tone, hyperreactivity to vasoactive substances, and decreased rate of relaxation.[61] Extracellular Ca^{2+}, on the other hand, not only serves as the activator Ca^{2+} to support contraction, but also acts as a membrane stabilizer to prevent further influx of Ca^{2+}, thus inhibiting the vasoconstriction and ultimately leading to relaxation.[62,63] Increased Ca sensitivity in contractile responses of vascular strips isolated from hypertensive animals has repeatedly been noted and recently reviewed.[64] However, such an abnormality was not observed in chemically skinned vascular preparations isolated from hypertensive animals.[25,65,66] This suggests that the contractile abnormalities found in vascular strips of hypertensive animals resides in the cell membranes.

A defective Ca^{2+}-extrusion mechanism in the PM of small muscular arteries[6,18,67] or a derangement of the Ca^{2+}-sequestration system in the ER of large elastic arteries[67–69] has been suspected to account for the functional abnormalities in hypertension. Several lines of evidence derived from the studies of PM from mesenteric arteries[70-74] collectively suggest that diminished ATP-driven Ca^{2+} transport could not be attributed solely to the consequence of hypertension and may, in fact, be related to the mechanisms initiating hypertension. These particular aspects have been reviewed quite adequately[70-72] and will not be discussed further.

Wei et al.[67] reported a significant decrease of Ca^{2+} binding to the PM fraction of mesenteric arteries from SHR at millimolar concentrations of Ca^{2+}. This finding seems to be consistent with the hypothesis that reduced binding of Ca^{2+} to the VSM cell membranes may be responsible for the decreased membrane stability that leads to increased constriction as a result of the influx of Ca^{2+} (*see* ref. 63) and reduced ability of the VSM to relax in hypertension.[61,63] However, Kwan et al.,[6,18,75] as well as others,[69,76] have shown that in the absence of ATP, the binding of Ca^{2+} at 20 μM free Ca^{2+} to the PM-enriched fractions isolated from the VSM of SHR was not different from that of WKY. They also demonstrated that the addition of A23187 prevented the ATP-supported Ca^{2+} transport and caused a rapid and substantial release of Ca^{2+} from the loaded membrane vesicles. However, the residual Ca^{2+} binding in the presence of A23187 was not different in fractions isolated from SHR and Wistar-Kyoto rats (WKY). It is possible that stabilization of cell membranes requires millimolar, not micromolar, concentrations of Ca^{2+}. Very few other abnormalities related to Ca^{2+}-handling properties of VSM membranes from small or resistant arteries in hypertension were noted. The Na–Ca exchange activity[30] and the permeability to Ca^{2+} of PM vesicles[71] isolated from VSM of mesenteric arteries of SHR and WKY were found to be not significantly different.

Membrane Receptors

The development of membrane-separation techniques in obtaining highly purified fragments of PM and the availability of new, highly selective radioligands has greatly facilitated the study of surface-membrane receptors in VSM. The radioligand binding technique using membrane fragments has contributed much of the quantitative information about some cell membrane receptors in the VSM of small arteries, such as adrenoreceptors and vasoactive peptide receptors, which have been adequately reviewed.[77,78] In the following sections, emphasis is placed on the progress in recent years using VSM from mesenteric vasculature.

Adrenoreceptors

In VSM, both postjunctional α- and β-adrenoreceptor subtypes have been identified by functional and radioligand-binding studies.[77,79,80] By functional criteria, with few exceptions, as the blood vessels become smaller, the α-adrenoreceptors become less dominant than the β-adrenoreceptors.[81] This has received experimental support from the radioligand-binding studies. For example, the maximal number of binding sites (B_{max}) of ^{3}H-prazosin (a selective antagonist for α_1-adrenoreceptor) in microsomal membranes was about fivefold higher in dog aorta than in dog mesenteric arteries, whereas that of ^{125}I-monoiodopindolol (ICYP) was substantially higher in mesenteric arteries than in aorta.[24,80]

Studies using selective antagonists for the subtypes of α- and β-adrenoreceptors in PM-enriched fractions of rat and dog mesenteric arteries indicated the dominance of β_2-adrenoreceptor subtype[80] and the presence of separate binding sites for ^{3}H-prazosin[82] (α_1-adrenoreceptor antagonist) and ^{3}H-rauwolscine[83] or ^{3}H-yohimbine[77] (α_2-adrenoreceptor antagonist). All these binding sites for adrenoreceptor ligands were positively correlated to the PM content. It is interesting to note that ^{3}H-prazosin binding differed between dog aorta and dog mesenteric arteries in both the B_{max} values and the K_d values (equilibrium dissociation constant) of the α_1-adrenoreceptor, thus suggesting the possible existence of α_1-adrenoreceptor subtypes.[82] It is also interesting to note that, in spite of the presence of ^{3}H-rauwolscine binding sites in PM-enriched fractions from dog aorta and mesenteric arteries, no evidence of functional α_2-adrenoreceptors was found (based on contractile responses elicited by B-HT 920). On the other hand, PM-enriched fraction from dog saphenous vein that responded to B-HT 920 showed a B_{max} value for ^{3}H-rauwolscine six- to eightfold higher than that from dog mesenteric arteries without significant changes in K_d values. These studies indicate that postjunctional α_2-adrenoreceptor is indeed present in mesenteric arteries in a functionally uncoupled state.

Alterations in VSM α_2- and β-adrenoreceptors in hypertension have long been suggested, but studies on the role of α- and β-adrenoreceptors on VSM of small arteries using a radioligand-binding technique are very limited. Agrawal et al.[77,79] reported a moderate increase in the B_{max} of ^{3}H-yohimbine binding to a PM-enriched fraction of mesenteric arteries isolated from SHR compared to that in WKY. The B_{max} and K_d values for ^{3}H-prazosin binding, however, were not altered in SHR. This is in contrast to the study by Schiffrin et al.,[84] who have reported an increase of B_{max} value in a crude membrane preparation from mesenteric arteries of 4-wk old SHR compared to a similar preparation from WKY. In SHR cerebral microvessels, an increased number of ^{3}H-binding sites was also reported. However, the lack of correlation studies between changes in binding and functional parameters in VSM adrenoreceptors make it difficult to relate these changes to hypertension. The studies of changes in postjunctional β-adrenoreceptors in small arteries in hypertension are also very scanty. A reduced β-adrenoreceptor density in hypertension has been reported in a crude membrane fraction isolated from mesenteric arteries of DOCA-salt hypertensive rats[85] and in unfractionated brain microvessels.[86] The interpretation of these results requires caution since ICYP binding to aortic muscle cells cultivated from 4-wk old SHR and WKY aortas indicated a four- to fivefold higher B_{max} in SHR than in WKY, whereas the crude microsomal fractions isolated from aortic muscle of SHR and WKY did not show significantly different B_{max} values.[87] The significance of different changes of β-adrenoreceptors in hypertension has recently been discussed in detail[24] and was related to both functional and structural changes of VSM associated with the progression of hypertension.

Vasoactive Peptide Receptors

Receptors for angiotensin II (AII), a potent vasoconstrictor, and a vasodilatory peptide, atrial natriuretic peptide (ANP), originally identified in the atria of mammals, have been identified and extensively characterized in VSM of mesenteric arteries.[88–93] The affinity of AII for its vascular binding sites is increased by mono-valent and divalent cations, but decreased by guanine nucleotides. McQueen et al.[94] and Grover et al.[88] reported that degradation of AII occurred when crude membrane fractions of mesenteric arteries were employed. Such a degradation of AII also occurred using membranes from large blood vessels.[95] This contributed artifactually higher K_d values. Degradation of ANP has also been noted recently using a PM-enriched fraction isolated from rat mesenteric artery.[96] The fact that these two peptide receptors coexist in the same vasculature with the opposite effects and that both receptors are under regulation by volume-dependent mechanisms led to the suggestion that ANP may be a physiological antagonist of AII.[78] Considerable information has become available in recent years on the changes of these peptide receptors in mesenteric arteries in hypertension using radioligand binding techniques. Readers interested in these specific aspects are referred to a recent comprehensive review by Schiffrin.[78] Vascular vasopressin (AVP) receptors have also been identified and characterized by radioligand-binding studies using crude microsomes derived from rat mesenteric arteries.[97,98] The mesenteric vascular AVP-receptor density was increased in sodium-depleted rats,[92] but decreased in SHR.[99] However, the tension development to AVP of the vascular muscle was enhanced in SHR compared to that in WKY. This was attributed to the enhanced activity of phospholipase C rendering increased production of IP_3 in spite of reduced AVP-receptor density.[78]

Membrane-Bound Enzymes

In VSM fractionation, several PM-bound enzymes, such as Na^+,K^+-ATPase, 5'-nucleotidase, basal ATPases, adenylate cyclase, and alkaline phosphatase, have been used as marker enzymes to monitor the PM content in isolated membrane fractions. Some of these enzymes are known to have important physiological roles. For an example, adenylate cyclase is a PM-bound enzyme known to be coupled to β-adrenoreceptors[100] and Na^+,K^+-ATPase is known to be the PM Na-pump. There is evidence that vascular muscle PM also contains an elastase-like activity[101–104] possibly involved in the catabolism of elastin.[105] Since many of these membrane-bound enzymes may have their functional and structural correlates, information on the pathophysiological consequences of these enzymatic changes in hypertension is important for a better understanding of the membrane changes of vascular muscle at a molecular level. However, very few systematic studies have been performed in VSM, particularly in small arteries. Identification and characterization of some of the PM enzymes have been made in the smooth muscle membranes isolated from mesenteric arteries. This includes 5'-nucleotidase,[6,10,17,18,75] alkaline and acid phosphatases,[6,75,106,107] elastase-like enzymes,[101] Mg^{2+}- or Ca^{2+}-activated ATPases[108] and K^+-activated p-nitrophenylphosphatase.[109] Changes of these PM-associated enzyme activities, as related to the development of hypertension, have also been briefly reviewed recently in large and small arteries.[72]

Membrane Phospholipids

It is conceivable that alterations in the composition of membrane lipids may cause physical, as well as functional, changes in the biological membranes. Although very little information is available on the properties of phospholipids of VSM cells in hypertension, it is not unreasonable to speculate that the aforementioned multiple abnormalities found in the cell membranes of VSM may be caused by a global membrane-lipid disturbance expressed in different forms. This is supported by some limited experimenal evidence. First, preincubation of certain phospholipids or phospholipases with the isolated smooth membrane vesicles resulted in altered Ca^{2+} binding and transport and ATPase activities.[110,111] Second, phosphatidylinositol turnover has now been shown to be associated with the Ca^{2+} release mechanism (*see* Section on Ca-Handling by Internal Membranes) as well as the control of Na^+,K^+-ATPase activity[112] in VSM. Third, the methylation of phospholipids has recently been studied in rat aortic microsomes.[113] The conversion of phosphatidylethanolamine by methylation to phosphatidylcholine, the principal product, caused a decrease in microsomal membrane fluidity (measured by a fluorescence polarization technique). It was also found that the aortic microsomes prepared from SHR showed substantially increased phospholipid methylation accompanied by increased fluidity compared to those prepared from WKY.[113] It is interesting to note that lipid methylation may influence the Ca^{2+} gating and transport[114,115] and β-adrenoreceptor responses.[116] Finally, oxygen radicals generated *in vitro* have been shown to cause lipid peroxidation, which increases phospholipid bilayer rigidity,[117] and to affect the contractile function of vascular smooth muscle.[118,119] Oxygen free radicals derived from hydrogen peroxide or from alloxan have recently been shown to inhibit the ATP-driven Ca^{2+}-transport[120,121] and ligand binding to adrenoreceptors[122] in VSM microsomes. Furthermore, Kostka and Kwan[123] have also noted that stimulation of lipid peroxidation of VSM microsomes by ascorbate and NADPH increased with prolonged cold storage (>1 wk at –20°C), which also caused diminished ATP-dependent Ca^{2+}-transport.

Conclusion

A wealth of information about the biochemical properties of VSM membranes from small arteries including resistance vessels, has become available by the way of an analytical subcellular-membrane approach. Such an approach enables dynamic studies of multiple ion-handling processes, including ion binding, ion pumps, ion channels, and ion exchangers, on the surface and internal membranes. It also allows biochemical dissection of the pharmacological events involving the recognition and transduction at the surface-membrane receptors. The least understood area of the membrane biochemistry of the small and resistance arteries appears to be the membrane-associated enzymes, many of which may be responsible for, and perhaps coupled to, the ion pumps, ion channels, and membrane receptors. Membrane abnormalities associated with pathophysiological processes often manifest themselves in the form of the above biochemical correlates, which may ultimately provide insight into the molecular control and prevention of cardiovascular diseases.

Acknowledgment

The cited work performed in the author's laboratory was supported by grants and personnel awards provided by the Heart and Stroke Foundation of Ontario, the Medical Research Council of Canada, and the Canadian Diabetes Association.

References

[1] Daniel, E. E., Grover, A. K., and Kwan, C. Y. (1983) *Biochemistry of Smooth Muscle* vol. 3., (Stephens, N. L., ed.), CRC, Boca Raton, FL, pp. 1–88.

[2] Daniel, E. E., Crankshaw, D. J., and Kwan, C. Y. (1979) *Trends in Autonomic Pharmacology* (Kalsner, S., ed.), Urban and Schwarzenberg, Baltimore, MD, pp. 443–484.

[3] Kwan, C. Y. (1987) *Sarcolemmal Biochemistry* (Kidwai, A. M., ed.), CRC, Boca Raton, FL, pp. 59–97.

[4] Daniel, E. E., Grover, A. K., and Kwan, C. Y. (1982) *Fed. Proc.* **41,** 2898–2904.

[5] Kwan, C. Y. (1985) *Biochim. Biophys. Acta* **819,** 148–152.

[6] Kwan, C. Y., Belbeck L., and Daniel, E. E. (1979) *Blood Vessels* **16,** 259–268.

[7] Kwan, C. Y., Lee, R. M. K. W., and Daniel, E. E. (1981) *Blood Vessels* **18,** 171–186.

[8] Kwan, C. Y., Triggle, C. R., Grover, A. K., Lee, R. M. K. W., and Daniel, E. E. (1983) *Preparative Biochem.* **13,** 275–314.

[9] Kwan, C. Y., Triggle, C. R., Grover, A. K., Lee, R. M. K. W., and Daniel, E. E. (1984) *J. Mol. J. Mol. Cell. Cardiol.* **16,** 747–764.

[10] Wei, J. W., Janis, R. A., and Daniel, E. E. (1976) *Blood Vessels* **13,** 293–308.

[11] Wuytack, F. and Casteels, R. (1980) *Biochim. Biophys. Acta* **595,** 257–263.

[12] Wuytack, F., Landon, E., Fleischer, R., and Hardman, J. G. (1978) *Biochim. Biophys. Acta* **540,** 253–269.

[13] Kwan, C. Y., Grover, A. K., Triggle, C. R., and Daniel, E. E. (1983) *Biochem. Int.* **6,** 713–722.

[14] Raeymaekers, L. and Casteels R. (1984) *Cell Calcium* **5,** 205–210

[15] Kwan, C. Y. (1986) *J. Bioenerg. Biomembr.* **18,** 487–505.

[16] Kwan, C. Y. (1986) *IRCS Med. Sci.* **14,** 433–434.

[17] Kwan, C. Y., Garfield, R. E., and Daniel, E. E. (1979) *J. Mol. Cell. Cardiol.* **11,** 639–659.

[18] Kwan, C. Y., Belbeck L., and Daniel, E. E. (1980) *Mol. Pharmacol.* **17,** 137–140.

[18a] Grover, A. K., Samson, S. E., and Lee, R. M. K. W. (1985) *Biochim. Biophys. Acta* **818,** 191–199.

[19] Kahn, A. M., Shelat, H., and Allen, J. C. (1986) *Am. J. Physiol.* **250**, H313–H319.

[20] Ford, G. D. and Hess, M. L. (1982) *Am. J. Physiol.* **242**, C242–C249.

[21] Nakayama, K., Suzuki, S., and Sugi H. (1986) *Jpn. J. Physiol.* **36**, 745–760.

[22] Grover, A. K., Kwan, C. Y., and Daniel, E. E. (1983) *Am. J. Physiol.* **244**, C61–C67.

[23] Kwan, C. Y., Grover, A. K., and Daniel, E. E. (1988) *Vasodilatation Vascular Smooth Muscle. Autonomic Nerves and Endothelium* (Vanhoutte, P.M., ed.), Raven, NY, pp. 21–28.

[24] Kwan, C. Y., Lee, R. M. K. W., and Daniel, E. E. (1989) *Essential Hypertension 2* (Aoki, K. and Buhler, F. R., eds.), Springer-Verlag, Tokyo, pp. 221–237.

[25] Mrwa, U., Guth, K., Haist, C., Troschka, M., Herrmann, R., Wojciechowski, R., and Gagelmann, M. (1986) *Life Sci.* **38**, 191–196.

[26] Kattenburg, D. M. and Daniel, E. E. (1984) *Blood Vessels* **21**, 257–266.

[27] Thorens, S. and Haeusler, G. (1978) *Biochim. Biophys. Acta* **512**, 415–428.

[28] Morel, N., Wibo, M. and Godfraind, T. (1981) *Biochim. Biophys. Acta* **64**, 812–822.

[29] Grover, A. K., Kwan, C. Y., and Daniel, E. E. (1981) *Am. J. Physiol.* **240**, C175–C182.

[30] Matlib, M. A., Schwartz, A., and Yamori, Y. A (1985) *Am. J. Physiol.* **249**, C166–C172.

[31] Matlib, M. A., Kihara, M., Farrell, C., and Dage, R. C. (1988) *Biochim. Biophys. Acta* **939**, 173–177.

[32] Matlib, M. A. and Reeves, J. P. (1987) *Biochim. Biophys. Acta* **904**, 145–148.

[33] Kahn, A. M., Allen, J. C., and Shelat, H. (1988) *Am. J. Physiol.* **254**, C441–C449.

[34] Grover, A. K., Kwan, C. Y., and Daniel, E. E. (1982) *Am. J. Physiol.* **242**, C278–C282.

[35] Blaustein, M. P. (1977) *Am. J. Physiol.* **232**, C165–C173.

[36] VanBreemen, C., Aaronson, P., and Loutzenhiser, R. (1979) *Pharmacol. Rev.* **30**, 167–205.

[37] Kahn, A. M., Allen, J. C., Cragoe, E. J., and Shelat, H. (1989) *Circ. Res.* **65**, 818–828.

[38] Kahn, A. M., Allen, J. C., Cragoe, E. J., Zimmer R., and Shelat, H. (1988) *Circ. Res.* **62**, 478–485.

[39] Mahnensmith, R. L. and Aronson, P. S. (1985) *Circ. Res.* **57**, 773–788.

[40] Triggle, D. J. (1984) *Hypertension, Physiological Basis and Treatment* (Ong, H. H. and Lewis, J. C., eds.), Academic, Orlando, FL, pp. 223–268.

[41] Triggle, D. J. and Janis, R. A. (1984) *Nitrendipine* (Scriabine, A., Vanor, S., and Deck, K., eds.), Urban and Schwarzenberg, Baltimore, MD, pp. 32–52.

[42] Morel, N. and Godfraind, T. (1988) *Br. J. Pharmacol.* **95**, 252–258.

[43] Barry, K. L., Mikkelsen, R. B., Shucart, W., Keouch, E. M., and Gavris, V. (1985) *J. Neurosurg.* **62**, 729–736.

[44] Kostka, P., Ahmad, S., Berezin, I., Kwan, C. Y., and Daniel, E. E. (1987) *J. Neurochem.* **49**, 1124–1132.

[45] Guan, W., He, H., and Chen, J. X. (1985) *Acta Pharmacol Sinica* **6**, 267–269.

[46] Guan, Y. Y., Kwan, C. Y., He H., and Daniel, E. E. (1988) *Blood Vessels* **25**, 312–315.

[47] Grover, A. K (1985) *Cell Calcium* **6**, 227–236.

[48] Wuytack, F., Raeymaekers, L., and Casteels, R. (1985) *Experientia* **41**, 900–905.

[49] Suematsu, E., Hirata, M., and Kuriyama, H. (1984) *Biochim. Biophys. Acta* **773**, 83–90.

[50] Kwan, C. Y., Grover, A. K., Daniel, E. E., Berenski, C. J., and Jung, C. Y. (1986) *Biochem. Arch.* **2**, 149–157.

[51] Chiesi M., Gasser J., Carafoli E. (1984) *Biochem. Biophys. Res. Comm.* **124**, 797–806.

[52] Suematsu, E., Hirata, M., Sasaguri, T., Hashimoto, T., and Kuriyama, H. (1985) *Comp. Biochem. Physiol.* **82A**, 645–649.

[52a] Yamamoto, H. and Van Breemen, C. (1985) *Biochem. Biophys. Res. Comm.* **130**, 270–274.

[53] Hwang, K. S. and van Breemen, C. (1987) *Pfluger's Arch.* **408**, 343–350.

[54] Julou-Schaeffer, G., Freslon, J. L. (1988) *Br. J. Pharmacol.* **95**, 605–613.

[55] Kanmura, Y., Missiaen, L., Raeymaekers, L., and Casteels R. (1988) *Pfluger's Arch.* **413**, 153–159.

[56] Bourreau, J. P., Gaspar V., Kwan, C. Y., and Daniel, E. E. (1990) Abstract presented at annual meeting of the Ontario chapter of the Canadian Hypertension Society. Kingston, Ontario; May.

[57] Bourreau, J. P., Zhang, Z. D., Low., A.M., Kwan, C. Y., and Daniel, E. E. (1990) *J. Pharmacol. Exp. Ther.* (1991) (in press).

[58] Losse, H., Zidek, W., and Vetter, H. (1984) *J. Cardiovasc. Pharmacol.* **6**, S32–S34.

[59] Sugiyama, T., Yoshizumi, M., Takaku, F., Urabe, H., Tsukakoshi, M., Kasuya, T., and Yazaki, Y. (1986) *Biochem. Biophys. Res. Comm.* **26**, 340–345.

[60] Zidek, W., Kerenyi, T., Losse, H., and Vetter, H. (1983) *Res. Exp. Med.* **183**, 129–132.

[61] Webb, R. C. and Bohr, D. F. (1981) *Am. Heart J.* **102**, 251–264.

[62] Bohr, D. F. (1963) *Science* **139**, 597–599.

[63] Holloway, E. T. and Bohr, D. F. (1981) *Circ. Res.* **33**, 678–685.

[64] Mulvany, M. J. (1986) *Essential Hypertension*(Aoki, K., ed.), Springer Verlag, Tokyo, pp. 51–65.

[65] McMahon, E. G. and Paul, R. J. (1985) *Circ. Res.* **56,** 427–435.

[66] Nghiem, C. X. and Rapp, J. P. (1983) *Clin. Exp. Hypertens.* **A5,** 849–856.

[67] Wei, J. W., Janis, R. A., and Daniel, E. E. (1976) *Circ. Res.* **39,** 133–140.

[68] Aoki, K., Ikeda, N., Yamashita, K., Tomita, N., Tazumi, N., and Hotta, K. (1974) *Jpn. Heart J.* **2,** 180,181.

[69] Moore, L., Hurwitz, L., Davenport, G. R., and Landon, E. J. (1975) *Biochim. Biophys. Acta* **413,** 432–443.

[70] Kwan, C. Y. (1985) *Can. J. Physiol. Pharmacol.* **63,** 366–374.

[71] Kwan, C. Y. (1986) *Essential Hypertension* (Aoki, K., ed.), Springer Verlag, Tokyo, pp. 135–144.

[72] Kwan, C. Y., ed. (1989) *Membrane Abnormalities in Hypertension,* vol. I, CRC, Boca Raton, FL, pp. 115–143.

[73] Kwan, C. Y. and Daniel, E. E. (1982) *Eur. J. Pharmacol.* **82,** 187–190.

[74] Kwan, C. Y., Sakai, Y., and Daniel, E. E. (1984) *Clin. Expt. Hypertension* **A6,** 1257–1265.

[75] Kwan, C. Y., Belbeck L., and Daniel, E. E. (1980) *Blood Vessels* **17,** 131–140.

[76] Webb, R. C. and Bhalla, R. C. (1976) *J. Mol. Cell. Cardiol.* **8,** 651–660.

[77] Agrawal, D. K., Crankshaw, D. J., and Daniel, E. E. (1987) *Sarcolemmal Biochemistry,* vol. 2 (Kidwai, A. M., ed.), CRC, Boca Raton, FL, pp. 99–152.

[78] Schiffrin, E. L. (1989) *Membrane Abnormalities in Hypertension* vol. II (Kwan, C. Y., ed.), CRC, Boca Raton, FL, pp. 59–90.

[79] Agrawal, D. K., Borkowski, K. R., and Daniel, E. E. (1989) *Membrane Abnormalities in Hypertension,* vol. II (Kwan, C. Y., ed.), CRC, Boca Raton, FL, pp. 23–57.

[80] Kwan, C. Y., Sipos, S. N., Osterroth A., and Daniel, E. E. (1987) *J. Pharamcol. Exp. Ther.* **243,** 1074–1081.

[81] Bevan, J. A., Bevan, R. D., and Laher, I. (1985) *Clin. Sci.* **68 (Suppl. 10),** 83s–88s.

[82] Shi, A. G., Kwan, C. Y., and Daniel, E. E. (1989) *J. Pharmacol. Exp. Ther.* **250,** 1119–1124.

[83] Shi, A. G., Ahmad S., Kwan, C. Y., and Daniel, E. E. (1990) *J. Cardiovasc. Pharmacol.* **15,** 515–526.

[84] Schiffrin, E. L. (1984) *J. Hypertens.* **2 (Suppl. 3),** S431,S432.

[85] Woodcock, E. A. and Olsson, C. A. (1980) *Biochem. Pharmacol.* **29,** 1465–1468.

[86] Maginoni, M. S., Koybayashi, H., Cazzaniga, A., Izumi, F., Spano, P. F., and Trabucchi, M. (1983) *Circulation* **67,** 610–613.

[87] Kwan, C. Y. and Lee, R. M. K. W. (1991) *Can. J. Physiol. Phamacol.* (in press).

[88] Grover, A. K., Kwan, C. Y., Kostka P., and Daniel, E. E. (1985) *Eur. J. Pharmacol.* **112,** 137–143.

89 Grover, A. K., Kwan, C. Y., Kostka P., Shephard P., and Daniel, E. E. (1984) *Can. J. Physiol. Pharmacol.* **62**, 1203–1208.

90 Gunther, S., Gimbrone, A. M., and Alexander, R. W. (1980) *Circ. Res.* **47**, 278–286.

91 Gunther, S., Gimbrone, A. M., and Alexander, R. W. (1980) *Nature (Lond.)*, **287**, 230–232.

92 St-Louis J. and Schiffrin, E. L. (1984) *Life Sci.* **35**, 1489–1495.

93 St-Louis J. and Schiffrin, E. L. (1988) *Can. J. Physiol. Pharmacol.* **66**, 951–956.

94 McQueen, J., Murray, G. D., and Semple, P. F. (1984) *Biochem. J.* **223**, 659–671.

95 Koziarz,P. and Moore, G. J. (1990) *Can. J. Physiol. Pharmacol.* **68**, 218–220.

96 Tamburini, P. P., Koehn, J. A., Gilligan, J. P., Charles, D., Palmesino, R. A., Sharif, R., McMartin, C., Erion, M. D., and Miller, M, J. S. (1989) *J. Pharmacol. Exp. Ther.* **251**, 956–961.

97 Lariviere, R. and Schiffrin, E. L. (1987) *Can. J. Physiol. Pharmacol.* **65**, 1171–1181.

98 Schiffrin, E. L. and Genest J. (1983) *Endocrinology* **113**, 409–411.

99 Lariviere, R., Baribeau, J., St-Louise J., and Schiffrin, E. L. (1989) *Can. J. Physiol. Pharmacol.* **67**, 232–239.

100 Lefkowitz, R. J., Stadel, J. M., and Caron, M. G. (1983) *Ann. Rev. Biochem.* **52**, 159–186.

101 Ito, H., Kwan, C. Y., and Daniel, E. E. (1986) *Am. J. Physiol.* **251**, H247–H252.

102 Kwan, C. Y., Lee, R. M. K. W., and Ito, H. (1987) *Biochem. Arch.* **3**, 217–221.

103 Kwan, C. Y., Wang, R. J., and Smeda, J. S. (1989) *Blood Vessels* **26**, 377–380.

104 Leake, D. S., Hornebesk, W., Brechmier, D., Robert, L., and Peters, T. I. (1983) *Biochim. Biophys. Acta* **761**, 41–47.

105 Ito, H., Yamamoto, K., and Okamoto, K. (1979) *Jpn. Heart J.* **20**(Suppl. I), 240–242.

106 Kwan, C. Y. (1983) *IRCS Med. Sci.* **11**, 690–691.

107 Kwan, C. Y. and Ito, H. (1987) *Comp. Biochem. Physiol.* **87B**, 367–372.

108 Kwan, C. Y. (1982) *Enzymes* **28**, 317–327.

109 Kwan, C. Y., Grover, A. K., and Daniel, E. E. (1984) *Arch. Int. Pharmacodyn. Ther.* **272**, 245–255.

110 Kutsky, P. and Weiss, G. B. (1983) *Arch. Int. Pharmacodyn. Ther.* **263**, 4–16.

111 Sakai, Y., Ichikawa, S., Yoshida, Y., Oouchi, M., and Miyagawa, M. (1982) *Jpn. J. Smooth Muscle Res.* **18**, 339–345.

112 Simmons, D. A., Kern, E. F. O., Winegrad, A. I., and Martin, D. B. (1986) *J. Clin. Invest.* **77**, 503–513.

[113] Jaiswal, R. K., Landon, E. J., and Sastry, B. V. R. (1983) *Biochim. Biophys. Acta* **735**, 367–379.

[114] Hirata, F. and Axalrod, J. (1980) *Science* **209**, 1082–1090.

[115] Landon, E. J., Owens, L., and Sastry, B. V. R. (1986) *Pharmacology* **32**, 190–201.

[116] Hirata, F., Strittmatter, W. J., and Axelrod, J. (1979) *Proc. Natl. Acad. Sci. USA* **76**, 368–372.

[117] Dobretsov, G. E., Borschevskaya, T. A., Petrov, V. A., and Vladimirov, Y. A. (1977) *FEBS Lett.* **84**, 125–128.

[118] Rosenblum, W. I. (1983) *Am. J. Physiol.* **245**, H139–H142.

[119] Wei, E. P., Christman, C. W., Kontos, H. A., and Povlishock, J. T. (1985) *Am. J. Physiol.* **248**, H157–H162.

[120] Kwan, C. Y. and Beazley, J. S. (1987) *Can. J. Physiol. Pharmacol.* **65**, 2346–2350.

[121] Kwan, C. Y. and Beazley, J. S. (1988) *J. Bioenerg. Biomembr.* **20**, 517–532.

[122] Kwan, C. Y., Sipos, S., and Gaspar, V. (1990) *Biochem. J.* **270**, 137–140.

[123] Kostka, P. and Kwan, C. Y. (1987) *Biochem. Arch.* **3**, 157–163.

Chapter 18

Regulation of Calcium Sensitivity in Vascular Smooth-Muscle

Ismail Laher and Cornelis van Breemen

Introduction

Calcium ions (Ca^{2+}) play a pivotal role in muscle contraction in that they regulate events that result in excitation–contraction coupling. The concept has emerged that signaling at the membrane of arteries and veins—for example, by receptor stimulation and depolarization—leads to a decrease in the approx 10,000-fold electrochemical gradient for Ca^{2+}, ostensibly by increasing the permeability to extracellular Ca^{2+} and by mobilizing intracellular stores of Ca^{2+}.[1–3] The mechanisms whereby activation of voltage, receptor, and stretch-operated Ca^{2+} channels lead to entry of extracellular Ca^{2+} is discussed elsewhere in this volume, as is the electrical coupling of vascular smooth-muscle cells. In this chapter we discuss recently obtained evidence on the regulation of intracellular sensitivity to Ca^{2+}, based primarily on evidence obtained in vascular preparations that may not be representative of resistance arteries.[4,5] Although specialization such as neurovascular transmission,[6] altered receptor pharmacology,[7] mechanical activation of myogenic characteristics[8] and regulation of transmitter release[9] occurs in

From *The Resistance Vasculature*, J. A. Bevan et al., eds. ©1991 Humana Press

arteries of diminishing size, it remains to be determined whether intracellular mechanisms regulating vascular tone are also differentially regulated in resistance arteries.

Voltage changes that accompany production of vascular tone influence the activation curves for Ca^{2+} entry owing to stimulation by neurotransmitter agonists.[10,11] Voltage-independent modulation of vascular tone resides in the biochemical sequelae following receptor stimulation in arteries. Signal transduction involving receptor stimulation involves coupling to G-proteins in such a way that upon occupation by an agonist, the receptor achieves a higher affinity for its G-protein subunit.[12–14] Multiple subtypes of G-protein have been linked to the activation of phospholipase C, an enzyme that cleaves membrane phospholipids.[14] Vasoconstriction by many agonists is accompanied by the generation of products derived from membrane phospholipids. Upon activation, phospholipase C leads to the subsequent formation of two key intracellular messengers—inositol 1,4,5-trisphosphate (IP_3) and diacylglycerol (DAG).[15,16]

Release of Intracellular Ca^{2+}

A second-messenger role proposed for IP_3 is in the release of intracellular Ca^{2+} from nonmitochondrial stores.[3,17,18] Although the details of the interaction of IP_3 with its receptor on the sarcoplasmic reticulum are unknown, several aspects of the possible modulatory role of IP_3 in the phasic component of agonist-mediated vasoconstriction have recently been described.[19] The importance of IP_3-mediated release of intracellular Ca^{2+} in resistance arteries is unclear because the direct evidence for a functional role of the sarcoplasmic reticulum in small arteries and arterioles is poorly described. A comprehensive study of the morphological characteristics and distribution of the sarcoplasmic reticulum in arteries of decreasing diameter has not been reported. The limitations imposed by current methodological considerations would partly account for the absence of such information.[20–22] Studies aimed at determining the relative importance of norepinephrine-induced phasic responses in arteries of diminishing size placed in Ca^{2+}-free solutions have been reported; such transient contractile effects represent intracellular events, most likely by the α-adrenoceptor-mediated generation of IP_3. The consensus is that the phasic contractile response to receptor stimulation in Ca^{2+}-free solutions diminish with arterial size.[23,24] This is in keeping with the greater susceptibility of smaller arteries to Ca^{2+}-entry inhibitors. [25,26]

In saponin-permeabilized vascular smooth-muscle, application of GTP enhanced the IP_3-induced Ca^{2+} release from the sarcoplasmic reticulum.[27] The enhancing effect of GTP or its nonhydrolyzable analog GTPγS on the IP_3-induced contraction is attenuated by GTPβS, a competitive inhibitor of the binding of GTP analogs to G-proteins.[28] Since heparin inhibited the IP_3, but not the GTPγS-mediated contraction, it was proposed that stimulation of G-proteins can result in intracellular Ca^{2+} release without activation of IP_3 receptors on the sarcoplasmic reticulum.[29] Direct measurements of Ca^{2+} movements in skinned rat aortic smooth-muscle cells indicate that the increased efflux of radiolabeled Ca^{2+} caused by IP_3 was specifically inhibited by heparin.[30] Heparin did not inhibit the caffeine-induced Ca^{2+} release. Heparin is presently the agent of choice for inhibiting IP_3-induced Ca^{2+} release; its potency in competing with IP_3 binding is similar to that for blocking release of Ca^{2+}. The use of heparin as an intracellular inhibitor of IP_3 binding is not without drawbacks, since (1) heparin does not freely cross membranes of intact arteries or cells and (2) heparin may have other effect on excitation–contraction coupling, e.g., heparin uncouples receptors from G-proteins linked to adenylate cyclase in nonmuscle cells.[31,32] In addition, supramaximal concentrations of heparin also inhibits Ca^{2+} influx.[30]

Additional support for an IP_3-independent G-protein-mediated release of Ca^{2+} derives from observations wherein GTPγS caused Ca^{2+} release and forced production in arteries depleted of an IP_3-sensitive Ca^{2+} pool.[29] The mechanism whereby GTPγS causes Ca^{2+} release is unknown, since the presence of a G-protein directly on the membranes of the SR has not been described and the possibility of a unique non-IP_3 mediator is not supported by observations made thus far.[33,35] In other cell types, GTP has also been shown to release Ca^{2+} from a component of endoplasmic reticulum (ER) that is insensitive to IP_3. This process appears to involve translocation of Ca^{2+} between ER components and may be mediated by a class of GTP-binding proteins.[36] Unlike the Ca^{2+} release in vascular tissue,[29] this release is inhibited by GTPs.[37]

It is apparent that upon agonist stimulation of receptor-coupled G-proteins, IP_3 causes a rapid release of Ca^{2+} stores; the latency for this response is on the order of 0.1 s as determined in flash-photolysis studies of caged IP_3.[19] Following release of intracellular Ca^{2+}, most notably by agonist stimulation, a number of investigators have noted that the concentration of Ca^{2+} remains elevated only transiently, since Ca^{2+} levels fall to near basal values in the face of maintained force production.[38–40] Although the declining fluorescence signal for free-Ca^{2+} concentration is explicable in a number of ways, one interpretation currently in vogue is that for force production to be maintained, intracellular modulation of Ca^{2+} sensitivity is a prerequisite *(discussed below)*. Clearly, consideration should be given to methodological aspects in the use of indicator dyes for Ca^{2+}, as well as in species and regional differences of vascular properties. Biochemical features of recorder dyes for Ca^{2+} may also contribute to methodological artifacts. As pointed out by Karaki[41] and Rasmussen et al.,[42] a better quantitative and temporal correlation exists between intracellular levels of Ca^{2+} and maintained vascular tone when using fura-2 as the indicator dye than when loading arteries with aequorin. One explanation proposed to account for the disparity in results obtained with fura-2 and aequorin measurements is the choice of HEPES or bicarbonate buffers;[42,43] this is unlikely to be the sole contributor to the variation observed in fluorescence-indicator dye studies of Ca^{2+}. Somlyo and Himpens[40] and Karaki[41] address key features of aequorin and fura-2, such as indicator affinity for Ca^{2+}, linearity of emission, interference by other ions, movement artifacts, intracellular distribution, and the like, as additional factors contributing to fluorescence profiles.

Fluorescence signals may also be influenced by differences attributable to regional specialization of vascular smooth-muscle, as well as to arterial vs venous characteristics.[38,44] When the α-adrenoreceptor agonist phenylephrine is added to the aorta precontracted with KCl, further contraction and a transient increase in the intracellular Ca^{2+} concentration are recorded; when similar experiments are made in the portal vein of the same species, the phenylephrine-stimulated transient Ca^{2+} increase is absent, though additional force production occurs. Though the relative roles of IP_3-generated Ca^{2+} release are not clear in such studies, the importance of stimulus additivity beyond Ca^{2+}-dependent components of the excitation–contraction coupling is evident in the venous preparation.

In addition to possible release of intracellular stores of Ca^{2+}, maintained tone production also requires the continued entry of extracellular Ca^{2+}. One intracellular target for Ca^{2+} is calmodulin; the Ca^{2+}–calmodulin complex activates phosphorylation of myosin light chain kinase (MLCK).[45] The prediction is that the extent of agonist-induced tonic contraction is related to the levels of Ca^{2+} and myosin light chain (MLC) phosphorylation. Attempts to support this hypothesis directly using recorder dyes for intracellular free Ca^{2+}, such as fura-2 and aequorin, and kinetic analysis of MLC phosphorylation patterns have not validated this.[42,46] It is apparent that high levels of maintained force occur in the face of nearly basal values of free Ca^{2+} and MLC phosphorylation levels. Moreover, it is clear that stimulus-specific patterns of Ca^{2+} fluorescence and MLC phosphorylation occur;[47] a consequence of this would be that uniform levels of agonist-stimulated force are produced with variants in the absolute availability or production of the key determinants of force, i.e., Ca^{2+} and MLC phosphorylation. Whereas myosin phosphorylation levels correlate to crossbridge cycling and shortening velocities, the maintained force that is developed is related to the myosin phosphorylation levels in a poorly defined manner. In keeping with this, a number of investigators have reported lower intracellular free Ca^{2+} for force production by agonist-mediated, compared to K^+-induced, tone (e.g., *see* refs. 44,47–49). These observations make apparent the participation of an additional regulator, the activity of which alters the complex relationship among steady-state force, Ca^{2+}, and MLC phosphorylation levels. According to Murphy and collaborators, MLC phosphorylation also regulates the rate of development of actomyosin crossbridges.[45] According to this theory, dephosphorylation of the phosphorylated MLC complex of an attached crossbridge will transform it to a longer lasting "latch"-type crossbridge. This represents another site of regulation for protein kinase C. For example, it has been reported that activation of protein kinase C by phorbol esters increases K^+-activated tone in mesenteric and coronary arteries, without significant elevations in cytoplasmic Ca^{2+} levels.[39] In skinned carotid arteries, activated protein kinase C augmented Ca^{2+}-induced contractions with minor increases in MLCK phosphorylation.[50]

Agonist-Mediated Changes in Intracellular Sensitivity to Ca^{2+}

In an elegant study of small arteries of the rat mesenteric circulation, Mulvany et al.[51] reported that the contraction induced by norepinephrine was considerably larger than that produced by potassium, even though both vasoconstrictors elicited the same magnitude of depolarization. Similar observations have been made in small arteries of the mesenteric circulation of the cat[52] and guinea pig.[53] At least three explanations can be offered in this regard. First, norepinephrine is able to release intracellular Ca^{2+} (IP_3-mediated) in addition to causing the influx of extracellular Ca^{2+}. As discussed earlier, the evidence for the presence and importance of intracellular Ca^{2+} release in small arteries is clearly not as convincing as in the case of larger, conduit vessels. Second, via a generation of DAG and subsequent activation of protein kinase C, intracellular sensitivity to Ca^{2+} is enhanced or the production of Ca^{2+}-independent processes are activated. It has been reported that phorbol esters increase the voltage-gated Ca^{2+} current in isolated vascular smooth-muscle cells,[54] as well as causing force production in intact artery segments in the face of constant or decreasing intracellular Ca^{2+} levels.[55] Third, norepinephrine is able to promote Ca^{2+} entry via both voltage-gated and receptor-mediated Ca^{2+} channels. In mesenteric resistance arteries, Ca^{2+}-sensitivity is greater when vessels are activated with norepinephrine than when activated by high K^+,[56] whereas the contractile responses to readmission of Ca^{2+} in the presence of norepinephrine are more sensitive to inhibition with dihydropyridines than when studied in a K^+-depolarizing solution.[57] In an attempt to reconcile these observations, Boonen and DeMey[58] suggest that norepinephrine modulates voltage-gated Ca^{2+} entry via production of second messengers, rather than by depolarization, resulting in fewer sites for high-affinity binding of dihydropyridines to voltage-inactivated Ca^{2+} channels.[59] In the studies by Mulvany et al.[51] and Hogestatt,[60] agonist-mediated tone was nearly maximal (80% of control) even though voltage-gated Ca^{2+}-entry was inhibited with nifedipine. These observations are at variance with those of Nelson et al., who reported that the affinity of the Ca^{2+}-channel for nisoldipine was optimized under depolarizing conditions.[61] Under such conditions, both agonist- (norepinephrine, serotonin) and potassium-induced tone and whole-cell Ca^{2+} currents were equally sensitive to dihydropyridines. Dose–response curves to Ca^{2+} in rat mesenteric resistance arteries are inhibited with a uniform rank order of potency by a number of organic inhibitors of Ca^{2+} entry, independent of activation of Ca^{2+} channels by either norepinephrine or potassium. The results of Nelson et al.[61] and Jolou and Freslon,[62] in contrast to those of Mulvany et al.[51] and Hogestatt,[60] suggest that depolarization and receptor activation by noradrenaline promote Ca^{2+} entry in resistance arteries via a common pathway.

The biochemical properties ascribed to DAG, notably that of converting protein kinase C to its Ca^{2+}-sensitive form,[63,64] make it a suitable candidate for a regulator of receptor-mediated production of tone. Current dogma supports the notion that activated protein kinase C reduces the Ca^{2+} requirement of a number of cellular processes, including that of force production.[42] In this context, it has recently been demonstrated that activation by protein kinase C, either by phorbol esters or by GTP-stimulation by analogs of GTP, causes the phosphorylation of the MLC 20-kDa proteins in arteries,[46,65] though it is unclear whether effects ascribed to phorbol esters mimic those caused by endogenous activation of diacylglycerol.[66]

The intracellular substrates, specifically light chain contractile proteins or other regulatory proteins of excitation–contraction coupling that are phosphorylated by endogenously activated protein kinase C, have yet to be identified; as such, the acceptance of a role for protein kinase C in maintained vascular tone, especially under physiologically relevant conditions, remains a contentious issue.[67] A number of studies have suggested a regulatory role for

Ca^{2+}–calmodulin-stimulated MLCK activity in force production, i.e., shortening velocity of arteries and veins.[45,48] The role of Ca^{2+}-independent MLC phosphatase activity during the maintained phase of vascular tone is not well documented, possibly because of the lack of specific inhibitors.[68]

Two credible suggestions have been offered to reconcile the low levels of Ca^{2+}-dependent MLC phosphorylation with high levels of maintained force. First, Ca^{2+} may alter the pattern of thin-filament regulation of vascular tone by binding to additional sites on contractile proteins.[45] Alternatively, other proteins may regulate actin–myosin attachment; for example, in the presence of Ca^{2+}, caldesmon is unable to inhibit binding of actin to myosin, permitting crossbridge formation.[69,70] Other actin-binding filaments may also be regulated by Ca^{2+} and protein kinase C, such as filamin,[71] gelsonin,[72] and profilin.[73]

The critical examination of modulation of intracellular sensitivity to Ca^{2+} has been aided by improvements in skinning procedures—notably by the application of agents like staphyloccocal α-toxin, which offer a number of advantages:[74,75] for example, α-toxin skinning allows the ionic composition of the cytoplasm to be controlled while maintaining the functional integrity of receptor-coupled second systems.[76] In such preparations, cellular proteins having mol wt >1000 are retained; included among these are calmodulin, diacylglycerol, protein kinase C, and the contractile proteins. The pores with a diameter of 2–3 nm created by α-toxin permit free movement of ions and ATP.

Alpha-adrenergic, as well as muscarinic, agonists were shown to induce contraction at constant EGTA-buffered Ca^{2+} concentration.[76,78] These contractions were enhanced by addition of GTP and abolished by GDPβS, indicating the involvement of a G protein. The G protein could be directly and irreversibly activated by GTPγS. Somlyo and collaborators[17] showed, in an elegant experiment using flash photolysis of cased GTPγS in an α-toxin-permeabilized fiber, that this activation has a delay of 10 s as a result of the slow dissociation of GDP from the G protein. As discussed before, the subsequent step in the agonist-initiated biochemical cascade is hydrolysis of PIP_2 by PLC with the production of IP_3 and DAG. In the Ca^{2+} clamped α-toxin permeabilized mesenteric smooth-muscle phorbol esters induce a relatively rapid contraction,[79] whereas the pKC inhibitor H-7 selectively relaxes the norepinephrine- and GTPγS-induced contractions. These results suggest that the component of agonist-induced contraction that is caused by enhancement of myofilament sensitivity to Ca^{2+} is attributable to DAG activation of PKC. The myofilament Ca^{2+} sensitivity could also be downregulated by adding cAMP or cGMP or by treating the α-toxin-permeabilized fibers with forskolin or sodium nitroprusside. These data support those reported by Karaki[41] in intact fibers, in which forskolin and sodium nitroprusside decreased the ratio of force over intracellular Ca^{2+} concentration measured with fura-2.

In addition to regulation by second messengers, smooth-muscle tone may also be affected by adenine nucleotides. Concentrations of substrate favoring accumulation of ADP enhanced force production at fixed $[Ca^{2+}]_i$.[80] As little as 100 μM ADPβS caused a significant shift to the left of the force–$[Ca^{2+}]_i$ curve. Since ADP concentrations as high as 400 μM have been measured in contracted smooth-muscle,[81] it is possible that this effect may contribute to physiological smooth-muscle tone. It has been suggested that elevated ADP would, by mass action, lead to accumulation of the strong binding crossbridge state AM–ADP,[82] which would lead to increased tension with decreased shortening velocity.

How the above states of contraction, which are characterized by high force-to-$[Ca^{2+}]_i$ ratios, are related to smooth-muscle latch remains to be resolved. Lash et al.[83] and Nishimura and van Breemen[80] favor regulation at the step of ADP release from the AM–ADP complex. This could possibly be effected by PKC-mediated phosphorylation by calmodulin or calponin. On the other hand, Somlyo and Somlyo[19] have obtained convincing evidence for cooperative attachment of nonphosphorylated (at the MLC site) crossbridges.

The observation of a dissociation among the kinetics of free Ca^{2+}, MLC phosphorylation and maintained force can be reconciled with the hypothesis that sensitivity to intracellular Ca^{2+} is dynamic and that the level of MLC phosphorylation has a limited predictive value on developed force. The implication, therefore, is that, although the rate and extent of MLCK activation correlated to shortening velocity of vascular tone production, maintained tone requires the cooperative activation of other proteins regulating contractile function. A number of such candidates, including caldesmon, have been proposed as factors governing the efficiency of the latch state of MLC. In support of this concept, it has recently been reported that vasoconstrictors inducing the latch state of vascular tone also stimulate phosphorylation of caldesmon.[84] Even though protein kinase C phosphorylates several contractile or regulatory proteins, such as MLCK and caldesmon,[42] in vitro, direct evidence implicating receptor-stimulated, mediated by phosphorylation of intact muscle G-protein (DAG, pKC) is absent, and thus represents an area of debate and intense research.[70,84–86]

Acknowledgments

Supported by funds from National Institutes of Health: NHLBI HL 42880 (I. Laher) and HL 35657 (C. van Breemen).

References

[1] Bolton, T. B. (1979) *Physiol. Rev.* **59,** 606–718.

[2] van Breemen, C., Aaronson, P., and Loutzenhiser, R. (1979) *Pharmacol. Rev.* **30,** 167–208.

[3] van Breemen, C. and Saida, K. (1989) *Annu. Rev. Physiol.* **51,** 315–329.

[4] Bevan, J. A. (1984) *Blood Vessels* **21,** 110–116.

[5] Bevan, J. A. (1987) *Clin. Invest. Med.* **10,** 568–572.

[6] Rowan, R. and Bevan, J. A. (1987) *Blood Vessels* **24,** 181–191.

[7] Owen, M. P., Quinn, C., and Bevan, J. A. (1985) *Am. J. Physiol.* **240,** H404–H414.

[8] Hwa, J. J. and Bevan, J. A. (1986) *Am. J. Physiol.* **250,** H87–H95.

[9] Bevan, J. A., Tayo, F. M., Rowan, R. A., and Bevan, R. D. (1984) *Blood Vessels. Fed. Proc.* **43,** 1365–1370.

[10] Bulbring, E. and Tomita, T. (1987) *Pharmacol. Rev.* **39,** 49–96.

[11] Nelson, M. T., Patlak, J. B., Worley, J. F., and Standen, N. B. (1990) *Am. J. Physiol.* C3–C18.

[12] Rosenthal, W., Hescheler, J., Trautwein, W., and Schultz, E. (1988) *FASEB J.* **2,** 72784–72790.

[13] Berridge, M. J., Cobbold, P. H., and Cuthbertson, K. S. R. (1988) *Philos. Trans. R. Soc. Lond. [Biol.]* **320,** 325–343.

[14] Freissmuth, M., Casey, P. J., and Gilman, A. S. (1989) *FASEB J.* **3,** 2125–2131.

[15] Berridge, M. J. (1987) *Annu. Rev. Biochem.* **56,** 159–193.

[16] Abdel-Latif, A. A. (1986) *Pharmacol. Rev.* **38**, 227–272.
[17] Somlyo, A. P., Walker, J. W., Goldman, Y. E., Trenthan, D. R., and Kobayashi, S. (1988) *Philos. Trans. R. Soc. Lond. [Biol]* **320**, 399–414.
[18] Suematsu, E., Hirata, M., Hashimoto, T., and Kuriyama, H. (1984) *Biochem. Biophys. Res. Commun.* **120**, 481–485.
[19] Somlyo, A. P. and Somlyo, A. V. (1990) *Annu. Rev. Physiol.* **52**, 857–874.
[20] Somlyo, A. P. (1985) *Circ. Res.* **57**, 497–507.
[21] Baker, J. R. J., Haigh, R. J., Warburton, G., Weston, A. H., and Williamson, I. H. M. (1987) *J. Pharmacol. Methods* **18**, 55–67.
[22] Waltz, B. and Baumann, O. (1989) *Prog. Histochem. Cytochem.* **20**, 1–47.
[23] Cauvin, C. and Malik, S. (1984) *J. Pharmacol. Exp. Ther.* **230**, 413–418.
[24] Owen, M. P., Joyce, E. H., and Bevan, J. A. (1987) *J. Pharmacol. Exp. Ther.* **243**, 27–34.
[25] Cauvin, C., Saida, K., and van Breemen, C. (1984) *Blood Vessels* **21**, 23–31.
[26] Loutzenhiser, R. and Epstein, M. (1985) *Am. J. Physiol.* **249**, F619–F629.
[27] Saida, K. and van Breemen, C. (1987) *Biochem. Biophys. Res. Comm.* **144**, 1313–1316.
[28] Kobayashi, S., Somlyo, A. P., and Somlyo, A. V. (1988) *J. Physiol.* **403**, 601–619.
[29] Kobayashi, S., Somlyo, A. V., and Somlyo, A. P. (1988) *Biochem. Biophys. Res. Commun.* **153**, 625–631.
[30] Yamamoto, H., Kanaide, H., and Nakamura, M. (1990) *Naunyn-Schmiedeberg's Arch. Pharmacol.* **341**, 273–278.
[31] Reches, A., Eldor, A., and Saloman, Y. (1979) *Thromb. Res.* **16**, 107–116.
[32] Willuweit, B. and Aktories, K. (1988) *Biochem. J.* **249**, 857–863.
[33] Jean, T. and Klee, C. B. (1986) *J. Biol. Chem.* **261**, 16414–16420.
[34] Henne, V., Puper, A., and Soling, H–D. (1987) *FEBS Lett.* **218**, 153–158.
[35] Nicchitta, C. V., Joseph, S. R., and Williamson, J. R. (1986) *FEBS Lett.* **209**, 243–248.
[36] Ghosh, T. R., Mullaney, J. M., Tarazi, F. I., and Gill, D. L. (1989) *Nature* **240**, 236–239.
[37] Chueh, S–H. and Gill, D. L. (1986) *J. Biol. Chem.* **261**, 13883–13886.
[38] Morgan, J. P. and Morgan, R. G. (1984) *J. Physiol.* **351**, 155–167.
[39] Itoh, T., Ranmura, Y., Kuriyama, H., and Sumimoto, K. (1986) *J. Physiol.* **375**, 515–534.
[40] Somlyo, A. P. and Himpens, B. (1989) *FASEB J.* **3**, 2266–2276.
[41] Karaki, H. (1989) *Tr. Pharmacol. Sci.* **10**, 320–325.
[42] Rasmussen, H., Haller, H., Takuwa, Y., Kelley, G., and Park, S. (1990) *Prog. Clin. Biol. Res.* **327**, 89–106.
[43] Ganz, M. B. and Rasmussen, J. (1990) *FASEB J.* **4**, 1638–1644.
[44] Ruzycky, A. L. and Morgan, K. G. (1989) *Br. J. Pharmacol.* **97**, 391–400.
[45] Hai, C.-M. and Murphy, R. A. (1989) *Annu. Rev. Physiol.* **51**, 285–298.

[46] Rasmussen, H., Takuwa, Y., and Park, S. (1987) *FASEB J.* **1,** 177–185.
[47] Jiang, M. J. and Morgan, K. G. (1989) *Pflugers Arch.* **413,** 637–643.
[48] Rembold, C. M. and Murphy, R. A. (1988) *Circ. Res.* **63,** 593–603.
[49] Somlyo, A. V., Kitazawa, T., Horiuti, K., Kobayashi, S., Trenthan, D., and Somlyo, A. P. (1990) *Prog. Clin. Biol. Res.* **327,** 167–182.
[50] Chatterjee, M. and Tejada, M. (1986) *Am. J. Physiol.* **251,** C892–C903.
[51] Mulvany, M. J., Nilsson, H., and Flatman, J. A. (1982) *J. Physiol.* **332,** 363–373.
[52] Skarby, T., Hogestatt, E. D., and Andersson, K–E. (1984) *Acta Physiol. Scand.* **123,** 445–456.
[53] Bolton, T. B., Lang, R. J., and Takewaki, T. (1984) *J. Physiol.* **351,** 549–572.
[54] Fish, R. D., Sperti, E., Colucci, W. S., and Clapham, D. E. (1988) *Circ. Res.* **62,** 1049–1054.
[55] Mori, T., Yanagisawa, T., and Taira, N. (1990) *Naunyn-Schmiedeberg's Arch. Pharmacol.* **341,** 251–255.
[56] Mulvany, M. J. and Nyborg, N. (1980) *Br. J. Pharmacol.* **71,** 585–596.
[57] Nyborg, N. C. B. and Mulvany, M. J. (1984) *J. Cardiovasc. Pharmacol.* **6,** 499–505.
[58] Boonen, H. C. M. and DeMey, J. G. R. (1990) *Eur. J. Pharmacol.* **179,** 403–412.
[59] Bean, B. P., Sturek, M., Puga, A., and Hermsmeyer, K. (1986) *Circ. Res.* **59,** 229–235.
[60] Hogestatt, E. D. (1984) *Acta Physiol. Scand.* **122,** 483–495.
[61] Nelson, M. T., Standen, N. B., Brayden, J. E., and Worley, J. F. (1988) *Nature* **336,** 382–385.
[62] Jolou, G. and Freslon, J.–L. (1986) *Eur. J. Pharmacol.* **129,** 261–270.
[63] Nishizuka, Y. (1984) *Nature* **308,** 693–697.
[64] Kaibuchi, K., Takai, Y., and Nishizuka, Y. (1981) *J. Biol. Chem.* **256,** 7146–7149.
[65] Fujiwara, T., Itoh, T., Kubota, Y., and Kuriyama, H. (1988) *Circ. Res.* **63,** 893–902.
[66] Fujiwara, T., Itoh, T., Kubota, Y., and Kuriyama, H. (1989) *J. Physiol.* **408,** 535–547.
[67] Rembold, C. M. and Weaver, B. A. (1990) *Circ. Res.* **15,** 692–698.
[68] Cohen, P., Holmes, C. F. B., and Tsukitani, Y. (1990) *Trends Biochem. Sci.* **15,** 98–102.
[69] Ngai, P. K. and Walsh, M. P. (1987) *Biochem. J.* **244,** 417–425.
[70] Marston, S. B. (1989) *J. Muscle Cell Motil.* **10,** 97–100.
[71] Small, J. V., Furst, D. O., and DeMeys, J. (1986) *J. Cell Biol.* **102,** 210–220.
[72] Yin, H. L. (1987) *BioEssays* **7,** 176–179.
[73] Lassing, I. and Lindberg, V. (1985) *Nature* **314,** 472–474.
[74] Ahnert-Hilger, G. (1988) *Trends Pharmacol. Sci.* **9,** 195–197.

[75] Ahnert-Hilger, G., Mach, W., Fohr, K. J., and Gratzl, M. (1989) *Meth. Cell Biol.* **31**, 63–90.

[76] Nishimura, J., Kolber, M., and van Breemen, C. (1988) *Biochem. Biophys. Res. Comm.* **157**, 677–683.

[77] Tobkes, N., Wallace, B. A., and Bayley, H. (1985) *Biochemistry* **24**, 1915–1920.

[78] Kitazawa, T., Kobayashi, S., Horiuchi, K., Somlyo, A. V., and Somlyo, A. P. (1989) *J. Biol. Chem.* **264**, 5339–5342.

[79] Nishimura, J. and van Breemen, C. (1989) *Biochem. Biophys. Res. Commun.* **163**, 929–935.

[80] Nishimura, J. and van Breemen, C. (1989) *Biochem. Biophys. Res. Commun.* **165**, 408–415.

[81] Krishanuda, J. M. and Paul, R. J. (1983) *Am. J. Physiol.* **244**, C385–C390.

[82] Hoar, P. E., Mahoney, C. W., and Kerrick, W. G. L. (1987) *Pflugers Arch.* **410**, 30–36.

[83] Lash, J. A., Sellers, J. R., and Hathaway, D. R. (1986) *J. Biol. Chem.* **261**, 16155–16160.

[84] Adam, L. P., Haeberle, J. R., and Hathaway, D. R. (1989) *J. Biol. Chem.* **264**, 7698–7703.

[85] Walsh, M. P. (1990) *Prog. Clin. Biol. Res.* **327**, 127–140.

[86] Khalil, R., Lodge, N., Saida, K., and van Breemen, C. (1987) *J. Hypertens.* **5**, 55–515.

Chapter 19

Hemodynamics
in Arteriolar Networks

Geert W. Schmid-Schönbein,
Richard Skalak, and Shu Chien

Introduction

The field of arterial hemodynamics is one of the most extensively studied areas in bioengineering and physiology today. Numerous books and papers have been published on this subject, [1-8] and this chapter will give a brief summary of the current research status. The venous circulation has received less attention, but the literature on this subject is growing; [9,10] it is in a sense more complex, because veins tend to be more distensible and may collapse. The arterial and venous systems are often characterized by the terms *resistance vessels* and *capacitance vessels*, respectively. These characterizations give predominant aspects correctly, but are oversimplifications in detail. In reality, all vessels in the circulation have some hemodynamic resistance resulting from their finite geometry and the viscosity of the blood; at the same time, all vessels have some volumetric capacitance as a result of their distensibility. In the present article on the artery–arteriolar circulation, the roles of capacitance and network topology will be discussed in connection with their predominant resistance function.

From *The Resistance Vasculature*, J. A. Bevan et al., eds. ©1991 Humana Press

One of the major functions of the arteriolar network is to distribute blood to the microcirculation: specifically, to the capillary network in accordance with the varying physiological functions and metabolic needs of different organs. These needs can change widely under various physiological conditions, such as sleep, exercise, and so on. The blood supplies to most organs of the body are in parallel, and regional blood-flow distribution is controlled by the arterial resistance-vessel system. Within each organ the arterial supply system needs to maintain a relatively uniform distribution of blood-flow, and hence, of blood cells, over the vast capillary network. These requirements are met in a complex manner by the artery–arteriolar system at the transition into the microcirculation, as discussed in this chapter. We will also point out several recent hemodynamic findings that go beyond the traditional resistance-capacitance model.

The Arteriolar-Network Topology

A basic requirement for developing a quantitative analysis of blood-flow in arterioles is a realistic picture of the vessel geometry and network branching pattern. Traditionally, the branching pattern of the arteries is depicted as an asymmetric branching tree, with some segments of the arteries and arterioles being accompanied by collateral blood vessels. This picture of the network topology of the small arteries and their transition into the arterioles of the microcirculation, however, is incomplete and often not in harmony with the hemodynamic data that have been obtained at the microcirculatory level.

In recent years, several organs in laboratory animals and in humans have been subject to detailed delineation of the vascular architecture.[11] We will focus on two organs, skeletal muscle in the rat and lung in the cat, in which the architecture has been elaborated.[10–14]

Let us first consider a general problem of the vascular architecture. One of the key functions of the artery–arteriolar network is to distribute the blood-flow to organs according to metabolic demands, despite considerable variation of the physical distance from the heart to the individual organs. Skeletal muscle may be just a few centimeters from the heart, as are the diaphragm and thoracic muscles, or may be remote, as are those in the extremities. It has been found that capillary velocities in skeletal muscle are remarkably uniform, irrespective of the position of the muscle relative to the heart or the location of the capillaries within the muscle. For example, typical capillary red-cell velocities in most skeletal muscles are about 2 mm/s, and a typical standard deviation about this value is about 0.3 mm/s. One does not find skeletal muscles that have capillary velocities an order of magnitude above or below this value. Considering the large number of capillaries in a typical capillary network, this is a remarkable situation. This uniformity of perfusion is important for tissue homeostasis and needs to be maintained during the action of autoregulatory controls.

Since the veins are more compliant and less muscular, it is to be expected that flow controls are built into the supply side of the capillaries, the artery–arteriolar network. If we postulate that the arterial network is entirely a tree-like structure with dichotomous branching, then the control mechanism for a uniform capillary perfusion would have to be entirely owing to diameter adjustments of the individual arteries and arterioles. Although this is possible, such a concept encounters considerable difficulties when one considers regulation at the arteriolar level. The flow through a tree-like vascular network is extraordinarily sensitive to small variations of diameter, as expected from Poiseuille's law. It is difficult to achieve the observed uniformity of capillary perfusion by adjustment of diameters within a tree-like artery–arteriolar network. Ultrafine adjustments of the arteriolar diameters would be required, and accurate coordination over large numbers of arterioles would be necessary, for example, during autoregulation. This is an unlikely flow-control mechanism. The available evidence indicates that, instead of such diameter control in a dichotomous tree, the arteriolar network has topological features that automatically produce a remarkably uniform capillary perfusion flow and pressure, even when all arterioles are fully dilated. This mechanism is illustrated by the spinotrapezius muscle in the rat (Figs. 1 and 2).

The blood supply to the rat spinotrapezius muscle originates at several locations along the thoracic aorta. The artery at the proximal end of the muscle originates from the brachial and thora-

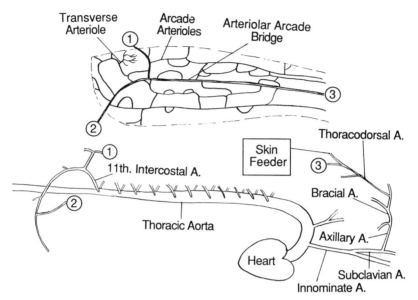

Fig. 1. Schematic drawing of the arterial/arteriolar blood supply to the spinotrapezius muscle in the mature rat. The two main feeders originate in the proximal portion of the muscle along the thoracodorsal artery and in the distal portion of the muscle along the 11th intercostal artery. Additional inflows vary from animal to animal, but usually include side branches to these main feeders as well as other intercostal arteries. The arcade arterioles in the spinotrapezius muscle are interconnected at various points (not shown) with the arcade arterioles of the underlying latissimus dorsi muscle, the overlying highly vascularized fascia, adipose tissue, lymph nodes, and the skin.

codorsal arteries, and the artery at the distal end of the muscle originates from the 11th intercostal muscle. In some animals there is also a third supply artery, originating from the 10th intercostal artery.[13] These supply arteries feed an arcade type of arteriolar network. The proximal and distal supply arteries are directly connected to each other by the vessels of the largest arcade arterioles (Fig. 2); this vessel is termed the arteriolar arcade bridge.[14] Thus, we have a complete loop formed by the aorta, the supply arteries, and the arcade bridge.

Besides the two main supply arteries, the spinotrapezius muscle has five or six additional smaller arterioles feeding into the

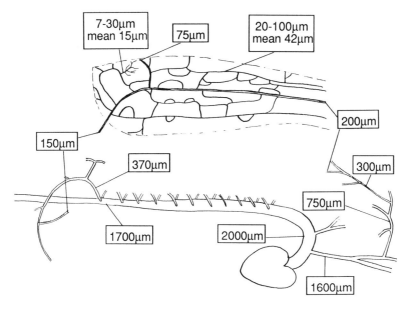

Fig. 2. Typical diameters of the artery/arteriolar blood vessels in the spinotrapezius muscle of the rat. For vessel designation, *see* Fig. 1.

arcade network and, further, there are five to ten interconnections with the arcade arterioles of neighboring skeletal muscles, connective tissue, adipose tissue, and skin (Fig. 1). The connections from the arcade arterioles to the capillaries are provided by the transverse arerioles, which are the terminal segment of the arteriolar network and have also been denoted as terminal arterioles.[5] Each terminal arteriole gives rise to a "tree" in the traditional sense; it has a single root at its origin along the arcades and branches asymmetrically into a multiple array of precapillaries. The arcade arterioles give rise to between five and ten transverse arteriolar trees/ mm[3] of muscle volume, and their terminal arterioles form a space-filling vessel system that reaches out into all regions of the muscle-capillary network.[16]

This arrangement of supply arteries, arcade arterioles, and transverse arterioles is also observed in other skeletal muscles of the rat[17] (Fig. 3), in the skeletal muscle of other species,[17,18] and in other organs.[19] The analog of the arcade arterioles is in other regions of the circulation, such as the array of arteries feeding the

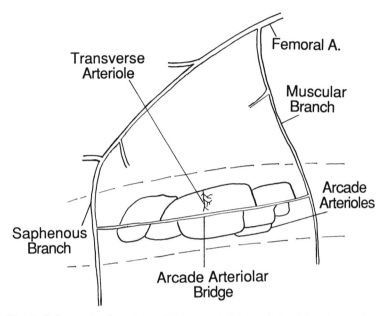

Fig. 3. Schematic drawing of the arterial/arteriolar blood supply to the gracilis muscle in the rat. Although its main supply originates from the femoral artery, the arcade arterioles are interconnected (not shown) with those of neighboring muscles and other tissues, the blood supply of which does not originate exclusively from the femoral artery. This situation contributes to the "zero-flow-pressure" phenomenon at steady perfusion in this muscle.[39]

hind leg or the canine coronary arteries, where the same network arrangements are usually called a collateral network. Collateral vessels and arcade arterioles are different terms for similar anatomical features, i.e., multiple, interconnected pathways by which blood can be transported from the heart to a capillary region. The presence of the arcade arterioles has important implications for the hemodynamics of the microcirculation, some of which will be discussed below.

It should be noted that not all microcirculations have arcade networks. Careful reconstruction of the vasculature in the lung shows no evidence for an arteriolar arcade network. Current vascular reconstructions of lungs in the cat[21,22] and in humans (S.S. Sobin, personal communication) show that the pulmonary arteries divide dichotomously until the capillary network (capillary sheets) are reached.[20] For example, in the cat, there are approx 11 branching orders. A single pulmonary artery gives rise to about 300,000 precapillary arterioles. Not one of them appears to anastomose. The capillary network itself is, of course, richly interconnected.

Pressure Distribution in Arteries and Arterioles

The mean blood pressure along the arterial and arteriolar network in most organs decreases progressively from its central value, about 100 mmHg, to between 20 and 30 mmHg at the entry to the capillary network. Direct measurements and theoretical predictions show that the average pressure reduction along the central arteries is relatively small (5–10 mmHg). In contrast, the pressure in arterioles at the microcirculatory level has been found to be significantly lower than that in the central arteries (*see* summaries in refs. 23–25), with organ-dependent values. Specifically, mean pressures in larger arterioles of skeletal muscle of the rat have been reported to be as low as 50–60 mmHg, or even lower.[26] This implies a steep pressure gradient between the central arteries and the larger arterioles of the microcirculation.

The values reported in the literature for the pressure gradient along the supply arteries are not consistent, sometimes even in the same species and organ. The differences may be attributed to several factors, e.g., the choice of the anesthetic agent and the surgical procedure used to expose the organ. We have therefore recently reevaluated this situation in skeletal muscle.[27] Systematic measurements of intravascular pressure were carried out in the feeding arter-ies at the entry to the muscle and at the midpoint of the arteriolar arcade bridge. Several different skeletal muscles in the rat, positioned close to and far from the heart, were selected, and pressures were measured after micropipet puncture using the servo-null technique.[28,29] Special precautions were taken to leave all feeders to the arteriolar arcades of the individual muscles intact and to minimize surgical interventions; the tissue was not exteriorized, because exteriorization and surgical trauma may cause an interruption of blood-flow and a disturbance of the pressure distribution in the arterioles.[17,27] The results revealed remarkably high

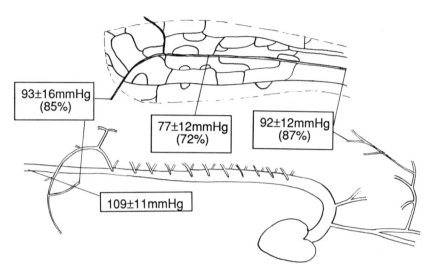

Fig. 4. Mean blood pressures in the distal segrnent of the central arteries feeding the spinotrapezius muscle of its proximal and distal ends, i.e., along the arcade bridge. Mean ± SD and percentile of simultaneous mean femoral pressure for four animals are shown. The values were obtained by micropuncture with minimal dissection of the surrounding tissue, to minimize disturbance of the blood flow around the measurement site and to the remainder of the muscle arcade arterioles. Data are summarizd from DeLano et al.[27] Note the relatively high blood pressures along the arcade arteriolar bridge.

feed pressures; values equivalent to 70–90% of the mean systemic pressure were found in the smallest arteries at their entry into the microcirculation (Figs. 4, 5). In the majority of the skeletal muscles, one finds only small pressure differences between the proximal and the distal feed arteries. Even at the midpoint of the arteriolar arcade bridge, located in the middle of the muscle within the arcade network, the diastolic pressures are >70% of the systemic diastolic pressure. The pressure becomes lower in the smaller arcade arterioles. The pulse pressure along the arcade arteriolar bridge is still >75% of the pulse pressure in the central arteries, and in some muscles it is almost the same as in the aorta. Since most arterioles are accompanied by lymphatic vessels,[30] the relatively high pulse pressure in the arcades serves to periodically expand the arterioles and compress the adjacent lymphatic vessels. This process may contribute to lymph transport out of the organ.

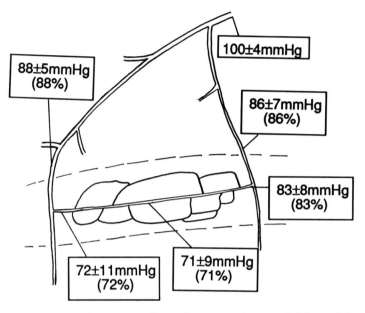

Fig. 5. Mean blood pressure along the two major arterial/arteriolar pathways into the arcade arterioles of the rat gracilis muscle. Mean pressures ± SD and percentile of simultaneous mean femoral pressures in five animals are shown. Note the relatively high blood pressures along the arcade arteriolar bridge, located in the middle of the muscle.

Although these observations need to be repeated in other species before general conclusions can be drawn, they lead to the following suggestions:

1. The pressure drop along the central arteries may be much smaller than presumed earlier, and there may be significant organ differences in this respect.
2. The pressure distribution in the arcade arterioles of skeletal muscle is relatively uniform, thus providing a nearly constant precapillary pressure at the entry to the transverse arterioles and capillaries. This may serve as one of the mechanisms for producing a relatively homogeneous capillary blood-flow.

Distensibility of Arterioles in Skeletal Muscle

Distensibility of the arteries is an essential element for wave propagation. In addition, distensibility plays a significant role in measurements of whole-organ capillary filtration coefficients involving the use of organ weight, since an expansion of the vascular weight would contribute to weight gain. The literature on arterial distensibility is vast (*see* refs. 31 and 32 for reviews). In this section, we focus on the passive properties of arteries in rat skeletal muscle, which were referred to above in discussions of the microanatomy. Arteries and arterioles exhibit viscoelastic properties, both in the active state with increased smooth-muscle tone and in the passive dilated state. These viscoelastic properties involve the presence of hysteresis, relaxation, and creep.

One of the ways to study the viscoelastic properties of arterioles is to fill the vascular system with a nondiffusible fluid and adjust the arterial and venous pressures to the same hydrostatic value. Elevating the pressures in either a step or an oscillatory fashion while recording the vessel diameter provides the basic data needed to derive the viscoelastic properties of the vessel. This experiment has been carried out on isolated arteries,[33,34] larger arterioles,[35] as well *in situ* on arterioles as small as the capillaries.[36] Several viscoelastic models have been proposed. In rat skeletal muscle, the diameter, d, is generally a nonlinear function of the transmural pressure, P. Rapid changes of P are accompanied by synchronous diameter changes, suggesting some elastic response. If the pressure is elevated and held for some time, however, d gradually creeps to a steady-state value and returns exponentially to its initial value on release of the pressure. This arterial-wall viscoelastic behavior, which is typical of biological tissues, can be closely approximated by a quasi-linear standard solid of the form

$$P + (\beta/\alpha_1)(\partial P/\partial t) = \alpha_2 E + \beta [1 + (\alpha_2/\alpha_1)](\partial E/\partial t) \tag{1}$$

In this expression, E is the diametral strain defined as

$$E = (1/2)[(d/d_0)^2 - 1] \tag{2}$$

E is quadratic in the diameter d. For larger arteries, E may have to be replaced by an exponential form, as proposed by Fung.[31] β, α_1, and α_2 are empirical coefficients whose values depend on the vessel diameter. Their values have been listed for all arcade and transverse arterioles[36] and were found to be constant for individual vessels. These values have been derived from step pressure

experiments, and they still need to be tested in harmonic oscillations for a range of frequencies. At zero transmural pressure ($P =$
0), d_o is the reference diameter. Experimental values of d_o were found
to be highly reproducible. For short-term, transient pressure
changes, the initial pressure amplitude $P(0)$ and the initial diameter $d(0)$ are related by

$$P(0) = (\alpha_1 + \alpha_2)[d(0)/d_0)^2 - 1]/2 \qquad (3)$$

and long-term changes by

$$\lim_{t \to \infty} /P(t) = (\alpha_2/2)[d(t)/d_0)^2 - 1] \qquad (4)$$

The advantage of the quasilinear standard solid, Eq. (1), is that
various solution techniques are available for a broad range of pressure histories. In the following, this model will be applied to the
case of blood flow in skeletal-muscle arterioles.

Flow Analysis

The general three-dimensional equations of motion and conservation of mass for blood, regarded as an incompressible, homogeneous non-Newtonian fluid, are

$$P(\partial v_i/\partial t) + v_j(\partial v_i/\partial x_j) = -(\partial p/\partial x_i) + (\partial/\partial x_j) [\mu_a(\partial v_i/\partial x_j)] + X_i \quad (5a)$$

$$(\partial v_i/\partial x_i) = 0 \qquad (5b)$$

where p is the local fluid pressure, v_i is the velocity, μ_a is the apparent viscosity, x_i are spatial coordinates, X_i is the body force
attributable to gravity, and X_i may be neglected in the rat, but plays
a significant role in humans or other large animals. The full problem of blood flow in the arteries requires solution of Eqs. (1, 2), and
(5a,b) with an appropriate set of initial and boundary conditions.
A constitutive equation for μ_a is also needed; for example, μ_a may
be a function of J_2 the second invariant of the velocity gradient
tensor. The general three-dimensional equations (1, 2, and 5a and
b) are seldom solved in full, although linearized equations for
Newtonian fluids can be solved for periodic motion in a straight,
circular vessel (e.g., Section 3.15 in Fung[4]). Usually, the general Eqs.
(5a and b) are integrated over the vessel cross-section to derive

one-dimensional equations with some additional assumptions concerning wall shear stress. Flow entry effects are neglected. Many such approximations have been derived for both rigid and flexible blood vessels.[2,3,6]

The general one-dimensional equations for unsteady blood-flow in a vessel are:

$$\rho[(\partial V/\partial t)+ V(\partial V/\partial x)] = -(\partial p/\partial x) + (f/A) \qquad (6a)$$

$$(\partial A/\partial t) + V(\partial A/\partial x) + A(\partial V/\partial x) = 0 \qquad (6b)$$

where A is the cross-sectional area, V is the mean velocity over a cross-section, x is the axial coordinate, and f is a frictional force at the tube wall. For Poiseuille flow,

$$f = -8\pi\mu V \qquad (7)$$

Equation (7) is often used as a first approximation.

The full equations (6a, b) give rise to a nonlinear wave theory, which yields realistic results in the large arteries.[37] The basic character of the wave propagation is a result of the interaction of the inertial term (left-hand side of Eq. (6a) and the elasticity of the wall, Eq. (1). The wave-propagation nature of the equations is preserved if equations 6a,b are linearized:

$$(\partial Q/\partial t) = (A_0/\rho)(\partial p/\partial x) + f/\rho \qquad (8a)$$

$$(\partial A/\partial t) + (\partial Q/\partial x) = 0 \qquad (8b)$$

where $Q = A V$ is the discharge and A_0 is the initial cross-sectional area. Equations (8a and b) are derived under the assumption that the area changes are small and that the convective term $V(\partial V/\partial x)$ in Eq. (6a) is negligible. Equations (8a and b) are the basis of the so-called transmission-line theory,[1] which reproduces many wave-propagation effects in large arteries for which the viscous term f is small.

As the diameter of blood vessels gradually decreases from the heart to the microcirculation, there comes a point at which the inertial terms in Eqs. (6 and 8) are relatively small compared to the viscous terms f. Table 1 shows the relative magnitudes of the inertial and viscous terms in Eq. (8a) for a number of vessels in the rat with diameters spanning three orders of magnitude. This is not the

Table 1
Ratio of the Unsteady Term $(\rho\partial Q/\partial t)$
to the Frictional Term f in Eq. (8a)

Vessel diameter, d (mm)	Ratio, $\rho\,(\partial Q/\partial t)/f$
10	654
1	6.54
0.1	0.0654
0.01	0.000654

[a]The flow is assumed to be sinusoidal plus a mean flow such that the minimum flow is exactly zero. The ratio $\rho\,(\partial Q/\partial t)/f)$ is based on the peak values of $\partial Q/\partial t$ and f. For the assumed sinusoidal flow, the ratio is equal to $N^2/16$, where N is the Womersley number $N = (d/2)\,(\omega/\upsilon)^{1/2}$, with ω being the circular frequency (rad/s).
[b]The values given here are based on a heart rate of 2 Hz and a kinematic viscosity of blood of $\upsilon = 0.03$ cm^2/s.

Reynolds number, because the inertial term here is the unsteady term $\partial Q/\partial t$. The Reynolds numbers are also small, as shown in Fig. 6. In small vessels, there is still a wave propagation, but it is highly damped, and a more revealing approximation may be derived by neglecting the inertial term in this case.

In rat skeletal muscle, the majority of small arteries and the arterioles have a Reynolds number <1 (Fig. 6), and the unsteady inertial term is also small (Table 1). Therefore, inertia forces can be neglected with good accuracy. The individual vessels are relatively long compared with their diameter, so that the flow field for a Newtonian fluid is fully developed within an entry length of about one vessel diameter. Thus, assuming blood to be an incompressible Newtonian fluid, as a first approximation, the equations of fluid motion and conservation of mass in the arteriolar network become, respectively,

$$A_o(\partial p/\partial x) + f = 0 \tag{9a}$$
$$(\partial A/\partial t) + (\partial Q/\partial x) = 0 \tag{9b}$$

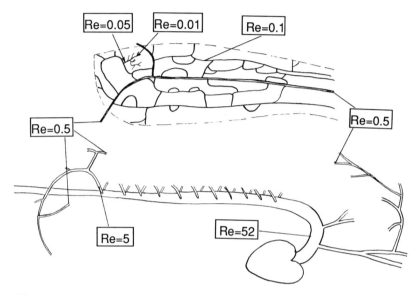

Fig. 6. Estimates of the Reynolds number along the arterial/arteriola network into the rat spinotrapezius muscle. The values are based on an average cardiac output of 50mL/min.

where f is given by Eq. (7). Equation (9)a predicts a linear relationship between pressure gradient, $\partial p/\partial x$ and velocity or flow rate, Q. In fact, in this approximation, Poiseuille's law for circular cylindrical vessels (with radius a) holds even though the flow is unsteady, i.e.,

$$Q = - (\pi\, a^4/8\mu)(\partial p/\partial x) \qquad (10)$$

Although Eq. (10) is linear in a $\partial p/\partial x$, a number of mechanisms exist that lead to nonlinearities in the pressure–flow relationships of single vessels and of the arteriolar network. Under physiological conditions, the main effects are (1) the active and the passive dependency of the vessel radius on local pressure and (2) the non-Newtonian flow properties of the blood. If we consider pathophysiological conditions, additional effects come into play: sclerotic narrowing of the blood vessels, aneurysm, thrombus and embolus formation, adhesion of leukocytes to the endothelium, and spasm of the smooth-muscle, such as during endothelial injury and lack of EDRF production. Each of these cases requires a separate analysis, and they are discussed to some extent in other chapters of this volume, we will limit the discussion here to physiological flow in the arterioles. In summary, flow in arterioles is a pulsatile flow at low Reynolds number, governed by Eqs. (7, 9a,b), and (10).

The typical heart rate of a rat is between 2 and 5 Hz. For such relatively rapid pulsations, the viscoelastic creep in the arteriolar wall can be neglected, so that the elastic limit expressed in Eq. (3) serves as a good approximation for the wall properties. Setting $\alpha_1 + \alpha_1 = \alpha$, Eq. 3 can be rewritten as

$$P = (\alpha/2)[(A/A_0) - 1] \tag{11}$$

where A_0 is the reference cross-sectional area at zero transmural pressure. We can combine Eqs. 7, 9a,b, and 10 in an equation governing the normalized pressure, $P' = (2P/\alpha) + 1$

$$(\partial P'/\partial t) = C^2(\partial^2 P'^3/\partial x^2) \tag{12}$$

with
$$C^2 = (A_0\alpha/48\pi\mu) \tag{13}$$

Thus, pulsatile flow in the arterioles is governed by a nonlinear diffusion equation (12), which has been referred to as dynamic viscous flow.[38] Some solutions of special situations are of interest. First, in steady flow, $\partial P/\partial t$ in Eq. 12 is zero, and integration of the right-hand side over the length, L, of an arteriole from its upstream position, $P'(0)$, to its downstream position, $P'(L)$, gives the steady flow Q_{ST}:

$$Q_{st} = (C^2/L)[P'(0)^3 - P'(L)^3] \tag{14}$$

Because of the distensibility of the vessel wall, the flow is not only a function of the upstream pressure, but also a nonlinear function of the downstream pressure. The third power is the consequence of the distensibility characteristics of the arterioles in skeletal muscle, Eq. (11). For the lung arterioles, the exponent is 5 instead of 3, because of the linear relationship between the diameter and transmural pressure.[4]

Under normal pulsatile perfusion, the left-hand side of Eq. (12) is not zero. To illustrate the effects of pulsatility, sinusoidal oscillations are considered. The magnitude of C^2 is about 95 cm^2/s in a 140-μm arteriole (one of the largest diameters in the arcades),

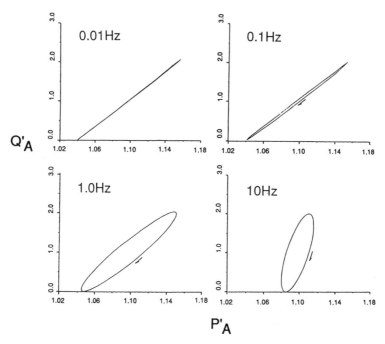

Fig. 7. The normalized arterial inflow Q'_A and P'_A of a single microvessel subjected to sinusoidal pressure oscillations. The data shown represent a solution to Eq. 12. Note that while the mean arterial inflow rate and its amplitude are maintained constant, the pressure amplitude decreases with increas-ing frequency of oscillation. Adapted from Lee et al.[40]

and it drops to about 0.9 in the transverse arterioles. Figure 7 shows some solutions to Eq. (12) in the form of pressure–flow curves for a single vessel subject to sinusoidal oscillations on the upstream end. The amplitude of the flow was stipulated to vary between a fixed maximum and zero for all frequencies. This graph shows that at physiological frequencies there exists a hysteresis in the pressure–flow curves and that the pressure at the instant of zero flow rises with frequency. Many other cases similar to this can be illustrated.[38] Dynamic viscous flow can lead to instantaneous reversal of the flow, although the mean velocity is positive. A step increase in pressure on the arterial side results in a flow overshoot, whereas a step increase in flow leads to a gradual rise in upstream pressure while the venous pressure is kept constant. Upstream and downstream pressures in these arterioles follow different time-courses, in general. All these features are closely related[39,40] and are similar to the flow phenomena observed in the Windkessel theory for the aorta. In the central arteries, however, inertia is present, and the well-known pressure wave propagation is possible also.

Another factor that can lead to a deviation from Poiseuille's law (Eq. 10) is variations in blood vessel cross-sections. Most blood-flow theories assume the vessel cross-sections to be circular. Although this may be a good first approximation, there are a number of situations for which this assumption is not completely satisfactory. A case in point is the transverse arterioles. These vessels are between 5–35 μm in the dilated state, and, when their smooth-muscle contracts, they become considerably narrower, to the point of complete lumen collapse.[41–43] In both the smaller arcade and the transverse arterioles, the endothelial cells are immured by a single coat of circumferentially oriented, spindle-shaped smooth-muscle cells. Contraction of the smooth-muscle cells leads to compression of the endothelial cells and protrusion of the nucleus into the lumen. Strong contractions lead to complete occlusion of the vessel lumen. In this way the transverse arterioles periodically interrupt blood-flow to their capillary networks during vasomotor cycles. In essence, this is the action of the precapillary sphincters.

The flow field in such arterioles deviates from Poiseuille's flow in circular vessels. Compared to that in a circular vessel of equal cross-sectional area, the hemodynamic resistance is elevated, and the ratio of maximum velocity to mean velocity is increased above the value of 2 for Poiseuille's flow. Furthermore, the wall shear stress is nonuniform: it is maximum at the tip of local endothelial-cell projections and near zero at the cell periphery. The average wall shear stress in noncircular vessels is less than that in circular vessels with the same lumen cross-sectional area and pressure gradient.[43] In the presence of blood cells, the effects of protrusions are currently unexplored, but in narrow arterioles the effects are expected to be amplifled, especially in the presence of relatively rigid leukocytes.

Cell Distribution in Small Arteries and Arterioles

The arteriolar topology influences microvascular hemodynamics via modulations of not only the pressure distribution, but also the cell distribution. The smaller the arterioles, the more pronounced is the influence of flow distribution on blood-cell distribution at bifurcations. The arterioles serve on one hand as a determinant of capillary hematocrit and oxygen transport to the tissue; on the other hand, they serve to direct leukocyte and platelet traffic in the microcirculation. Consequently, they affect the efficiency of the immune system and the degree of organ injury by inflammatory cells.

The cell distribution in the cross-section of an arteriole is nonuniform, with a tendency for the smaller cells to be displaced toward the vessel periphery by larger cells or cell aggregates. This phenomenon can be observed when individual platelets in whole blood are tracked along the arterioles.[44,45] Similarly, white-cell margination is enhanced when red cells form aggregates that exceed the size of the white cells at low flow states,[46–48] and this effect can be reversed when the red cells are disaggregated by high flows. Individual platelets or white cells follow an average path that is parallel to the arteriolar wall, but superimposed on the average cell path are radial and longitudinal displacements resulting from random interactions with neighboring cells.[49] The random dispersions are larger in the vicinity of the arteriolar endothelium because of high shear rates, and smaller in the center of the vessel, where the shear rate is low. The radial dispersions lead to frequent encounters between leukocytes or platelets and the arteriolar endothelium, a process that may be important in the deposition of monocytes on the arterial wall and the initiation of arteriogenesis.[50,51]

At the divergent bifurcations of the arteries and arterioles, the blood stream is separated into two daughter vessels. However, the divisions of plasma and of blood cells into the daughter vessels do not occur in a homogeneous fashion. In the region of bifurcation, blood cells and plasma may be separated in different proportions into the daughter vessels.[52–61] The unequal division of cells and plasma is negligible in the large bifurcations of the central arteries but becomes progressively more pronounced in the arcade arterioles and dominates in the transverse arterioles. The origin of this phenomenon lies in the fact that blood cells are separated into the daughter vessels in accordance with their position relative to a dividing streamline in the parent vessel of the bifurcation. The dividing streamline separates the blood cells that will flow into the right and the left daughter vessel. Cells positioned in immediate proximity to the dividing streamline, or actually positioned on it, tend to flow into the daughter vessel with the higher flow. The daughter vessel with the higher flow exerts a higher plasma shear stress and a steeper pressure gradient on the blood cells entering its lumen.[62]

Plasma and cell divisions are readily described by the blood cell separation function at divergent bifurcations,[52]

$$\phi_i = \phi_i(\psi_i \ldots) \tag{15}$$

which relates the fractional cell flux ϕ_i into daughter vessel i ($i = 1$, 2) to the fractional bulk flow of blood ψ_i (cells plus plasma). These functions always satisfy the relations

$$\phi_1 + \phi_2 = 1 \text{ and } \psi_1 + \psi_2 = 1 \tag{16}$$

In large arteries, $\phi_i = \psi_i$, but in arterioles the cell-separation function becomes nonlinear. For symmetric bifurcations, such as T or Y junctions, with a symmetric distribution of blood cells in the entry region of the bifurcation, the cell distribution function is symmetric about the position of equal cell flux and bulk flow into the daughter vessels, i.e., $\phi_i = \psi_i = 0.5$. In this specific case, the shape of the dividing streamline in a cross-section of the parent vessel is essentially a straight line.[58] If the diameters of the daughter vessels are unequal, as is generally the case at the junctions between the arcade and transverse arterioles,[14] the cell-distribution functions become asymmetric, since the dividing streamlines in the parent vessel assume a curved shape.[61] If the daughter vessels have equal flow rates and a uniform cross-sectional distribution of blood cells at the entry to the bifurcation, the daughter vessel with the *smaller* diameter may actually collect the *larger* number of blood cells. This is because, under this circumstance, the smaller daughter arteriole draws a larger proportion of its blood from the center of the parent vessel. If preferential shunting of blood cells continues to repeat itself at consecutive branches, the main channels of the arcade and transverse arterioles will continue to receive the higher cell fraction. This effect leads to shunting of leukocytes, as has been demonstrated in the transverse arterioles of skeletal muscle.[63,64]

If the blood cell distribution in the parent vessel becomes asymmetric as a result of the presence of upstream arteriolar bifurcations,[57] then the cell-distribution functions may be asymmetric irrespective of the diameters of the daughter vessels. The branching angles have relatively small effects on the cell distribution in the small arteries and arterioles, because their influence on the flow field is weak at low Reynolds numbers. A case of special significance for the microcirculation is when the flow distribution between arteriolar daughter vessels becomes unequal. This occurs regularly during arteriolar vasomotion[65] or in resting skeletal muscle when the transverse arterioles constrict, thereby reducing capillary flow. Under these conditions, a point can be reached at which the daughter arteriole with the lower flow will collect cells at progressively lower concentrations, until a small, but still finite, flow rate is reached at which no red or white cells can enter this daughter vessel. This can lead to nearly complete washout of red and white cells from the capillaries downstream of the transverse arterioles, although usually platelets can still be observed in such low-hematocrit microvessels with the use of sufficiently high optical magnifcation.

The sequence of events can be illustrated at the bifurcations between the arcade and transverse arterioles. During regular vasomotor control, the rate of transverse arteriolar constriction is sufficiently slow[66] to cause a gradual washout of blood cells from the transverse arteriole, as well as a major portion of the capillary network that it supplies. As the flow rate into the transverse arteriole is reduced, fewer cells enter, until zero flow is reached with closure of the arteriolar lumen and attainment of near-zero hematocrit (Fig. 8). At the same time, the other daughter vessel, i.e., the arcade arteriole, transiently collects blood with a hematocrit higher than that in the parent vessel of the bifurcation. Since, under conditions of such local autoregulation, the flow through the arcade arterioles continues, this bolus of blood with an elevated hematocrit is rapidly carried away, either into neighboring transverse arterioles or through vascular connections between the arcade arter-

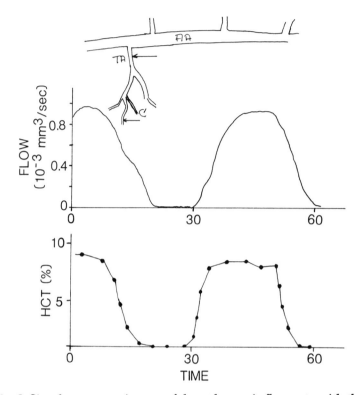

Fig. 8. Simultaneous estimates of the volumetric flow rate with the dual-slit method and microtube hematocrit (by direct cell counting) in a third-order branch of a transverse arteriole in rat spinotrapezius muscle. The transverse arteriole exhibits marked vasomotor activity. Note that with each reduction of flow rate, the microhematocrit drops and is then maintained close to zero (at a reduced, but still open, vessel lumen). During such hematocrit reduction, a transient bolus with elevated hematrocrit is being carreied away along the arcade arterioles into neighboring transverse arterioles or into the arterioles of neighboring organs.

ioles and blood vessels in neighboring organs.[15] If several transverse arterioles along an arcade constrict simultaneously, the bolus may be enlarged, but it is still carried away by the continuing blood-flow through the arcade arterioles. Thus, hematocrit adjustments caused by arteriolar bifurcations, as an addition to the traditional Fahraeus effect caused by radial redistribution of blood cells in the vessels, can be caused by variations in flow rates attributable to arteriolar constrictions. The lower the flow rate, the lower is the hematocrit in transverse arterioles. In fact, near-zero hematocrit occurs at essentially zero-bulk flow rates.

It may be mentioned in this regard that recent attempts to demonstrate low-discharge hematocrits from small arterioles or capillaries by aspiration into a micropipet[67] require that the low flow rates in the vessels are not disturbed. Thus, collection of low-discharge hematocrit may require near-zero aspiration flow rates, a condition that sets extraordinary experimental requirements, which may be difficult to meet. If the flow rates in the vascular network during the blood aspiration are disturbed, the flow ratios at the upstream arteriolar bifurcations are altered, thereby disturbing the cell distribution through the microcirculation and, thus, the tube and discharge hematocrits. Large-network models may be useful in demonstrating these effects, but at the present time, a detailed accounting of low capillary hematocrits is still elusive.

Closure

The arteriolar system regulates the division of the total cardiac output by controlling the resistance of each organ to blood-flow. This is accomplished by active diameter changes of the arterioles.

The arteriolar system also provides for uniform distribution of blood-flow to all parts of each muscle or organ. This uniform distribution is attributable in large measure to an arcade system of feeder arterioles from which the final generations of arterioles arise. This pressure in the arcade arterioles is relatively uniform and equal to 70% or more of central arterial pressure.

The flow in the arteries is governed by viscous and inertial forces and driven by the pressure gradient. There is a progressively decreasing importance of the inertial forces towards the arterioles and capillaries. In small arteries and arterioles, the flow is governed by a nonlinear diffusion equation reflecting a balance between vessel wall stress and viscous flow in the lumen. Nonlinear flow phenomena, similar to those seen in a Windkessel model are present.

The distribution of blood cells is not necessarily the same as the distribution of total blood at any bifurcation. A vessel receiving a smaller portion of total flow generally receives an even smaller fraction of the blood cells. This mechanism operates more strongly in the small arterioles and capillaries. Such an effect contributes to the low capillary hematocrits observed, especially in capillaries near stasis. Arterioles constrict rhythmically at most levels of autoregulation, leading to hematocrit fluctuations.

Acknowledgments

This work was supported by NSF grant DCB-88-19346 and US Public Health Service grants HL-10881 and HL-44147.

References

[1] McDonald, D. A. (1974) *Blood Flow in Arteries,* 2nd Ed., Williams and Wilkin, Baltimore, MD.

[2] Pedley, T. J. (1980) *The Fluid Mechanics of Large Blood Vessels,* Cambridge University Press, Cambridge, MA.

[3] Patel, D. J. and Vaishnav, R. N., eds. (1980) *Basic Hemodynamics and its Role in Disease Process,* University Park Press, Baltimore, MD.

[4] Fung, Y. C. (1984) *Biodynamics: Circulation,* Springer-Verlag, New York.

[5] Caro, C. G., Pedley, T. J., Schroter, R. C., and Seed, W. A. (1978) *Mechanics of the Circulation* Oxford University Press, Oxford.

[6] Bergel, D. H., ed. (1972) *Cardiovascular Fluid Dynamics,* Academic, New York.

[7] Burton, A. C. (1965) *Physiology and Biophysics of the Circulation,* Year Book Medical Publishers, Chicago, IL.

[8] Wetterer, E. and Kenner, T. (1968) *Grundlagen der Dynamik des Arterienpulses,* Springer-Verlag, Berlin.

[9] Guyton, A. C. (1963) *Handbook of Physiology, Section 2: Circulation, vol. 2* (Hamilton, W. F. and Down, P., eds.), American Physiological Society, Washinton, D.C.

[10] Kamm, R. D. (1987) *Handbook of Bioengineering* (Skalak R. and Chien, S., eds.), McGraw-Hill, New York, pp. 23.1-23.19.

[11] Wiedeman, M. P. (1984) *Handbook of Physiology, Section 2: The Cardiovascular System, vol. 4: Microcirculation* (Renkin, E. M. and Michel, C. C., eds.), Amierican Physiological Society, Betheseda, MD, pp. 11–40.

[12] Spalteholz, W. (1988) *Abh. Sächs. Ges. Wiss. Math. Phys. Kl.* **14,** 509–528.

13 Schmid-Schönbein, G. W., Firestone, G., and Zweifach, B. W. (1986) *Blood Vessels* **23,** 34–49.

14 Schmid-Schönbein, G. W., Skalak, T. C., and Firestone, G. (1987) *Microvasc. Res.* **34,** 385–393.

15 Lindbom, L (1984) *Acta Physiol. Scand.* **122,** 225–233.

16 Engelson, E. T., Skalak, T. C., and Schmid-Schönbein, G. W. (1985) *Microvasc. Res.* **30,** 29–44.

17 Hill, M. A., Simpson, B. E., and Meininger, G. A. (1990) *Microvasc. Res.* **39,** 349–363.

18 Myrhage, R. (1977) *Acta Orthop. Scand..* **168(Suppl.),** 1.

19 Frashier, W. A. and Wayland, H. (1972) *Microvasc. Res.* **4,** 62–76.

20 Sobin, S. S., Fung, Y. C., Lindal, R. G., Tremen, H. M., and Clark L (1980) *Microvasc. Res.* **19,** 217–233.

21 Yen, R. T., Zhuang, F. Y., Fung, Y. C., Tremer, H., and Sobin, S. S. (1984) *J. Biomech. Eng.* **106,** 131–136.

22 Sobin, S. S. and Kremer, H. M. (1966) *Fed. Proc.* **25,** 1744–1752.

23 Zweifach, B. W. and Lipowsky, H. H. (1984) *Handbook of Physiology,* Section 2: *The Cardiovascular System, vol. 4: Microcirculation* (Renkin, E. M. and Michel, C. C., eds.), American Physiological Society, Bethesda, MD, pp. 251–307.

24 Renkin, E. M. (1984) *Handbook of Physiology, Section 2: The Cardiovascular System, vol. 4: Microcirculation* (Renkin, E. M. and Michel, C. C., eds.), American Physiological Society, Bethesda, MD, pp. 627–687.

25 Schmid-Scönbein, G. W. and Chien, S. (1986) *Handbook of Hypertension, vol. 7: Pathophysiology of Hypertension: Cardiovascular Aspects* (Zanchetti, A. and Tarazi, R. C., eds.), Elsevier Science Publishers, Amsterdam, pp. 465–489.

26 Roy, J. W. and Mayrowitz, H. N. (1984) *Hypertension* **6,** 877–886.

27 Delano, F. A., Schmid-Schönbein, G. W., and Zweifach, B. W. (1990) *FASEB J.* **3,** 1383.

28 Wiederhielm, C. A., Woodbury, J. W., Kirk, S., and Rushmer, R. F. L. (1964) *Am. J. Physiol.* **207,** 173–176.

29 Intaglietta, M., Pawula, R. F., and Tompkins, W. R. (1970) *Microvasc. Res.* **2,** 212–220.

30 Skalak, T. C., Schmid-Schönbein, G. W., and Zweifach, B. W. (1984) *Microvasc. Res.* **28,** 95–112.

31 Fung, Y. C. (1981) *Biomechanics,* Springer-Verlag, New York, pp. 261–301.

32 Canfield, T. R. and Dobrin, P. B. (1987) *Handbook of Bioengineering* (Skalak, R. and Chien, S., eds.), McGraw-Hill, New York, pp. 16.1–16.28.

33 Bauer, R. D., Busse, R., and Schabert, A. (1982) *Biorheology* **19,** 409–424.

34 Bauer, R. D., Busse, R., and Schabert, A. (1985) *Pflügers Arch.* **403,** 308–311.

[35] Jackson, P. A. and Duling, B. R. (1989) *Am. J. Physiol.* **257**, H1147–H1155.

[36] Skalak, T. C. and Schmid-Schönbein, G. W. (1986) *J. Biomech. Eng.* **108**, 193–200.

[37] Stettler, J. C., Niederer, P., and Anlicker, M. (1987) *Handbook of Bioengineering* (Skalak , R. and Chien, S., eds.) McGraw-Hill, New York, 17.1–17.26.

[38] Schmid-Schönbein, G. W., Lee, S. Y., and Sutton, D. W. (1989) *Biorheogy* **26**, 215–227.

[39] Sutton, D. W. and Schmid-Schönbein, G. W. (1989) *Am. J. Physiol.* **257**, H1419–H1427.

[40] Lee, S. Y. and Schmid-Schönbein, G. W. (1990) *J. Biomech. Eng.* **112**, 437–443.

[41] Van Citters, R. L. (1966) *Circ. Res.* **18**, 199–204.

[42] Walmsley, J. G., Gore, R. W., Dacey, R. A., Damon, D. N., and Duling, B. R. (1982) *Microvasc. Res.* **24**, 249–271.

[43] Schmid-Schönbein, G. W. and Murakami, H. (1985) *Int. J. Microcirc. Clin. Exp.* **4**, 311–328.

[44] Palmer, A. A. (1967) *Bibl. Anat.* **9**, 300–303.

[45] Tangelder, G. J., Teirinck, H. C., Slaaf, D. W., and Reneman, R. S. (1985) *Am. J. Physiol.* **248**, H318–H323.

[46] Phibbs, R. H. (1966) *Am. J. Physiol.* **210**, 919–925.

[47] Nobis, U., Pries, A. R., Cokelet, G. R., and Gaehtgens, P. (1984) *Microvasc. Res.* **29**, 295–304.

[48] Goldsmith, H. L. and Spain, S. (1984) *Microvasc. Res.* **27**, 204–222.

[49] Goldsmith, H. L. and Karino, T. (1977) *Ann. N.Y. Acad. Sci.* **293**, 241–255.

[50] Gerrity, R. G. (1981) *Am. J. Pathol.* **103**, 181–190.

[51] Back, M. R. (1986) Master's Thesis, University of California, San Diego, CA.

[52] Schmid-Schönbein, G. W., Skalak, R., Usami, S., and Chien, S. (1980) *Microvasc. Res.* **19**, 18–44.

[53] Kranzow, G., Pries, A. R., and Gaehtgens, P. (1982) *Int. J. Microcirc. Clin. Exp.* **1**, 67–79.

[54] Øfjord, E. S. and Clausen, G. (1983) *Amer. J. Physiol.* **145**, H429–H436.

[55] Dellimore, J. W., Dunlop, M. J., and Canham, P. B. (1983) *Am. J. Physiol.* **244**, H635–H643.

[56] Klitzman, B. and Johnson, P. C. (1982) *Am. J. Physiol.* **242**, H211–H219

[57] Fenton, B.M, Carr, R. T., and Cokelet, G. R. (1985) *Microvasc. Res.* **29**, 103–126.

[58] Chien, S., Tvetenstrand, C. D., Farrell-Epstein, M., and Schmid-Schönbein, G. W. (1986) *Am. J. Physiol.* **248**, H568–H576.

[59] Perkkiö, J., Hokkanen, J., and Keskinen, R. (1986) *Med. Phys.* **13**, 882–886.

[60] Pries, A. R., Key, K., Claassen, M., and Gaehtgens, P. (1989) *Microvasc. Res.* **38**, 81–101.

[61] Rong, F. W. and Carr, R. T. (1990) *Microvasc. Res.* **39,** 186–202.

[62] Fung, Y. C. (1973) *Microvasc. Res.* **5,** 34–48.

[63] Blixt, A., Braide, M., Myrhage, R., and Bagge, U. (1987) *Int. J. Microcirc. Clin. Exp.* **6,** 273–286.

[64] Ley, K., Meyer, J. U., Intaglietta, M., and Arfors, K. E. (1989) *Am. J. Physiol.* **256,** H85–H93.

[65] Meyer, U., Borgstrom, P., Lindbom, L, and Intaglietta, M. (1988) *Microvasc. Res.* **35,** 193–203.

[66] Wetter, T., Schmid-Schönbein, H., Johnson, P. C., and Klitzman, B. (1981) *Bibl. Anat.* **20,** 237–241.

[67] Dejardins, C. and Duling, B. R. (1987) *Am. J. Physiol.* **257,** H494–H503.

Chapter 20

Autoregulation and Resistance-Artery Function

Gerald A. Meininger, Jeff C. Falcone, and Michael A. Hill

Introduction

Autoregulation is defined as the tendency for blood flow to remain constant despite fluctuations in arterial perfusion pressure. This phenomenon is a local property of the vasculature that exists in the absence of neural influences and humoral factors.[1] The efficiency of autoregulation varies from tissue to tissue, with those organs most vital for survival, e.g., brain, heart, and kidney, demonstrating the most marked autoregulatory capacity. Autoregulatory mechanisms conceivably also contribute to the local control of other circulatory parameters, such as capillary pressure, fluid exchange, and vascular resistance. The vascular elements responsible for these autoregulatory events are considered to reside in the arterial section of the circulation, in particular, in those vessels responsible for the largest fraction of tissue vascular resistance. Traditionally, the terminal arterioles of the microcirculation have been viewed as the principal site of these local vasoregulatory events. However, more recent studies aimed at quantitating the distribu-

From *The Resistance Vasculature*, J. A. Bevan et al., eds. ©1991 Humana Press

tion of vascular resistance have suggested that small resistance arteries proximal to the microcirculation may also be important sites of local regulation. This chapter represents our attempt to summarize some of the issues that may be important to consider regarding autoregulation and resistance-artery function. This topic has been largely neglected, compared to the attention given the role of microcirculation in autoregulation.

Local Regulatory Mechanisms

A number of local mechanisms have been proposed to account for the homeostatic properties of autoregulation. Three principal mechanisms of intrinsic vascular control include metabolic, myogenic, and flow-dependent regulation. Other, more specialized mechanisms include the tissue pressure hypothesis[2] and renal tubuloglomerular feedback.[3] Acting alone or in concert, these mechanisms are believed to form the basis for reactive and functional hyperemia and blood-flow autoregulation. The three main autoregulatory mechanisms are briefly defined and considered in more detail below.

Metabolic Mechanism

The metabolic hypothesis for local blood-flow regulation suggests that the resistance of arterial vessels is modulated by direct or indirect effects of oxygen availability. Thus, in the face of declining oxygen levels or increasing concentrations of vasoactive metabolites (e.g., lactate, adenosine, K^+), arterial resistance decreases with a resultant increase in blood flow. The increased blood flow acts to either promote the washout of vasoactive substances accumulating in the interstitium or correct the deficit in nutrient supply. Restoration of resting metabolic conditions removes the vasoactive stimulus and returns vascular resistance towards normal.

Myogenic Mechanism

"Myogenic regulation" refers to the ability of arterial vessels to respond to alterations in intravascular pressure in such a way that at increased transmural pressure there is a tendency for resistance vessels to actively constrict, whereas at decreased pressure these vessels dilate. In the case of an increase in arterial pressure, a myogenic vasoconstriction would act to limit hyperperfusion and transmission of the elevated pressure into the exchange vessels. The sensed variable and feedback signals that trigger and regulate the myogenic response are, at present, uncertain. It is, however, unlikely that the controlled variable is simply vascular diameter or blood flow, since, during a maintained myogenic constriction, the diameter and flow in many vessels will not return to basal values. Some evidence exists to support the propsition that mechanical tension within the vascular wall may be the controlled factor,[4,5] whereas other investigators have suggested that the controlled variable may be either wall stress[6] or vascular smooth muscle membrane tension.[7]

Flow-Dependent Mechanism

The third major regulatory mechanism is that of flow-dependent regulation. It has been suggested that alterations in blood-flow velocity within a vessel are coupled to changes in that vessel's caliber.[8] Such adjustments in vascular diameter may be attributable to vasoactive substances (e.g., endothelial-dependent relaxing factor, eicosanoids) released from the vessel wall as a result of alterations in shear stress. The endothelial-derived vasoactive substances released from larger resistance vessels, following an increase in shear stress, could act both as paracrine factors (acting locally) and as endocrine factors, being released upstream and then acting on downstream microvascular sites. Thus, larger resistance vessels could in effect act as "endocrine vessels." In support of this latter concept, it has been shown in bioassay studies that endothelial factors released from isolated arterial vessels can exert vasoactive influences on downstream endothelium-denuded arterial segments.[9] Thus, the endothelial factors are transferable. In an elegant set of recent experiments, Kon et al. have provided some intriguing in vivo support for this concept.[10] They demonstrated that acetylcholine infused into the proximal renal artery increased the glomerular filtration rate and renal plasma flow. These renal effects were abolished by endothelial denudation of the renal artery, suggesting that the endothelium of the renal artery was producing vasoactive factors capable of acting on the downstream renal microvasculature (Fig. 1).

Mechanism Interaction

A key question that is yet to be fully resolved is whether local regulatory mechanisms act together to regulate a single microcirculatory variable (e.g., local blood flow or capillary pressure) or whether each mechanism of local regulation is selectively recruited for the control of separate microvascular parameters. Although it is evident under certain conditions (e.g., increased or decreased arterial pressure) that metabolic and myogenic mechanisms can synergistically act to autoregulate blood flow, it is also apparent that these regulatory mechanisms can be forced into opposition. An illustration of this is the vasoconstrictor response elicited following elevation of venous pressure. In this situation a predominant metabolic mechanism would trigger vasodilation, whereas the predominance of the myogenic or the flow-dependent mechanism would cause net vasoconstriction. Under other conditions it appears that one mechanism can selectively operate to override another. For example, in resting skeletal muscle, a pressure-induced myogenic constriction of arterioles can be maintained despite marked decreases in local blood flow (Fig. 2) and tissue oxygen content (Fig. 3). Thus, the myogenic mechanism can, under resting conditions, act as a more dominant form of local control in skeletal muscle.

On the basis of such arguments regarding selectivity, it is possible to conceive of a control scheme to describe local vasoregulation that involves recruitment of the metabolic and myogenic mechanisms for control of different variables. For example, the metabolic mechanism may act principally to regulate blood flow to match metabolic needs, and the myogenic mechanism may relate more to the control of intravascular pressure in exchange vessels. Accordingly, one might expect that at low perfusion pressures metabolic mechanisms might predominate to protect against tissue hypoxia and relative hypoperfusion, whereas, at high perfusion pressure,

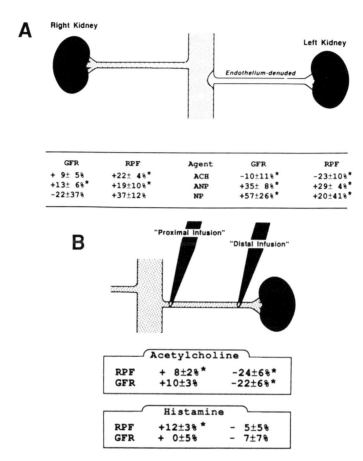

Fig. 1. Demonstration that an upstream arterial vessel can function as an "endocrine" source of vasoactive substances, which can exert an effect in the more distal segments of the arteriolar circulation.[10] Panel **A** provides data suggesting that the endothelial cells of the renal artery produce vasodilator factors, in response to acetylcholine, (ACH), that lower intrarenal vascular resistance, as assessed by the glomerular filtration rate (GFR) and renal plasma flow (RPF). Panel **B** demonstrates that arterial administration of endothelial-dependent vasodilators at the distal end of the renal artery does not lower renal vascular resistance. This further supports the suggestion that the endothelial cells of the renal artery itself are required for the production of a vasodilator substance. In both panels, results are expressed as percent change in GFR and RPF. ANP, atrial natriuretic peptide; NP, nitroprusside. (Reproduced from the *Journal of Clinical Investigation*, 1990, **85**, 1728–1733, by copyright permission of the American Society for Clinical Investigation.)

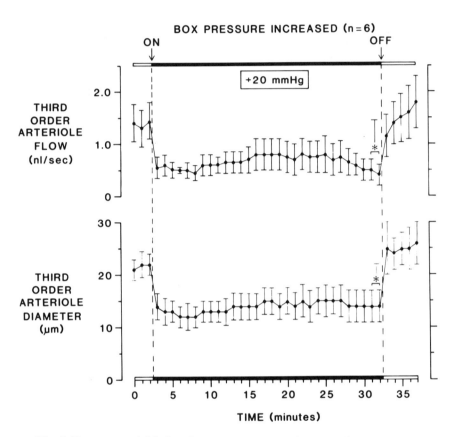

Fig. 2. Response of third-order cremaster muscle arterioles to a sustained (30-min) 20-mmHg increase in intravascular pressure. The myogenic vasoconstriction (lower panel) is maintained well during the pressure step, despite a marked reduction in arteriolar blood flow (upper panel). These data indicate that in resting skeletal muscle the myogenic mechanisms dominates metabolic influences. (Reproduced from *Circulation Research*, 1987, **60(6)**, 861–870, by permission of the American Heart Association.)

the myogenic mechanism would dominate local control to protect against excess fluid filtration and edema. In such a scheme, the flow-dependent mechanism would be recruited to act as a secondary control mechanism. It would not become involved unless the metabolic or the myogenic mechanism precipitated an alteration in vascular resistance that would cause flow to change in vessels

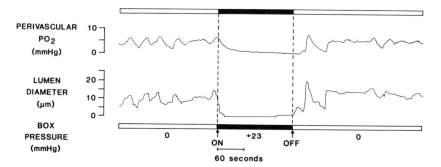

Fig. 3. Example tracing of a third-order cremaster muscle arteriole, which demonstrates that a myogenic vasoconstriction can be maintained despite a marked fall in perivascular oxygen levels. In support of Fig. 2, this data indicates that under resting conditions the myogenic response overrides metabolic factors. (Meininger and Bohlen, unpublished data.)

sensitive to shear stress. Once recruited, the flow-dependent mechanism would reinforce either a vasoconstrictive or a vasodilatory response, since decreases in flow would stimulate constriction, and increases in flow, dilation.

Vascular Sites Involved in Autoregulation

An established view is that the primary vascular elements contributing to autoregulatory phenomena are the precapillary arterioles. This view has been based largely on studies demonstrating that these vessels constitute a major site of vascular resistance and exhibit marked reactivity to a variety of vasoactive stimuli. However, direct measurements of intravascular pressure in a number of tissues have suggested that a substantial fraction of vascular resistance can reside upstream from the vessels that are classically considered as the microcirculation (i.e., arterial vessels >100 μm in diameter). For example, in brain,[11] heart,[12] small intestine,[13] and certain skeletal muscles,[14,15] some 30–50% of vascular resistance is reported to reside in such arterial vessels. It is, therefore, important to establish whether the resistance of the larger arterioles is accessible to local regulatory mechanisms or whether it represents an element of structural resistance caused by such factors as path length. In support of a possible role in local regulation, in vitro studies have demonstrated that these larger vessels are capable of exhibiting spontaneous tone and myogenic reactivity. This has been shown in middle cerebral vessels (approx 400μm) of the cat,[16] and renal interlobular arterioles (approx 125 μm)[17] and the first-order arteriole (approx 120μm) of the rat cremaster muscle.[18]

There is often use of the term resistance artery without due care being given to the location of the vessel within the intact vascular network and to whether the vessel truly occupies a site of high resistance within the network. Several criteria should be used in identifying sites of vascular resistance and in their consideration as sites of resistance control. A standard technique for identification of sites of vascular resistance has been to localize within a network the vascular sites responsible for the major pressure drops. However, it has become apparent that caution should be exercised in ascribing a resistance role to the larger arterioles and small arteries based only on in vivo servo-null pressure measurements or calculations of resistance without careful attention being given to the vascular design of the network. It is evident that the surgical preparation required for microvascular study of some tissues compromises the local circulation by disrupting arterial supply vessels. This has been shown to result in both a fall in pressure in remaining feed vessels and overestimation of the upstream component of vascular resistance. A particular example of this is the rat cremaster muscle, in which the standard preparation involves ligation of one of two major supply arterioles and extensive disruption of the arcade network.[19,20] As such, in networks with multiple feed vessels and network interconnections, it is difficult to define vascular resistance in a meaningful manner based on the pressure distributions.

An alternative to the pressure-distribution approach for arcade networks is to use power dissipation as a measure of resistance within the vascular network.[21] Power dissipation does not require assumptions to be made about the network structure. This approach is based on identifying the vascular sites of major energy loss as blood flows through a network. As implied above, another important factor to assess is whether the suspected resistance vessels contain a significant amount of vascular tone and are capable of exhibiting vasoreactivity. Thus, in the intact network, verification of both the existence of a pressure drop and the presence of significant vascular tone should increase the reliability of the estimates of resistance distribution.

Coupling Activity of Resistance Arteries and Arterioles During Autoregulation

As discussed above, a role for small resistance arteries in autoregulation is attractive especially in vascular networks, in which they contribute significantly to overall network resistance. In fact, this appears to be the most compelling argument favoring a role for these vessels in autoregulation. This can be illustrated by considering a vascular network in which 50% of the vascular resistance resides in the arterial segment proximal to the microcirculation. In such a network it can be calculated that reducing microvascular resistance to zero without involving the small resistance arteries can bring about only a twofold increase in blood flow. In comparison to these theoretical calculations, evidence exists in skeletal muscle to indicate that during exercise there may be increases in blood flow that typically reach four- to sixfold above basal levels. Thus, a reduction in resistance involving the microcirculation alone would be inadequate to account for blood-flow increases of this magnitude. This reasoning has been used to support the necessity for small resistance arteries located proximal to the microcirculation to be involved in autoregulation.[15,22–24] Arguments concerning a role of small arteries in autoregulation for those organs in which these vessels do not contain a significant fraction of the overall network resistance are not as convincing.

A conceptual problem facing the inclusion of resistance arteries into a model of global network behavior during autoregulation is that these vessels, as a class, are not necessarily located in close proximity to the tissue parenchymal cells. Thus, metabolic or other tissue-derived vasoactive signals would not necessarily be free to directly influence the behavior of these vessels. This spatial separation from the more active terminal arterioles invites the question of how these two vascular segments coordinate their activity during autoregulation. Several schemes have been propsed to explain how small-artery vascular behavior could be enlisted and coordinated with the vascular events occurring in the terminal arterioles. These schemes include (1) propagated vasodilation or vasoconstriction; (2) vein-to-artery diffusion of vasoactive substances; and (3) series-distributed vascular control. Experimental evidence exists to supprt some of these control schemes, whereas others remain yet to be verified.

Propagated Vasodilation and Vasoconstriction

The concept that vasomotor respnses could be conducted longitudinally along the arterial tree has been proposed as a possible mechanism to coordinate small artery and microvessel behavior. Propagated or conducted vasomotor responses have been described in both large[25-27] and small arterial vessels.[28] Muscle contraction in the innervated and denervated dog limb has been shown to result in dilation of the femoral artery.[25,27] An ascending vasodilation of the femoral artery has also been induced by localized injections (downstream of the femoral artery) of acetylcholine, histamine, and bradykinin.[26] This dilation could be blocked with cocaine. In the microcirculation of hamster cheek pouch, microiontophoretic application of acetylcholine was observed to produce a localized vasodilation that spread upstream and downstream from the point of application.[28] In more recent studies,[23,24,29] application of acetylcholine to the distal vasculature of the hamster cremaster muscle was reported to produce a dilation that spread upstream to the large feeder arterioles (Fig. 4). According to this concept, vascular dilation responses initiated in the terminal portion of the vascular bed by autoregulatory events are conducted along the arterial wall via cell-to-cell electrical coupling.

The conduction would act to spread the vasomotor changes throughout the vascular bed and to upstream resistance segments. Thus, a vasodilation initiated in a terminal arteriole could spread upstream to recruit small arteries.

Vein-to-Artery Diffusion of Vasoactive Factors

Another mechanism that may be involved in coordinating vascular responses is chemical communication between veins

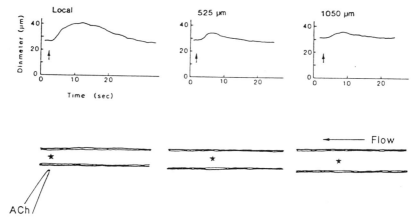

Fig. 4. Local application of acetylcholine to a hamster cheek pouch arteriole results in a local dilation that is propagated upstream from the site of
application. Arrows in each panel indicate time of drug application; stars
indicate sites of observation. (Reproduced from *Circulation Research*, 1987,
61(Suppl. II), II-20–II-25, by permission of the American Heart Association.)

and nearby arteries. An attractive mechanism by which this communication could occur is through diffusion of vasoactive substances from the vein to an adjacent artery. Most blood vessels of
the body are arranged in parallel artery–vein pairs. This relationship is found especially in small arteries and veins and extends
into the microcirculation to include large arterioles and venules. In
addition to the vessels being paired in parallel, other examples of
close spatial arrangements include arteries and veins twining
around one another and smaller arteriole branches that cross or
come into close contact with a venule of similar or higher branch
order (i.e., a third-order arteriole crossing a first-order venule).[23,24,30,31] Tigno et al. recently reported that approx 62% of the
venules draining a specific section of capillaries in the mesenteric
vasculature cross the arteriole supplying those same capillaries.[32]
Therefore, based on the anatomic layout, the spatial characteristics
necessary to support diffusion between the arterial and venous sides
of the circulation exist.

Several types of substances are likely candidates for this diffusional communication scheme. Examples include the dissolved
gases involved in tissue respiration. These gases, such as oxygen
and carbon dioxide, are extremely lipid-soluble, suggesting that

they could readily cross the vascular wall. Early investigations tested various hypotheses related to these diffusional possibilities. Kampp et al. designed studies that specifically investigated the countercurrent movement of oxygen across the vasculature.[33] The experiments examined whether oxygen shunting existed between arteries and veins by comparing the time until appearance of oxygen in veins with both oxygenated hemoglobin and methemoglobin (metHb) red cells after their simultaneous injection into the superior mesenteric artery. Their findings showed that oxygen appeared 1–2 s earlier in the intestinal vein than the red cells containing methemoglobin. From this data, Kampp et al. concluded that oxygen was diffusing from artery to vein, short-circuiting the capillaries.

With oxygen-sensitive microelectrodes,[34] studies have been able to look more specifically at oxygen and the possibility of countercurrent diffusional shunting.[35–37] Oxygen-sensitive microelectrodes have been utilized by Lash and Bohlen to measure perivascular and tissue oxygen tension during contraction of the rat spinotrapezius muscle.[37] Their findings demonstrated that periarteriolar and tissue P_{O_2} decreased at the onset of skeletal-muscle contraction but recovered rapidly during the stimulation period (Figs. 5 A,B). In comparison, perivenular oxygen tension decreased and remained low throughout the contraction period (Fig. 5B). Concurrent with these changes, arteriolar diameter increased in response to the contraction-induced hyperemia and remained dilated throughout the stimulation period, recovering only after the muscular contraction ceased. Thus, the arteriolar vasodilation was more closely correlated to the perivenous changes in oxygen tension.

Another observation made in this study was that resting perivenous oxygen tension was higher than capillary or tissue oxygen tension. This suggested that the venule might be gaining oxygen—perhaps by diffusion from arterioles, which would act as an oxygen source. As a corollary, Duling[35] has observed a longitudinal periarteriolar oxygen gradient that indicates diffusional loss of oxygen along the arteriolar tree. Thus, one possible control scheme could involve diffusional movement of oxygen. During increased metabolic activity, increased oxygen consumption and higher blood

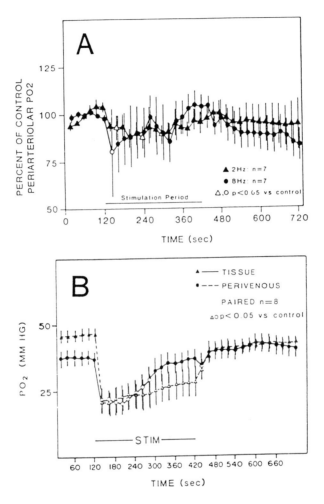

Fig. 5. Work from Lash and Bohlen.[37] Panel **A** illustrates periarteriolar P_{O_2} (mean ± SEM) during stimulation of skeletal muscle at 2 and 8 Hz. Open symbols represent significant ($p < 0.05$) decreases in P_{O_2} relative to the resting state. During the fourth and fifth minute of stimulation, periarteriolar P_{O_2} was not significantly different from control. Panel **B** shows tissue/capillary bed (triangles) and perivenular (circles) P_{O_2} during 12-Hz skeletal-muscle stimulation. Open symbols represent significant ($p < 0.05$) decreases in P_{O_2} relative to the resting state. Note that capillary bed P_{O_2} recovered to control levels during stimulation whereas perivenous P_{O_2} remained decreased from control. (Reproduced from the *American Journal of Physiology*, 1987, **252**, H1192–H1202, by permission of the American Physiological Society.)

flow would reduce venular oxygen tension. The decrease in venular oxygen could then act to precipitate dilation of upstream small arteries that are in close proximity to the veins.

It is also possible that metabolic or endocrine factors other than gases are free to move from vein to artery. Hester[31] examined the hypothesis that increases in venous adenosine arising from tissue metabolic activity could diffuse from venules to cause arteriolar vasodilation. In support of this idea, he found that adenosine infused into venules caused dilation in adjacent arterioles and in other arterioles at points of crossing the infused venule (Fig. 6). Recently, Tigno et al. infused norepinephrine into precapillary arterioles of the rat mesentery.[32] In arterioles upstream from the site of infusion, they observed vasoconstriction at locations where these arterioles were crossed by venules that drained the region supplied by the infused arteriole. The arteriolar constriction was observed to propagate in both upstream and downstream directions and across arterio-arteriolar arcades. Therefore, these studies suggest that the accumulation of vasoactive tissue metabolites, such as aden-osine, or neurohumoral factors in the venous blood could lead to vasomotor changes in upstream resistance vessels. This mechanism could provide a means for feedback control in which vascular resis-tance of the larger arterioles and small arteries is coupled to downstream tissue metabolism.

Another possibility that has been considered is that venules may contain a sensor mechanism for monitoring venous blood chemisty and/or the parenchymal conditions of their immediate surroundings. The vein would then release vasoactive factors that result in either arteriolar vasoconstriction (e.g., to low levels of metabolites and/or increased nutrients) or vasodilation (e.g., to high levels of metabolites and/or decreased nutrients). Falcone and Bohlen investigated this idea in the intestinal and skeletal-muscle circulations.[30] In their study, they considered the venule as a sensor and hypthesized that the venules might release endothelium-derived relaxing factor (EDRF) in response to changes in venous blood chemistry. The EDRF would then diffuse from the venule to produce dilation of paired arterioles. This was tested by localized iontophoretic application of acetylcholine, an endothelium-dependent dilator, onto a large venule. They observed an arteri-

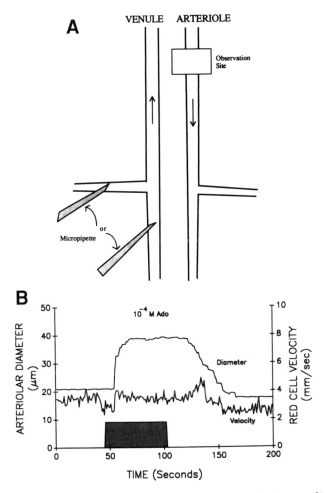

Fig. 6. Panel **A** depicts the experimental scheme that Hester[31] used for infusing adenosine into a venule while observing the diameter of a nearby arteriole. The arrows denote the normal direction of blood flow. Panel **B** illustrates arteriolar diameter and red cell velocity during 1 min of venular perfusion with $10^{-4}M$ adenosine. The top trace represents diameter and the lower is velocity. The shaded square denotes the perfusion period. The venular–arteriolar distance was 35 μm and the distance from the micropipet to the observation site was 495 μm. During the infusion of the adenosine there was a significant vasodilation of the adjacent arteriole, apparently caused by vein-to-artery diffusion of the infused adenosine. (Reproduced from the *American Journal of Physiology,* 1990, **258,** H1918–H1924, by permission of the American Physiological Society.)

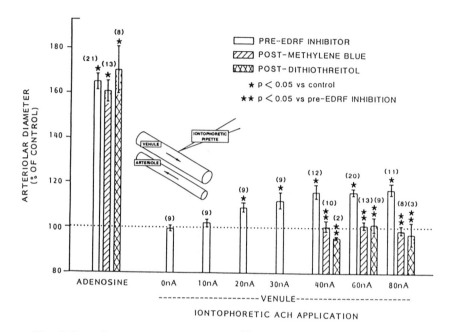

Fig. 7. Data from Falcone and Bohlen[30] indicating that venular release of EDRF may act to dilate paired arterioles. Maximal dilation of arterioles with adenosine (left-most columns), an endothelium-independent dilator, was unaffected by EDRF inhibition with methylene blue (hatched) or dithiothreitol (cross-hatched). In comparison, localized iontophoretic application of acetylcholine, an endothelium-dependent dilator, to the distal side of a paired large venule (*see inset*) induced a concentration-dependent arteriolar dilation. The dilation was abolished by the EDRF inhibitors methylene blue and dithiothreitol. The arteriolar dilation could not be accounted for by tissue diffusion of acetylcholine through the tissue to the arteriole. (Reproduced from the *American Journal of Physiology*, 1990, **258**, H1515–H1523, by permission of the American Physiological Society.)

olar dilation that could be inhibited by either methylene blue or dithiothreitol, two EDRF-inhibiting agents (Fig. 7). The dilation could not be accounted for by simple diffusion of acetylcholine from the site of venular application, by local neural reflexes, or by local generation of prostaglandins. Thus, their studies suggest that venules may have an endocrine influence on the behavior of resistance arteries through release of vasoactive factors.

Series-Distributed Vascular Control

Another possible scheme that could act to coordinate vascular regulation is through a series-coupled arrangement of vasoregulatory mechanisms along the arterial tree. This series distribution of vascular control could involve a single vascular control modality or several control modalities. Johnson[38] and Lang and Johnson[39] proposed that autoregulation could be accounted for on the basis of series distribution of the myogenic mechanism. In this scheme it is suggested that upstream arteriolar segments are more sensitive to small alterations in intravascular pressure than are the more distal arterioles. Thus, as arterial pressure is reduced, the larger arterioles respond by dilating, which acts to preserve intravascular pressure in the distal arterioles. A myogenic vasodilation would therefore not be seen in the smaller arterioles unless arterial pressure was reduced to a level at which the necessary compensation exceeded the dilation capacity of the upstream vessels. The authors suggested this coupling mechanism based on their observations of cat mesentery and Oien and Aukland's studies of renal blood flow.[40] This hypothesis does not appear to be applicable to all tissues, since studies of rat skeletal muscle or bat wing indicated that small arterioles (<20 μm) respond more rapidly than large arterioles following small alterations in intravascular pressure.[6,41]

Another type of series-distributed control scheme has been theoretically considered by Granger et al.[42–45] using a mathematical model of a vascular network in which the metabolic, myogenic, and flow-dependent mechanisms were differentially distributed along the arterial tree. In the modeled vascular network, metabolic control was located in the most terminal portion of the microvascular bed, myogenic control in intermediate-sized arterioles upstream from the terminal vessels, and flow-dependent control in the most proximal arterial segment. In a vascular bed with this arrangement, an increase in metabolic rate or tissue hypoxia would lead to metabolic vasodilation of the terminal arterioles. This vasodilation would reduce upstream intravascular pressure, providing a stimulus for myogenic vasodilation of the intermediate-sized vessels. At the same time, the fall in terminal arteriolar resistance would increase blood flow, which would act as the stimulus to di-

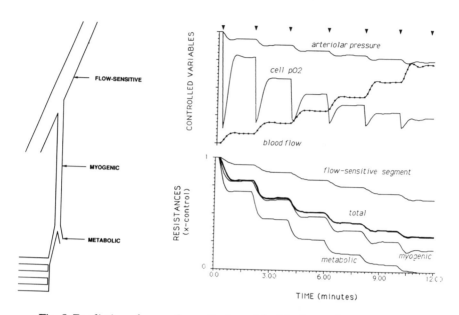

Fig. 8. Predictions from a theoretical model of the hemodynamic results that would occur following an increase in tissue metabolic rate for an arterial network in which terminal vessels are under metabolic control, intermediate-sized vessels are under myogenic control, and larger proximal vessels are under flow-dependent control. Following each step increase in metabolic rate (doubling; arrows), tissue oxygen tension (upper trace; cell P_{O_2}) falls, leading to a decrease in resistance for the metabolically controlled terminal arterioles (lower trace). The fall in resistance in this segment causes intravascular pressure (upper trace) upstream in the myogenically active arterial segment to fall, leading to a decrease in the resistance of this segment (lower trace). The fall in resistance of the metabolic segment also causes blood flow to increase (upper trace) through the vessels of the proximal arterial segment that are controlled by a flow-sensitive mechanism, which leads to reduced resistance in this segment (lower trace). Thus, the metabolic dilation is effectively able to recruit additional vasodilation from upstream arterial segments. (Unpublished results, courtesy H. J. Granger.)

late the flow-sensitive segment of the arterial tree (Fig. 8). According to this scheme, the proximal resistance-artery vessels are coupled to the metabolic segment by means of physical factors (i.e., changes in intravascular pressure and blood flow). Experimental verification of this control scheme will require demonstration of the localized predominance of these mechanisms along the arterial tree.

In summary, there are a number of possible mechanisms that could act to involve resistance arteries in blood-flow autoregulation. It would be premature to speculate on which of the control schemes discussed, if any, are physiologically significant until such time as experimental evidence becomes available to establish their importance. Understanding the functional integration of the vascular network will help to provide new insight into the phenomenon of autoregulation of blood flow and the role of resistance arteries.

Significance of Autoregulation in Resistance-Artery Control: Interaction with Neurohumoral Mechanisms

As discussed above, there are numerous local regulatory mechanisms that are important to tissue homeostasis. The operation of these local mechanisms during control of the systemic circulation is often difficult to discern because of the vasomotor influence of neurohumoral regulation. Nonetheless, these local vasoregulatory mechanisms operate in conjunction with neural and hormonal mechanisms of cardiovascular control. Consequently, the resistance vasculature plays a dual physiological role and is governed by the balance between these two groups of control mechanisms. Assessing the impact of autoregulation on this balance is important to establish the physiological significance of autoregulation in cardiovascular control and to develop a unified concept of resistance-vessel control. A goal of the remainder of this chapter will be to discuss some of the various forms that an interaction between local and neurohumoral control might take.

There are numerous types of interplay that can be envisioned to develop between local and neurohumoral control. In general, the quantitative contribution of autoregulation to changes in vascular resistance can be based on how neurohumoral factors affect the stimuli for autoregulation and/or the autoregulatory process itself. Consequently, alterations in local arterial pressure, blood flow, and/or autoregulatory gain form a convenient categorical basis for discussing the interplay between local and neurohumoral mechanisms of vascular control.

Pressure-Mediated Interactions

Pressure is a physical force that acts to distend the vascular wall thereby increasing wall tension or stress. Several explanations have been advanced that could account for the ability of this force to modify the vascular effects of neural or circulating vasoconstrictors. These include (1) pressure-induced shifts along the length–tension curve, (2) length-dependent sensitivity changes, and (3) alterations in myogenic vascular tone. The basis for each of these possibilities appears to be in the ability of pressure as a mechanical force to act as a determinant of vascular-smooth-muscle function.

In studies of the reactivity of small frog mesenteric arteries and arterioles the contractile responsiveness to norepinephrine was shown to be maximized when wall stress was within an optimal range.[46] Accordingly, increases in intravascular pressure would augment responsiveness in resistance vessels, which are normally understressed, and would decrease responsiveness in resistance vessels, which are normally within the optimal range or over-stressed. Presumably, these observations are related to the length–tension curve. Alterations in length of vascular smooth-muscle also appear to affect the sensitivity of vascular smooth muscle to contractile agonists. Price et al.[47,48] found that arterial rings of the dog tibial artery showed a progressive increase in sensitivity to norepinephrine as the preparation circumference was increased. The sensitivity continued to increase as length was increased and did not demonstrate a maximum that could be correlated to the length for maximum force development. Thus, this relationship appears to depend on a mechanism other than that responsible for the length–tension relationship. In addition to altering these mechanical properties, changes in pressure also have the ability to activate vascular smooth muscle via the myogenic mechanism.[1] The contribution of the myogenic response to vascular tone will be dependent on the intrinsic myogenic responsiveness of the vasculature and proportional to intravascular pressure. Therefore, any time intravascular pressure is altered by neural or circulating vasoconstrictors, there should be a predictable change in myogenic vascular tone.

Examples demonstrating a role for pressure as a determinant of vascular resistance have been described in previous studies of animals with experimental hypertension[49,50] or by infusion of vaso-constrictors.[51-55] In these studies, a significant fraction of the vasoconstriction accompanying hypertension or vasoconstrictor infusion was dependent on the increase in arterial pressure. When the vasculature was selectively protected from the increase in arterial pressure so local perfusion pressures were maintained at normal levels, there was a significant reduction in local vasoconstriction (e.g., Fig. 9). Under some conditions it is possible to demonstrate that this pressure-dependent mechanism is capable of accounting for all of the change in local vascular resistance during changes in arterial pressure. This was apparent in experiments in which a carefully defined dose of phenylephrine, which was subthreshold for intestinal vasoconstriction, was infused into the systemic circulation. This same dose, however, was capable of elevating mean arterial pressure and causing constriction in skeletal muscle. Under these conditions, it was found that intestinal vasoconstriction occurred solely as a result of an autoregulatory response to the rise in systemic arterial pressure.[52,53] Collectively, these studies have been interpreted to indicate that a pressure-dependent form of autoregulation plays an important role as a determinant of vascular resistance.

Flow-Mediated Interactions

Flow is an additional variable that should be considered in any analysis of the interaction between autoregulation and neurohumoral factors. Conceivably, flow could influence vascular tone by altering the delivery or washdown of diffusible metabolic factors that are vasoactive. Also, flow-induced shear stress should be considered as an additional physical force that can act on the endothelium of the vascular wall through a flow-mediated mechanism. The ability to predict which of these two types of local control act to determine the outcome of an interaction would depend on knowledge of the predominant mechanism.

In situations in which neural or circulating vasoconstrictors increase peripheral vascular resistance and reduce local blood flow, the decrease in flow would result in reduced oxygen delivery and/or reduced washout of vasodilator substances and metabolites.

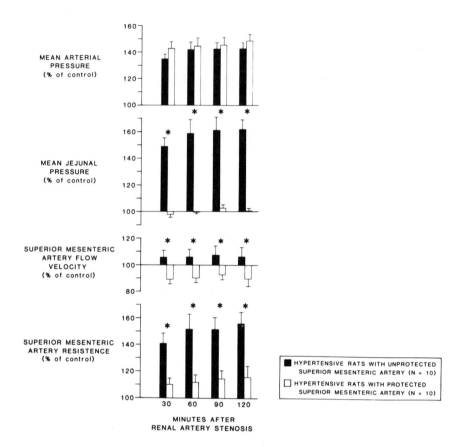

Fig. 9. Renal hypertension was induced in a group of rats by acutely reducing renal blood flow by 50%. After 2 h, mean arterial pressure and intestinal vascular resistance were increased by approx 40 and 50%, respectively
(black bars). A vascular occluder around the superior mesenteric artery was
periodically inflated (30, 60, 90, and 120 min) to restore normal intestinal
perfusion pressure (white bars). This maneuver prevented transmission of
the elevated arterial pressure into the intestinal circulation. Protection of the
intestinal vasculature from the elevated intravascular pressure significantly
reduced the vascular resistance compared to that present when local pressure was elevated. The data suggest that pressure acts as a stimulus to increase the intensity of the vasoconstriction associated with acute hypertension. Data are mean ± SEM; indicates significant ($p < 0.05$) difference
between unprotected and protected conditions. (Reproduced from *Hypertension*, 1985, **7(3)**, 364–373, by permission of the American Heart Association.)

Thus, a net dilation that would counteract the vasoconstriction would be generated locally. This locally derived dilator influence could modulate vasoconstriction throughout the network if the dilation signals were spread by propagation[23,24,29] or through countercurrent diffusion from veins to parallel resistance arteries.[30,31] Regardless, this mode of interplay would reduce the effectiveness of the neural or humoral vasoconstriction. In contrast, the reduction in flow through vessels with a flow-dependent control mechanism would lead to further vasoconstriction of the vessels. As such, this process would act in a synergistic fashion with the neural and circulating vasoconstrictors. The net result would be an amplification of the vasoconstriction.

A flow-mediated autoregulatory process based on the delivery/washdown of vasoactive factors has been proposed as a mechanism of vasoconstriction in hypertension resulting from volume expansion.[56-58] This theory, called the "whole-body autoregulation theory," proposes that during volume expansion the initiating hemodynamic event is an elevated cardiac output that leads to tissue overperfusion. The overperfusion in turn causes vasoconstriction. The increase in vascular resistance then reduces cardiac output back to normal levels. A selective overperfusion of organs could also occur if the initial increases in vascular resistance that occur during the development of hypertension are not uniformly distributed. As a result, arterial pressure would increase proportionally more than vascular resistance for some organs. Consequently, these organs would experience overperfusion and flow could act to produce vasoconstriction in these organs.

Gain-Mediated Interactions

Gain-mediated interactions are envisioned to occur if neural or hormonal vasoconstrictors are capable of altering the intrinsic vasoregulatory properties of the vasculature. In this type of interplay, the contribution of autoregulation would be amplified if neurohumoral agonists increased the efficiency of autoregulation or be diminished if autoregulatory capacity is reduced. The importance of vasoregulation in establishing basal vascular tone emphasizes the need to understand how these local mechanisms are modulated.

Recent evidence indicates that myogenic reactivity can be influenced by a number of factors. In large skeletal-muscle arterioles (100–120 μm in diameter), preconstriction to varying degrees with norepinephrine or with selective α_1 and α_2 agonists significantly enhanced the myogenic reponsiveness of the vessels to changes in intravascular pressure.[51,55,59,60] Thus, the adrenergic tone acted to amplify myogenic vasoconstrictor responses. In comparison, these studies indicated that α_2-mediated tone was more susceptible to inhibition by myogenic withdrawal of tone during decreases in pressure. An enhanced responsiveness could not be duplicated by constriction to similar degrees with potassium chloride, but myogenic responses were facilitated when vascular tone was increased with Bay K8644, thromboxane, indolactam, or cyclooxygenase inhibition.[59,61,62] Enhanced blood-flow autoregulation has also been reported for the cat skeletal muscle following sympathetic-nerve stimulation[63] and in dog skeletal muscle following norepinephrine infusion.[64] The ability of various agonists to interact with the myogenic mechanism very likely results from similarities in the cellular mechanisms leading to smooth-muscle activation. Functionally, these types of interactions may act to optimize the effectiveness of local and extrinsic regulation of the vasculature.

Summary

In summary, autoregulatory control of large arterioles and small resistance arteries may be very important in tissues in which a significant fraction of the total network vascular resistance resides in these vessels. Understanding the mechanisms of autoregulation that are operative in this vascular segment will help to clarify how vascular behavior in this segment is coordinated with local vasoregulation and with mechanisms of blood pressure regulation.

Acknowledgments

The work presented was in part supported by the National Institutes of Health, HL33324, RR05814, and HL08398 and the American Heart Association with a Grant-in-Aid, and a CIBA-GEIGY Established Investigator Award to G. A. Meininger. M. A.

Hill was supported by grants from the Juvenile Diabetes Foundation International and the American Heart Association, Texas Affiliate.

References

[1] Johnson, P. C. (1986) *Circ. Res.* **59**, 483–495.

[2] Hinshaw, L. B. (1964) *Circ. Res.* **14–15** (Suppl 1), I–120–I–129.

[3] Hall, J. E. (1982) *International Review of Physiology*, vol 26. (Guyton, A. C. and Hall, J. E., eds.), University Park Press, Baltimore, pp. 243–321.

[4] Bouskela, E. and Wiederhielm, C. A. (1979) *Am. J. Physiol.* **237**, H59–H65.

[5] Burrows, M. E. and Johnson, P. C. (1981) *Am. J. Physiol.* **241**, H829–H837.

[6] Davis, M. J. (1987) *Microcirculation—an Update, vol. 1* (Tsuchiya, M., et al., eds.), Elsevier Science, Amsterdam, pp. 239, 240.

[7] Fleming, B. P. (1990) *FASEB J.* **4**, A1252.

[8] Koller, A. and Kaley, G. (1990) *Am. J. Physiol.* **258**, H916–H920.

[9] Harder, D. R., Sanchez-Ferrer, C., Kauser, K., Stekiel, W. J., and Rubanyi, G. M. (1989) *Circ. Res.* **65**, 193–198.

[10] Kon, V., Harris, R. C., and Ichikawa, I. (1990) *J. Clin. Invest.* **85**, 1728–1733.

[11] Shapiro, H. M., Stromberg, D. D., Lee, D. R., and Wiederhielm, C. A. (1971) *Am. J. Physiol.* **221**, 279–283.

[12] Chilian, W. M., Eastham, C. L., and Marcus, M. L. (1986) *Am. J. Physiol.* **251**, H779–H788.

[13] Meininger, G. A., Fehr, K. L, and Yates, M. B. (1986) *Hypertension* **8**, 66–75.

[14] Bohlen, H. G, Gore, R. W., and Hutchins, P. M. (1977) *Microvasc. Res.* **13**, 125–130.

[15] Meininger, G. A., Fehr, K. L, and Yates, M. B. (1987) *Microvasc. Res.* **33**, 81–97.

[16] Harder, D. R. (1984) *Circ. Res.* **55**, 197–202.

[17] Harder, D. R, Gilbert, R., and Lombard, J. H. (1987) *Am. J. Physiol.* **253**, F778–F781.

[18] Falcone, J. C., Davis, M. J., and Meininger, G. A. (1991) *Am. J. Physiol.* **260**, H130–H135.

[19] Delano, F. A., Schmid-Schönbein, G. W., and Zweifach, B. W. (1989) *FASEB J.* **3**, 1383.

[20] Hill, M. A., Simpson, B. E., and Meininger, G. A. (1990) *Microvasc. Res.* **39**, 349–363.

[21] Borders, J. L. and Granger, H. J. (1986) *Hypertension* **8**, 184–191.

[22] Meininger, G. A. (1987) *Microvasc. Res.* **34**, 29–45.

[23] Segal, S. S. and Duling, B. R. (1986) *Circ. Res.* **59**, 283–290.

[24] Segal, S. S. and Duling, B. R. (1987) *Circ. Res.* **61**(Suppl. II), II-20–II-25.

[25] Fleisch, A. (1935) *Arch. Int. Physiol.* **41**, 141–167.

[26] Hilton, S. M. (1959) *J. Physiol.* **149**, 93–1.

[27] Schretzenmayr, A. (1933) *Pflugers Arch. Gesamte Physiol.* **232**, 743–748.

[28] Duling, B. R. and Berne, R. M. (1970) *Circ. Res.* **26**, 163–170.

[29] Segal, S. S. and Duling, B. R. (1986) *Science* **234**, 868–870.

[30] Falcone, J. C. and Bohlen, H. G. (1990) *Am. J. Physiol.* **258**, H1515–H1523.

[31] Hester, R. L. (1990) *Am. J. Physiol.* **258** (in press).

[32] Tigno, X. T., Ley, K., Pries, A. R., and Gaehtgens, P. (1989) *Pflugers Arch.* **414(4)**, 450–456.

[33] Kampp, M., Lundgren, O., and Nilsson, N. J. (1967) *Experientia* **23**, 197,198.

[34] Whalen, W. J., Nair, P., and Ganfield, R. A. (1973) *Microvasc. Res.* **5**, 254–262.

[35] Duling, B. R. (1974) *Am. J. Physiol.* **227(1)**, 42–49.

[36] Duling, B. R. and Berne, R. M. (1970) *Circ. Res.* **27**, 669–678.

[37] Lash, J. M. and Bohlen, H. G. (1987) *Am. J. Physiol.* **252**, H1192–H1202.

[38] Johnson, P. C. (1980) *Handbook of Physiology, Section 2: The Cardiovascular System Vascular Smooth Muscle.* vol. 2 (Bohr, D. F., Somlyo, A. P., and Sparks, H. V., Jr., eds.), American Physiological Society, Bethesda MD, pp. 409–422.

[39] Lang, D. J. and Johnson, P. C. (1986) *Microvascular Networks: Experimental and Theoretical Studies* (Popel, A. and Johnson, P. C., eds.), Karger, Basel, pp. 112–122.

[40] Oien, A. H. and Aukland, K. (1983) *Circ. Res.* **52**, 241–252.

[41] Meininger, G. A., Mack, C. A., Fehr, K. L., and Bohlen, H. G. (1987) *Circ. Res.* **60**, 861–870.

[42] Granger, H. J. (1989) *FASEB J.* **3(4)**, A1387.

[43] Granger, H. J., Barnes, G. E., Meininger, G. A., and Goodman, A. H. (1984) *Int. J. Microcirc.* **3**, 222.

[44] Granger, H. J. and Guyton, A. C. (1969) *Circ. Res.* **25**, 379–388.

[45] Granger, H. J., Meininger, G. A., Borders, J. and Morff, R. J. (1984) *Physiology and Pharmacology of the Microcirculation,* vol. 2 (Mortillaro, N. A., ed.), Academic, New York, pp. 181–265.

[46] Gore, R. W. (1972) *Am. J. Physiol.* **222**, 82–91.

[47] Price, J. M., Davis, D. L., and Knauss, E. B. (1983) *Am. J. Physiol.* **245**, H379–H384.

[48] Price, J. M., Davis, D. L., and Knauss, E. B. (1981) *Am. J. Physiol.* **241**, H557–H563.

[49] Hinojosa-Laborde, C., Greene, A. S., and Cowley, A. W. (1988) *Hypertension.* **11**, 685–691.

[50] Meininger, G. A., Routh, L. K., and Granger, H. J. (1985) *Hypertension* **7**, 364–373.

51 Meininger, G. A., Faber, J. E., and Hard, S. D. (1989) *FASEB J.* **3(4),** A1395.

52 Meininger, G. A. and Trzeciakowski, J. P. (1990) *Am. J. Physiol.* **258,** H1032–H1041.

53 Meininger, G. A. and Trzeciakowski, J. P. (1988) *Am. J. Physiol.* **254,** H709–H718.

54 Metting, P. J., Stein, P. M., Stoos, K. A., Kostrzewski, K. A., and Britton, S. L. (1989)*Am. J. Physiol.* **256,** R98–R105.

55 Nilsson, H. and Sjoblom, N. (1985) *Acta Physiol. Scand.* **125,** 429–435.

56 Borst, J. G. G. and Borst-DeGeus, A. (1963) *Lancet* **1,** 677–682.

57 Coleman, T. G., Granger, H. J., and Guyton, A. C. (1971) *Circ. Res.* **29**(Suppl. II), 76–86.

58 Ledingham, J. M. and Cohen, R. D. (1963) *Lancet* **1,** 887, 888 (letter to the Editor).

59 Faber, J. E. and Meininger, G. A. (1990) *Am. J. Physiol.*(in press).

60 Meininger, G. A. and Faber, J. E. (1989) *FASEB J.* **3,** A257.

61 Hill, M. A., Davis, M. J., and Meininger, G. A. (1990) *Am. J. Physiol.* **258,** H127–H133.

62 Hill, M. A., Falcone, J. C., and Meininger, G. A. (1990) *Am. J. Physiol.* **259,** H1586–H1594.

63 Ping, P. P. and Johnson, P. C. (1989) *FASEB J.* **3,** 270.

64 Goodman, A. H., Einstein R., and Granger, H. J. (1978) *Circ. Res.* **43,** 769–776.

Chapter 21

The Microvascular Consequences of Diabetes Mellitus and Hypertension

H. Glenn Bohlen and Julia M. Lash

Introduction

The two major diseases of the microvasculature in man are essential hypertension and the secondary complications of diabetes mellitus. The prevalence of the disease is estimated to be approx 20% for hypertension[1] and 3–6% for various forms of diabetes mellitus.[2] In this chapter, we will be primarily concerned with microvascular consequences found in animal models of the diseases and where possible, correlate these changes to known problems in hypertensive and diabetic humans.

Hypertension as a Microvascular Disorder

Format of the Increased Resistance

One of the major characteristics of the form of hypertension in the spontaneously hypertensive rat (SHR) is that vascular resistance in all major organ systems is elevated almost proportionally to the mean arterial pressure.[3–6] However, as will be subsequently

From *The Resistance Vasculature*, J. A. Bevan et al., eds. ©1991 Humana Press

discussed, the hemodynamic format used to elevate vascular resistance is somewhat different between organ systems, and may even be different between various skeletal muscle groups.

The hemodynamic mechanisms that cause increased resistance of the microvasculature have been, and are, major topics in clinical and basic research. An obvious possibility would be vasoconstriction of all or part of the vasculature. For example, a 10–12% constriction of all arterial and arteriolar vessels would increase vascular resistance about 50–60%. However, as judged from studies of the microvasculature in anesthetized and very recently, unanesthetized SHR, uniform vasoconstriction of all arterioles does not seem to occur. Constriction of the largest and smallest arterioles in SHR is known to occur in the intestinal and cerebral vasculatures.[7–10] Excessive vasoconstriction of small arterioles in skeletal muscle is probable, based on a much steeper pressure dissipation in hypertensive than normal rats.[11] However, microvascular hemodynamics are sufficiently complex that finding or not finding vasoconstriction alone may not be adequate to predict altered vascular resistance. For example, even slight constriction of some of the larger arterioles and smallest arteries may have profound effects on vascular resistance based on Borders and Granger's[12] analysis of power dissipation in the vasculature. Their calculations and measurements indicate that the most efficient location to increase power dissipation, which to an external observer would have the practical effect of increasing resistance, is in those vessels with the highest individual blood flows, the larger resistance arterioles. As shown in Fig. 1 for the pressure distributions in the intestinal and cerebral cortical vasculatures of both adult normal and SHR rats, a large pressure decrease occurs prior to the large arterioles.[7,9] This pressure reduction is difficult to account for based solely on the size of the vessels involved and may well reflect to some extent the process of power dissipation. There is also the possibility that the branching of small arteries into many large arterioles of much smaller diameter dissipates a substantial amount of energy at branch points, having the same effect as a large resistance. The theoretical basis for such large amounts of pressure dissipation across branching locations of daughter vessels with small diameters relative to the parent vessel has been modeled by Vawter

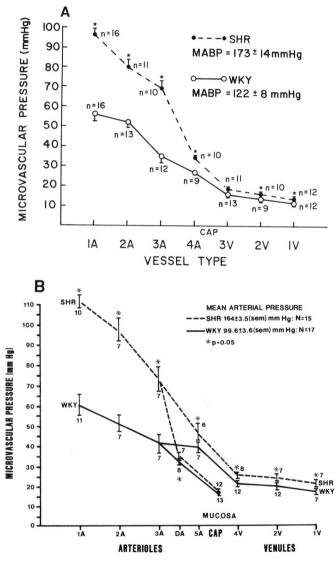

Fig. 1. Panel **A** presents cerebral cortical microvascular pressures and panel **B** displays pressures in the small intestine microvasculature of adult normal (WKY) and spontaneously hypertensive (SHR) rats. Note the higher arteriolar pressures in first (1A) to third (3A) branch generation arterioles in both vasculatures of SHR but near normal to slightly elevated pressures in all succeeding vessel types. Data from refs. 7 (Panel B) and 9 (Panel A) reprinted with the permission of the American Heart Association.

et al.,[13] and the model fits pressure declines across branch points of arterioles.[14] The point to be made is that the relative diameters of vessels arranged in series is potentially as, or perhaps more, important than the absolute vessel diameters in the overall process of determining vascular resistance.

A potentially important vascular alteration that increases vascular resistance in hypertension was described by Short[15,16] for the small intestine of hypertensive humans and confirmed by Hutchins and Darnell[17] for the cremasteric muscle in SHR. This mechanism is both temporary and permanent closure of arterioles. Based on studies of lower body muscles in SHR, approx twice as many small arterioles are closed, either temporarily or permanently, in SHR as in normal rats.[17–20] Computer modeling of vessel closure indicates up to a 20% increase in vascular resistance could occur with typical levels of rarefaction found in SHR.[21] Such an increase in resistance potentially could account for the total increase in resistance in mild hypertension and approx one–third to one–half of the increased resistance in advanced disease. Meininger[22] has demonstrated that renal vascular hypertension also causes anatomical rarefaction of intermediate diameter arterioles in the small intestine. Therefore, anatomical rarefaction may not be unique to the form of hypertension in SHR. However, vessel closure or rarefaction either by physiological means for temporary closure or anatomical processes for permanent closure has only been consistently found for lower body muscle and intestinal vasculatures in SHR[7,17–19,23] and small intestine,[15,16] conjunctiva,[24] and skeletal muscle,[25] of hypertensive man. At this time, anatomical rarefaction seems to be virtually nonexistent in renal[26] and cerebral vasculatures,[9] as well as the spinotrapezius muscle,[27] an upper body muscle, in SHR. In addition, physiological rarefaction has not been found to occur in the cerebral[19] and spinotrapezius muscle[28,29] vasculatures of SHR. However, the importance of physiological rarefaction may be substantially underestimated if anesthesia and surgical preparation suppresses vascular tone. Therefore, the resolution of the issue of physiological rarefaction depends in part on the expanded use of microvascular observations in unanesthetized animals as well as means to test this issue in hypertensive humans.

Hypertensive Microvascular Pressures
and Their Consequences

Measurements of microvascular pressures indicate that, with the exception of the smallest arterioles in most tissues, vessel wall tension should be increased at least 30–60%, depending on the stage of hypertension and degree of vasoconstriction. Folkow[30,31] has proposed that the overload on the vasculature would induce hypertrophy of the vascular smooth muscle and presented results from pressure–flow studies that are consistent with encroachment on the vessel lumen through wall hypertrophy. There is ample evidence that even the smallest arteries in hypertensive SHR and humans with essential hypertension have enlarged vessel walls through a combination of vascular smooth muscle cell hypertrophy and some degree of cellular hyperplasia.[32-36] However, it is difficult to demonstrate directly vessel wall hypertrophy in terms of more vascular muscle mass, either by cellular hyperplasia or hypertrophy opposed to the commonly reported increased wall thickness to lumen ratio for arterioles. This difficulty may arise either because hypertrophy does not appreciably occur or because the histological processing of tissue distorts tissues sufficiently, through dehydration and shrinkage, that wall dimensions are underestimated. The vast majority of arterioles have essentially a monolayer of vascular smooth muscle cells, as depicted in Fig. 2, and at their thickest radial point, e.g., the nucleus, the cells are about 2–2.5 μ in width. Miller et al.[37] have counted the number of vascular smooth muscle cells per 10 μ longitudinal length of arteriole and found no difference for comparable branch orders of intestinal arterioles in young or adult normotensive and SHR rats. Measurements of the cell length and average width from scanning electron micrographs of these intestinal vessels as well as morphometric analysis of just the muscular component of sectioned vessels indicated that vascular smooth muscle cell hypertrophy is only likely to exist for the largest of intestinal arterioles.[37] Therefore, hyperplasia and hypertrophy in this particular vasculature is limited up to the early adult phase of life when hypertension is well established.

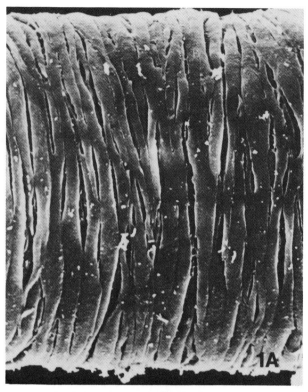

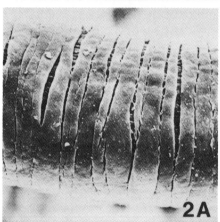

Fig. 2. Scanning electron micrographs of normal intestinal arterioles demonstrate the monolayer of vascular smooth muscle cells that are circumferentially wrapped around the various sizes of arterioles. The insets for third and fourth generation vessels are the same magnification as for the larger arterioles. Images are from ref. 108, reprinted with permission.

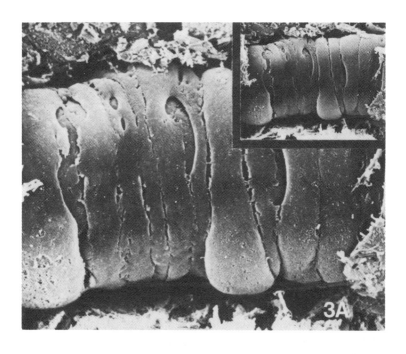

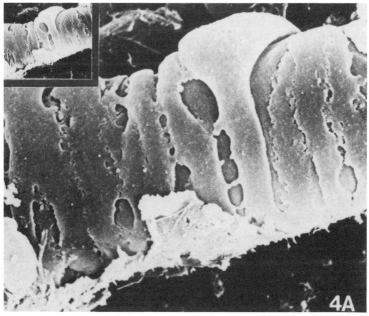

Whether or not appreciable cellular hypertrophy occurs for the arterioles in hypertensive animals, there is sufficient evidence to conclude that the passive mechanical properties of these arterioles are altered. For example, in virtually every microvascular study of a wide variety of organs, the passive diameter of the arterioles are equal to or smaller than their normal counterparts, even though microvascular pressures in SHR[7,9,37] or renal vascular hypertensive rats[22] are substantially higher than normal. Baumbach et al.[38] have shown that passive cerebral arterioles in stroke-prone SHR are smaller than normal at virtually all pressures from about 10 mmHg to the prevailing pressure during vasodilation at natural resting arterial pressures. However, their data also reveal that the relative change in diameter for a given increase in pressure in SHR is greater than normal. In effect, these vessels behave as if their capacitance is decreased but their compliance is increased. This would have the functional effect of a higher than normal passive resistance for a given intravascular pressure but a greater than normal relative change in passive resistance for a given change in pressure. Studies of the passive perfusion pressure–blood flow characteristics of the hind limb in normal and SHR rats[30,31] also predict such changes in the pressure–diameter relationship of resistance vessels. However, similar results would be obtained in the presence of vessel wall hypertrophy of a large fraction of the resistance vessels as a result of encroachment on the vessel lumen or alterations of the passive wall characteristics.[30,31] As all of these possibilities probably exist simultaneously to various extents, the practical importance of the observed data is that, for hypertensive rats to achieve normal blood flows during vasodilation events, higher than normal intravascular pressures are required to obtain expansion of the vessels.

Genetic vs Pressure Dependent Abnormalities

An issue that is far from being resolved is, which of the changes in microvascular structure and function reflect the genetic abnormalities of inherited hypertension or the adaptation to increased intravascular pressure? In various renal and genetic hypertensive models, a chronic reduction of the hind quarters arterial pressure to about normal by partial or total aortic occlusion has shown that substantial, but not total, protection from microvascular modification can be achieved.[39-42] Such studies indicate that simply limiting the increase in intravascular pressures has a protective effect, independent of circulating agents or sympathetic activity. However, it is fair to ask the questions of whether increased microvascular pressures are possibly required for full expression of genetic abnormalities or whether limiting arterial vascular pressures invokes protective mechanisms, such that perfusion is maintained. For example, Hogan and Hirschmann[43] found that reduction of the perfusion pressure to the cremasteric muscle of normal young rats caused the formation of a greater than normal number of arterioles. If something similar were to happen in SHR, it might counteract the anatomical and physiological rarefaction usually found in this particular vasculature. Pharmacological intervention to maintain relatively normal total body arterial pressure in SHR[44-47] and renal vascular hypertensive rats[48] also limits, and in some cases reverses, vascular problems and suppresses the process of rarefaction. However, whereas such improvements are presumably in the best clinical interests for the hypertensive animal or human, whether a given pharmacological intervention is influencing a specific genetic or induced abnormality of hypertension or actually creating a hopefully benign abnormality that happens to lower arterial pressure remains to be determined.

Functional Characteristics
of the Hypertensive Microvasculature

Whether a given microvascular consequence of hypertension is of genetic or adaptative origin, there is ample evidence to conclude that overall microvascular function to maintain the host tissue is remarkably preserved. For example, edema formation, impaired wound healing, and tissue pain or pathology, consistent with inadequate microvascular perfusion are very rare complications of hypertension during most of the life-span in the hypertensive human or SHR. Such admittedly anecdotal observations are in part the impetus for many of the studies of how the microvasculature of hypertensive animals regulate microvascular

behavior. As shown in Fig. 1 for intestinal and cerebral vasculatures, directly measured pressures in capillaries and the smallest
venules of all organs studied thus far,[7–11,49] except the glomerular
capillaries of the kidney,[26] tend to be essentially normal or only very
slightly increased in SHR rats with fully developed hypertension.
Therefore, the risk of edema formation, at least at rest, is minimal.
Whether or not any of the resistance vessels specifically change
their function to maintain capillary pressures in a limited range is
at best speculative. However, a common finding in studies of
microvascular pressure is that pressures in intermediate diameter
arterioles are hypertensive but the smallest arterioles of SHR dissi-
pate such a large fraction of the arterial pressure that inflow pressures to the capillaries are approx normal.[17–11] Furthermore, whether
studied at the whole organ level or for individual resistance vessels in in vivo situations, the resistance vessels of SHR demonstrate
a much greater than normal constrictor response to norepinephrine, even at their naturally elevated intravascular pressures.[18,50]
This increased responsiveness is such that iontophoretic application of norepinephrine onto denervated small arterioles in skeletal
muscle of SHR will cause constriction to near closure at dose levels
that are just above threshold for normal vessels.[18] This type of
behavior could partially explain the temporary rarefaction of small
arterioles and acute denervation does cause temporarily closed
arterioles to open in some but not all skeletal muscles.[18,20]

One of the best examples of maintenance of normal "nutritive" microvascular function during hypertension is preservation
of cerebral blood flow autoregulation in both humans[51,52] with
essential hypertension and the SHR.[8–10] Not only is cerebral blood
flow and exchange vessel pressure approx normal at rest, it is normal despite arterial and arteriolar pressures that can cause trauma
in normal vessels. Werber and Heistad[10] have shown that whereas
acute autoregulatory events in intact SHR do not rely on the sympathetic nervous system, except possibly at very high arterial pressures, chronic sympathetic denervation impairs, but does not
eliminate, the enhanced cerebral vascular regulation in hypertension. Other important contributors to increased cerebral regulation
which have been demonstrated, are altered vessel wall mechanical
properties as previously mentioned,[38] some degree of vessel wall

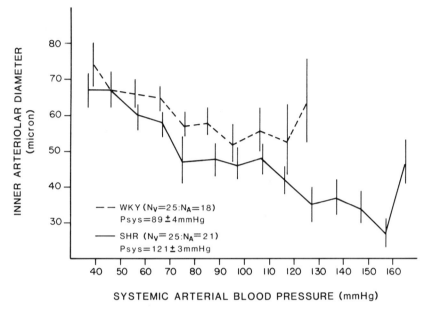

Fig. 3. Average inner diameters of large arterioles in 4–5-wk-old normal and spontaneously hypertensive rats over a wide range mean arterial pressures. These data from ref. 8 (reprinted with permission) indicate that the arterioles of SHR are usually constricted relative to normal and can maintain constriction at much higher pressures than in normal animals.

hypertrophy,[9,10,32] and increased myogenic responses of small arteries.[53] The net effect is that the upper and lower limits of arterial pressure at which cerebral blood flow remains virtually constant are increased 30–70 mmHg in SHR[8–10,44] and hypertensive humans.[51,52] This overall change in regulation begins very early in life because 4–6 wk SHR can maintain cerebral arteriolar constriction at pressures very similar to those in adult SHR, as shown in Fig. 3.[8] However, when the mean arterial pressure was increased to 110–120 mmHg in young normal rats, arteriolar dilation occurred, presumably owing to the overpowering intravascular pressure, and tissue swelling rapidly developed in open-skull preparations.

It would be inappropriate to conclude that all aspects of nutritive vascular regulation are normal in hypertensive animals. Boegehold and Bohlen[29] have shown that, whereas the average tissue PO_2 in the spinotrapezius muscle of young adult SHR is statistically normal, the coefficient of variation was much greater in SHR (51%) than normal rats (36%). In effect, the SHR had a much greater incidence of unusually low and high tissue PO_2 than in normal rats. Although not confirmed, this could indicate capillary regions with abnormally low and high perfusion. This interpretation is consistent with the current understanding of physiological rarefaction of terminal arterioles because other investigators[54] have found minimal, if any, anatomical rarefaction in this muscle. When the muscle contracted at frequencies of 1–8 Hz, all sizes of arterioles studied demonstrated both a normal relative and absolute increase in diameter.[29] However, tissue PO_2 in hypertensive rats initially decreased to 21.3 ± 2.3% (SEM) and recovered to 41.2 ±13% after 3 min of 8 Hz muscle contractions compared to an initial decrease to about 50% of control and a recovery to greater than 80% of control in normal rats. Boegehold and Bohlen[29] proposed, but did not test, that impaired capillary perfusion or inadequate dilation of small arteries secondary to wall hypertrophy[55] could explain the inappropriate regulation of tissue PO_2 during muscle contractions in the presence of normal functional dilation and vessel density. In any event, the PO_2 data for both resting and contraction conditions do indicate that overall vascular performance is potentially compromised during hypertension.

The Microvascular Consequences of Diabetes Mellitus

The Questionable Roles of Insulin and Glucose in Diabetic Microangiopathy

Although the majority of fatalities secondary to vascular disease during diabetes are usually caused by cerebral, cardiac, or renal macro- and microvascular disturbances, there is very little question that all organ microvasculatures are impaired. In humans and some animal models, diabetes also accelerates the atherosclerotic process, which not only threatens survival but also may complicate microvascular alterations during diabetes. Exactly what constitutes the most important microvascular abnormalities during diabetes can not be precisely defined, in part because there are so many abnormalities that can occur simultaneously. An equally important issue for which there is very limited information is whether any given vascular abnormality is caused by inadequate production or actions of insulin, some factor related to hyperglycemia and disturbed carbohydrate metabolism, or secondary complications of diabetes, such as increased glucagon or growth hormone.

Glucose uptake and utilization by fragments of microvessels isolated from normal animals occurs independently of insulin, although vessel fragments from diabetic animals have a subnormal glucose consumption.[56] The normal endothelial cell internalizes insulin through a receptor-mediated process, and subsequently releases unaltered insulin.[57] This is unusual because most cells degrade insulin after it is internalized. What fraction of the plasma insulin enters the interstitium by first passing through endothelial cells vs simple diffusion through the extracellular pore system of the capillary wall is not known. Studies by Yang et al.[58] suggest a delay in insulin passage from the blood plasma to the lymphatic fluid, which includes a time factor related to transit of insulin across the vessel wall. They also found that total body glucose utilization in normal dogs is much better correlated to lymph insulin concentration than plasma insulin concentration, probably because lymph composition more closely approximates the insulin concentration presented to target cells after passage through the vascular wall. The extent to which endothelial cell "transport" of insulin is a component of insulin resistance in diabetes is currently speculative, and could be related to both the numbers of endothelial cells available and potential alterations of their insulin transport function.

Structural and Functional Damage to the Diabetic Microvessels

As shown in Fig. 4, within 3–5 wk after insulin-dependent hyperglycemia in rats is initiated with streptozotocin, the vascular smooth muscle cells of the cerebral vasculature lose their spindle shape and become stellate in appearance. Simultaneously, the endo-

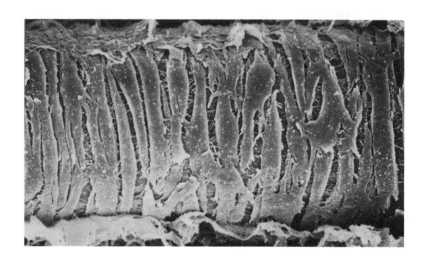

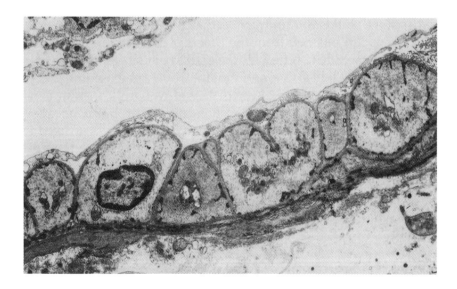

Fig. 4. Scanning and transmission electron micrographs of cerebral cortical arterioles from normal and streptozotocin diabetic rats demonstrate the
major changes in vascular smooth muscle cell external and internal morphology, which occur within 4–5 wk after severe hyperglycemia begins.
From ref. 59, reprinted with permission.

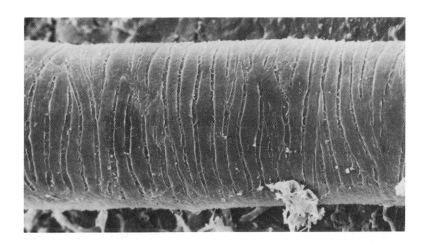

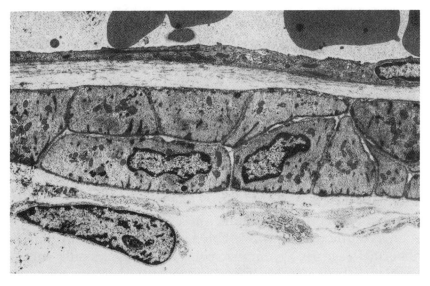

thelial cells become morphologically abnormal and about 20% of the cells appear to be dying. These abnormalities demonstrated by Moore et al.[59] have been observed in other tissues of rats (unpublished observations), and somewhat similar morphology has been described for microvessels in severely diabetic humans.[60] In addition, Mayhan[61] has shown that some facet of cerebral endothelial cell or vascular smooth muscle cell function is severely impaired in diabetic rats because adenosine diphosphate and serotonin cause minimal vasodilation. The dilator action of these agents is known to be caused by *endothelial derived relaxing factor* (EDRF), which relaxes smooth muscle. The cerebral vessels of diabetic rats will dilate by approx normal amounts in response to endothelial independent dilators.[61,62] Therefore, vascular smooth muscle cells may retain a substantial fraction of the normal ability to relax when appropriately challenged .

Based on the morphological changes in diabetic vessels, one would expect substantial impairment of cerebral vascular regulation. However, Rubin and Bohlen[62] found that cortical arteriolar blood flow was autoregulated over approx the same hypotensive and hypertensive range of arterial pressures (Fig. 5, upper panel), the tissue PO_2 at rest was normal, and as shown in the lower panel of Fig. 5, tissue PO_2 was maintained similar to normal as the arterial pressure was altered. These data were obtained after 8–10 wk of hyperglycemia in young adult rats (24–26 wk-old) when plasma glucose was over 600 mg/100 mL and body wt was about 75% of normal. There were, however, clear indications that the vasculature was abnormal in some respects. Although arteriolar density (number of vessels/mm² surface) was normal during maximum vasodilation, venular density was reduced about 20%, which may indicate a permanent loss of venules.[62] When arterial pressures were increased to above about 160 mmHg (norepinephrine infusion), both the arterioles of normal and hyperglycemic rats developed alternating areas of constriction and relative dilation. This pattern reverted to a normal appearance and resting diameter almost as soon as normotension was restored in the normal animal but the abnormal vascular appearance persisted for hours after such an event in diabetic animals. If the "sausage string" pattern represents a form of cerebral vasospasm, then cerebral vessels of hyperglycemic rats are predisposed to this problem.

The previously described study of normal cerebral tissue PO_2[62] and subsequent studies that indicated normal intestinal muscular and mucosal tissue PO_2[63] in diabetic rats are germane to the question of whether or not hypoxia, either intermittent or sustained, possibly contributes to the development of diabetic microvascular disease. Dietzel[64] has proposed that disturbances of 2,3-diphos-

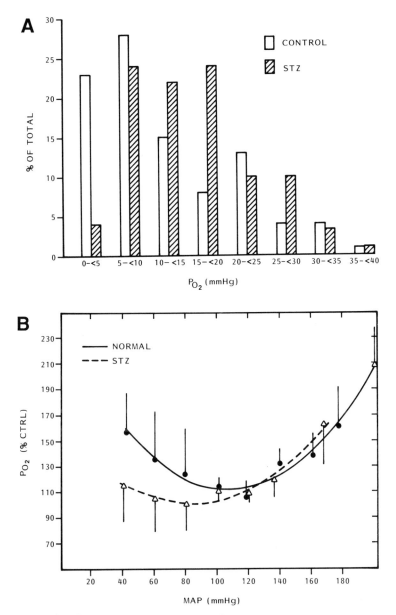

Fig. 5. The data in panel **A** indicate that whereas the average cerebral cortical tissue PO_2 is normal in diabetic rats, the distribution is skewed to higher values. The data in panel **B** indicate that maintenance of cerebral cortical tissue PO_2 as the arterial pressure is modified is similar in normal and diabetic rats. From ref. 62, reprinted with permission.

phoglycerate and glycosylation of hemoglobin simultaneously contribute to intermittent hypoxia and eventual vessel damage by impairing oxygen disassociation. Studies of vascular reperfusion injury indicate that brief periods of hypoxia followed by restoration of normoxia causes endothelial cell damage owing to oxygen radical formation.[65] In the studies conducted in Bohlen's laboratory in which an oxygen sensitive microelectrode remained at one point for often an hour, intermittent bouts of relative hypoxia did not occur in the brain or small intestine.[62,63] However, that such bouts are possible is easy to defend. Spontaneous formation of what appear to be platelet clots occurs in the arterioles of hyperglycemic rodents[66,67] and white blood cells constantly stick to the lining of the arterioles: these events rarely occur in normal animals except after mechanical or reperfusion injury.[65] Another problem of concern is that hyperglycemia causes chronic decreases in red cell deformabilty, which could further complicate perfusion in general and, for partially obstructed vessels,[68,69] in particular. Based on these observable and routine problems in diabetic animals, intermittent bouts of hypoxic damage probably do occur and their cause is as likely related to occlusive problems in arterioles and capillaries as abnormal tissue gas exchange caused by altered hemoglobin function.

Microvascular Rarefaction and Growth During Diabetes

Assuming vascular damage occurs during the bouts of mechanical obstruction and that other factors related to diabetes cause vessel impairments, an expected long-term outcome would be loss of microvessels. The loss of cerebral venules in streptozotocin treated rats has been mentioned in this context.[70] To determine if the cause of hyperglycemia influenced whether or not anatomical rarefaction occurs, vascular density of the cremasteric skeletal muscle of DbDb, ObOb, and streptozotocin (STZ) treated mice were compared at equal ages and for approx equal durations and severities of hyperglycemia.[71] The DbDb and ObOb mice were hyperglycemic at the ages studied because of excessive food intake and were hyperinsulinemic. The number of smallest arterioles per mass of tissue was decreased relative to normal animals a minimum of

25% in DbDb, ObOb, and STZ treated animals, but numbers of large to intermediate diameter arterioles (1A–3A) were normal, or even elevated, resulting in decreased skeletal muscle vascular resistance relative to normal animals.[71] Subsequent studies[72,73] of the cremasteric muscle in the DbDb and streptozotocin models indicated increased intercapillary distances, such that about 25–30% of the capillaries were absent. A loss of capillaries in skeletal muscle is not unique to diabetic mice because morphological studies of skeletal muscle tissue from diabetic humans indicates a loss of capillaries.[74,75]

The extent to which any or all of the vascular abnormalities described thus far can be ascribed to specific abnormalities related to insulin, carbohydrate metabolism or hyperglycemia is very difficult. However, there is substantial circumstantial evidence that some aspect of hyperglycemia is associated with microvascular damage. Several studies have indicated that a very high carbohydrate diet[76–78] for rats can initiate microvascular disturbances consistent with those in diabetes. Human infants born to diabetic mothers are assumed to be occasionally hyperglycemic, are known to be hyperinsulinemic, and have cellular morphological alterations in the umbilical vessels consistent with those of a diabetic vasculature.[79] Normal rats exposed to twice daily intraperitoneal hyperglycemia(400 mg/100 mL, isotonic glucose mixed into isotonic Normasol-R, Abbott Labs.) for 4–5 wk qualitatively demonstrate intestinal vascular abnormalities.[80] The mass of glucose given had little effect on the plasma glucose concentration. As shown in Fig. 6, the distribution of intercapillary distances in the intestinal longitudinal muscle layer of such animals is very similar to that of streptozotocin treated rats which were hyperglycemic for 4–5 wk. In effect, the diabetic and ip hyperglycemic rats both appeared to lose capillaries and thereby increase their average intercapillary distances.

Whereas the study[80] just mentioned may indicate that hyperglycemia can cause capillary rarefaction in the intestinal microvasculature, additional studies have shown that the intestinal tract has a number of abnormalities during diabetes that greatly complicate definitive conclusions. Rats with untreated diabetes mellitus experience a profound hypertrophy of the small intestine. Whether

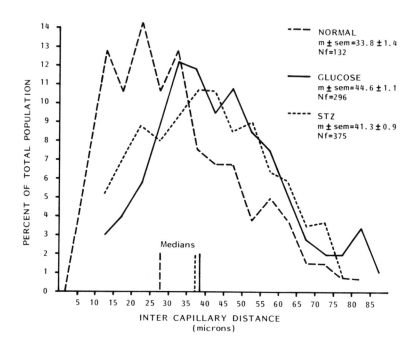

Fig. 6. The mean and distributions of inter capillary distances for the intestinal muscle layers of normal, diabetic (STZ), and intraperitoneally hyperglycemic (glucose) rats indicate that whole body and regional hyperglycemia both cause increased separation distances between capillaries, probably because capillaries are lost. From ref. 80, reprinted with permission.

this is secondary to hyperphagia, some aspect of hyperglycemia, or the excessive glucagon concentrations, a hormone that induces bowel growth, is debatable. To better understand how the microvasculature of a diabetic animal copes or fails to cope with the excessive growth of the bowel, a defined region of ileum was studied for microvascular density at age 4–5 wk and restudied at age 9–10 wk in normal and streptozotocin diabetic rats.[81] The same arterioles present at age 4–5 wk were present at age 9–10 wk in both normal and diabetic animals. However, as documented in Table 1, the greater growth of bowel in diabetic than normal rats caused the number of arterioles per mass of tissue to be about half of normal. This decrease in numbers of vessels per mass of tissue was

Table 1
Comparison of Physical, Intestinal, and Microvascular Growth
in Normal and Diabetic Rats

	Normal		Diabetic		Control diabetic	
Age						
T1 (days)	39 ± 1.0	(8)	34 ± 0.5	(15)	NA	
T2 (days)	76 ± 1.1	(6)	73 ± 1.5	(8)	76 ± 3.1	(5)
Body weight						
T1(grams)	137 ± 6.9	(8)	108 ± 2.0	(15)	NA	
T2(grams)	329 ± 9.0	(6)	$250 \pm 6.9^*$	(7)	286 ± 23.4	(5)
Blood glucose at T2						
(mg/100 mL)	106 ± 2.3	(6)	$528 \pm 13.5^*$	(10)	501 ± 55.3	(5)
Intestinal mass per unit length at T2						
(mg/mm)	8.52 ± 0.76	(5)	$13.97 \pm 1.06^*$	(7)	11.81 ± 0.58	(5)
Intestinal mass per arteriole at T2						
mg/1A	11.49 ± 0.48	(10)	$21.87 \pm 1.25^*$	(5)	22.53 ± 1.73	(4)
mg/2A	0.47 ± 0.04	(9)	$0.758 \pm 0.20^*$	(6)	0.729 ± 0.51	(4)
mg/3A	0.095 ± 0.22	(4)	0.169 ± 0.024	(6)	0.158 ± 0.029	(4)
Intercapiillary distances						
T1 longitudinal	59 ± 4.0	(5)	55 ± 4.8	(5)	NA	
T2 longitudinal	61 ± 5.7	(5)	65 ± 3.4	(5)	66 ± 1.8	(3)
T1 radial	91 ± 9.1	(5)	97 ± 3.0	(5)	NA	(3)
T2 radial	$78 \pm 6.8†$	(5)	$75 \pm 5.4†$	(5)	76 ± 2.3	(3)

Intercapillary distances in the longitudinal bowel axis are measuremnents of the distance between walls of adjacent capilaries of the inner muscle layer and intercapillary distances in the radial bowel axis are for capillaries of the outer muscle layer. T1 is the first observation and T2 is the second observation of the same tissue region.

*Dotted diabetic ≠ dotted normal;$p \leq 0.01$.

†Intercapillary distance T2 ≠T1;$p \leq 0.05$.

From ref. 81, with permission of the American Heart Association, Inc.

strictly caused by tissue growth and not a loss of arterioles. Studies of intercapillary distances in the visceral muscle layers indicated the average capillary spacing was virtually normal in diabetic rats despite abnormally large tissue growth (Table 1). In effect, additional capillaries must have been added to support both the normal juvenile growth of the intestine plus the additional tissue expansion in diabetic animals. This is an important observation because it clearly indicates that capillary angiogenesis in excess of that which occurs in normal juvenile rats can occur in juvenile diabetic animals.[81] In addition, it would appear that arteriolar angiogenesis was not required to support the enlarged diabetic bowel, possibly because of enhanced capillary formation.

One would expect that the capillaries of the enlarged diabetic intestine might be more permeable than normal. This assumption is now known to be only partially correct based on studies by Korthuis et al.[82] They found that the capillary filtration coefficient was increased about 75%, which would indicate that water permeability was elevated after 4 wk of streptozotocin induced hyperglycemia in rats. However, the osmotic reflection coefficient for plasma proteins was normal in the diabetic animal and suggests permeability to large molecules is preserved despite diabetes and the growth of large numbers of capillaries. Korthuis et al.[82,83] have found that blood flow/100 g of tissue was increased about 20% in the duodenum to almost doubled in the ileum, yet gastric and colonic blood flows were normal. When a diabetic rat was used to perfuse a normal segment of bowel, intestinal blood flow increased an average of 30%. In normal animals, blood hyperosmolarity (glucose or sucrose) and elevated plasma glucagon concentrations equivalent to those in diabetic animals produced intestinal hyperemia of about the same magnitude as naturally occurs in diabetic animals.[83] When glucagon antiserum was given to diabetic rats, blood flow in the stomach, duodenum, jejunum, and kidney decreased by up to one-fourth and the antiserum had no appreciable effect on these blood flows in normal rats.[84] These various observations[84,85] indicate that circulating vasodilators, including both hyperosmolarity and glucagon, in diabetic animals may in part explain chronic intestinal hyperemia.

Effects of Exercise
on Hypertensive and Diabetic Vasculatures

In the clinical treatment of hypertension and diabetes, aerobic exercise is often suggested as an adjunct to dietary and pharmacological treatment.[86] From the perspective of microvascular physiology, the cellular and structural microvascular adaptations to exercise, as well as changes in total body homeostasis, have the potential to substantially benefit both hypertensive and diabetic individuals. In normotensive subjects, aerobic exercise training slightly decreases resting arterial pressure[87] and decreases the minimum vascular resistance of cardiac[88] and skeletal muscle by 20–40%,[89,90] probably by enlargement of resistance vessels. Exercise training increases skeletal muscle capillary density[91] and total body insulin sensitivity,[92] most likely through changes in skeletal muscle metabolism and perfusion. In addition, baroreceptor gain is decreased in aerobically trained humans[93] and rats,[94] and neural vascular regulation is altered.[95]

Specific Benefits of Exercise in Hypertension

Aerobic training decreases, but does not normalize, arterial blood pressure in SHR[96] or in humans with mild essential hypertension.[97,98] In addition, clinical studies indicate that the normal increase in arterial pressure during exercise is reduced in hypertensive and normotensive subjects after aerobic training.[99] A reduced pressure response to environmental stress is also observed in aerobically trained rats concurrent to a decrease in plasma norepinephrine,[100] and resting heart rate is significantly decreased after training of mildly hypertensive humans.[97] At a structural level, myocardial capillary density increases in trained SHR[101] but the effects on skeletal muscle capillarization have not been thoroughly evaluated. Additionally, the maximum force generation of in vitro arterial smooth muscle is decreased by training in the SHR, but the hypersensitivity to norepinephrine is unaltered.[102]

Exercise Benefits in Diabetes

It has been hypothesized that delivery of insulin to parenchymal cells is diffusion limited as diabetic arteriolar and capillary

rarefaction progresses,[103] producing increased insulin resistance at
the organ level. This hypothesis is supported by the proportionality of insulin action to capillary density in normal skeletal muscle.[74]
While the potential to increase capillary density in diabetes has
been questioned,[104] appropriate aerobic training programs cause
an approx normal increase in capillary density in insulin resistant
diabetic animal models.[105] However, the severity and stage of the
diabetic process may influence the beneficial effects of aerobic
training. Sedentary insulin–resistant obese Zucker rats (OZR) have
a lower capillary density and higher hind limb vascular resistance
than lean Zucker rats (LZR). Training the OZR from age 6–11 wk
does not increase capillary density but does reduce hind limb vascular resistance and increase submaximal glucose uptake to near
normal. In contrast, longer durations of exercise training for OZR
from age 6–18 wk or 11–18 wk increases capillary density by 40%,
concurrent with improved hind limb minimum resistance and
glucose uptake. However, glucose uptake was not normalized in
the older rats (18 wk) despite attaining capillary density and vascular resistance conditions equal to or better than those in normal
lean rats. These various observations indicate that whereas aerobic
training improves the vascular status in diabetes, training alone
can not achieve complete suppression of diabetic complications.

Nonspecific Effects of Exercise

A complicating factor for the effects of exercise training on
hypertensive and diabetic subjects is the frequent simultaneous
decline in body wt[87] and change in dietary composition. For
example, sedentary obese subjects on a diet–based wt loss program
exhibit comparable declines in blood pressure as those in training.[87] In addition, sedentary OZR fed a high carbohydrate diet demonstrate increases in submaximal glucose uptake similar to that
seen in trained rats on a high–fat diet.[106] However, exercise training does have some unique effects. Trained OZR demonstrate
decreased vascular resistance, increased capillary density, and
increased insulin sensitivity without a significant decrease in body
wt.[105,107] In a similar context, glucose tolerance and capillary density were improved in the absence of changes in body wt or body
composition in obese women.[92] Furthermore, dietary manipula-

tion and exercise training were found to have synergistic effects on glucose uptake and disposal in the OZR.[106] The extent to which the metabolic adaptations involve changes in microvascular structure or function is not known. However, an important consideration for diabetes and hypertension research may be that the animals used, both normals and those with disease, are usually totally sedentary, and probably more sedentary than typically occurs in the human population. This raises the distinct possibility that results obtained with sedentary animals reflect the worse case scenario for microvascular disturbances and should be viewed in this context.

Summary

The study of the microvascular consequences of diabetes mellitus and the forms of hypertension in human essential hypertension and the spontaneously hypertensive rat tends to lead to two major conclusions. First, and somewhat surprisingly, the overall functions of the microvasculature in terms of tissue support are remarkably normal until the hypertension is highly advanced. It is difficult to predict whether this reflects that the microvasculature simply compensates for a bad situation, or that some abnormal genetic characteristics of the cells in the vessel wall require an elevated intravascular pressure to accomplish relatively normal microvascular function, or that both occur to some extent. Second, diabetes mellitus, unlike hypertension, causes a relatively easily documented degeneration of both the structure and function of the endothelial and vascular smooth muscle cells, such that the regulation and range of microvascular performance is progressively altered. However, clinical experience with various approaches to restore normal carbohydrate metabolism indicate that diabetic microvascular disease can be downgraded from an almost foregone conclusion of an eventual life threating abnormality in the untreated state to a mild to moderate nusiance. This observation may reflect the typical research laboratory findings in untreated diabetic man and animals that at least 60–70%, and often more, of the normal microvascular response to a given pertubation is preserved, except in the worse case situations. Therefore, although performance is neither normal nor dangerously abnormal, even modest improvement might provide a relatively safe level of microvascular reserve.

References

[1] Purice, S., Harnagea, P., Damsa, T., Schioiu, L., Georgescu, M., Cucu, F., Mitu, S., and Voiculescu, M. (1986) *Med. Int.* **24,** 253–261.

[2] Harris, M. I., Hadden, W. C., Knowler, W. C., and Bennett, P. H. (1987) *Diabetes* **36,** 523–534.

[3] Tobia, A. J., Lee, J. Y., and Walsh, G. M. (1974) *Cardiovasc. Res.* **8,** 758–762.

[4] Tobia, A. J., Walsh, G. M., Tadepalli, A. S., and Lee, J. Y. (1974) *Blood Vessels* **11,** 287–294.

[5] Hsu, C. H., Slavicek, J. H., and Kurtz, T. W. (1982) *Am. J. Physiol.* **242,** H961–H966.

[6] Ferrone, R. A., Walsh, G. M., Tsuchiya, M., and Frohlich, E. D. (1979) *Am. J. Physiol.* **236,** H403.

[7] Bohlen, H. G (1983) *Hypertension* **5,** 739–745.

[8] Bohlen, H. G (1987) *Hypertension* **9,** 325–331.

[9] Harper, S. L. and Bohlen, H. G. (1984) *Hypertension* **6,** 408–419

[10] Werber, A. H. and Heistad, D. D. (1984) *Circ. Res.* **55,** 286–294.

[11] Zweifach, B. W., Kovalcheck, S., De, Lano, F., and Chen, P. (1981) *Hypertension* **3,** 601–614.

[12] Borders, J. L. and Granger, H. J. (1986) *Hypertension* **8,** 184–191.

[13] Vawter, D., Fung, Y. C., and Zweifach, B. W. (1974) *Microvasc. Res.* **8,** 44–52

[14] Gore, RW. and Bohlen, H. G (1977) *Am. J. Physiol.* **233,** H685–H693.

[15] Short, D. (1966) *Br. Heart J.* **28,** 184–191.

[16] Short, D. S. and Thomson, A. D. (1959) *J. Path. Bact.* **78,** 321–334.

[17] Hutchins, P. M. and Darnell, A. E. (1974) *Circ. Res.* **34,** 1–161.

[18] Bohlen, H. G (1979) *Am. J. Physiol.* **236,** HI57–H164.

[19] Prewitt, R. L. and Dowell, R. F. (1979) *Bibl. Anat.* **18,** 169–173.

[20] Prewitt, R. L., Chen, I. I. H., and Dowell, R. (1982) *Am. J. Physiol.* **243,** H243–H251.

[21] Greene, A. S., Tonellato, PJ., Lui, J., Lombard, J. H., Cowley, A. W., Jr. (1989) *Am. J. Physiol.* **256,** H126–H131.

[22] Meininger, G. A., Fehr, K. L., Yates, M. B., Borders, J. L., and Granger, H. J. (1986) *Hypertension* **8,** 66–75.

[23] Cox, R. H. (1979) *Am. J. Physiol.* **237,** H597–H605.

[24] Sullivan, J. M., Prewitt, R. L., and Josephs, J. A. (1983) *Hypertension* **5,** 844–851.

[25] Henrich, H. A., Romen, W., Heimgartner, W., Hartung, E., and Baumer, F. (1988) *Klin. Wochenschr.* **66,** 54–60.

[26] Azar, S., Tobian, L., and Johnson, M. S. (1974) *Am. J. Physiol.* **227**, 1045–1050.

[27] Englson, E. T., Schmid-Schönbein, G. W., and Zweifach, B. W. (1986) *Micro-vasc. Res.* **31**, 356–374.

[28] Gray, S. D. (1982) *Clin. Sci.* **63**, 383s–385s.

[29] Boegehold, M. A. and Bohlen, H. G. (1988) *Hypertension* **12**, 184–191.

[30] Folkow, B., Hallback, M., Lundgren, Y., and Weiss, L. (1970) *Acta Physiol. Scand.* **79**, 373–378.

[31] Folkow, B., Hallback, M., Lundgren, Y., Sivertsson, R., and Weiss, L. (1973) *Circ. Res.* **32**, 1,2.

[32] Nordborg, C., Fredriksson, K., and Johansson, B. B. (1985) *Stroke* **16**, 313–320.

[33] Mark, A. L. (1984) *Hypertension* **6**, III-69–III-73.

[34] Lee, R. M. K., Garfield, R. E., Forrest, J. B., and Daniel, E. E. (1983) *Blood Vessels* **20**, 57–71.

[35] Mulvany, M. J., Baandrup, U., and Gundersen, H. J. G. (1985) *Circ. Res.* **57**, 794–800.

[36] Furuyama, M. (1962) *Tohoku J. Exp. Med.* **76**, 388–414.

[37] Miller, B. G., Connors, B. A., Bohlen, H. G., and Evan, A. P. (1987) *Hypertension* **9**, 59–68.

[38] Baumbach, G. L., Dobrin, P. B., Hart, M. N., and Heistad, D. D. (1988) *Circ. Res.* **62**, 56–64.

[39] Bell, D. R. and Overbeck, H. W. (1979) *Hypertension* **1**, 78–85.

[40] Stacy, D. L. and Prewitt, R. L. (1989) *Am. J. Physiol.* **256**, H213–H221.

[41] Folkow, B., Garevich, M., Hallback, M., Lundgren, Y., and Weiss, L. (1971) *Acta Physiol. Scand.* **83**, 532–541.

[42] Folkow, B., Hallback, M., Lundgren, Y., and Weiss, L. (1973) *Acta Physiol. Scand.* **87**, 10A,11A.

[43] Hogan, R. D. and Hirschmann, L. (1984) *Microvasc. Res.* **27**, 290–296.

[44] Harper, S. L. (1987) *Circ. Res.* **60**, 229–237.

[45] Dusseau, J. W. and Hutchins, P. M. (1979) *Am. J. Physiol.* **236**, H134–H140.

[46] Owens, G. K. (1985) *Circ. Res.* **56**, 525–536.

[47] Hutchins, P. M., Marshburn, T. H., Maultsby, S. J., Lynch, C. D., Smith, T. L., and Dusseau, J. W. (1988) *Hypertension* **12**, 74–79.

[48] Wang, D.-H. and Prewitt, R. L. (1990) *Hypertension* **15**, 68–77.

[49] Bohlen, H. G., Gore, R. W., and Hutchins, P. M. (1977) *Microvasc. Res.* **13**, 125–130.

[50] Folkow, B., Hallback, M., Lundgren, Y., Weiss, L., Albrecht, I., and Julius, S. (1974) *Acta Physiol. Scand.* **90**, 654–656.

[51] Strandgaard, S. (1976) *Circulation* **53**, 720–726.

[52] Strandgaard, S., Oleson, J., Skinhoj, E., and Lassen, N. A. (1973) *Br. Med. J.* **1**, S07–511.

53 Harder, D. R., Smeda, J., and Lombard, J. (1985) *Circ. Res.* **57**, 319–322.

54 Gray, S. D. (1984) *Microvasc. Res.* **27**, 39–50.

55 Schmid-Schönbein, G. W., Firestone, G., and Zweifach, B. W. (1986) *Blood Vessels* **23**, 34–49.

56 Kern, T. S. and Engerman, R. L. (1986) *Metabolism* **355**, 24–27.

57 King, G. L. and Johnson, S. M. (1985) *Science* **227**, 1583–1585.

58 Yang, Y. J., Hope, I. D., Ader, M., and Bergman, R. N. (1984) *J. Clin. Invest.* **84**, 1620–1628.

59 Moore, S. A., Bohlen, H. G., Miller, B. G., and Evan, A. P. (1985) *Blood Vessels* **22**, 265–277.

60 Angervall, L. and Save-Soderbergh, J. (1966) *Diabetologia* **2**, 117–122.

61 Mayhan, W. G. (1989) *Am. J. Physiol.* **256**, H621–H625.

62 Rubin, M. J. and Bohlen, H. G. (1985) *Am. J. Physiol.* **249**, H540–H546.

63 Bohlen, H. G. (1983) *Circ. Res.* **52**, 677–682.

64 Ditzel, J. (1976) *Diabetes* **25**, 832–838.

65 Granger, D. N. (1988) *Am. J. Physiol.* **255**, h1269–h1275.

66 Rosenblum, W. I., El-Sabban, F., and Loria, R. M. (1981) *Diabetes* **30**, 89–92.

67 Rosenblum, W. I. and El-Sabban, F. (1983) *Microvasc. Res.* **26**, 254–257.

68 Schmid-Schönbein, H. and Volger, E. (1976) *Diabetes* **25**, 897–902.

69 Ernst, E. and Matrai, A. (1986) *Diabetes* **35**, 1412–1415.

70 Christensen, N. J. (1972) *Acta Med. Scand.* **541**, 3–60.

71 Bohlen, H. G. and Niggl, B. A. (1979) *Circ. Res.* **45**, 390–396.

72 Bohlen, H. G. and Niggl, B. A. (1979) *Blood Vessels* **16**, 269–276.

73 Bohlen, H. G. and Niggl, B. A. (1980) *Microvasc. Res.* **20**, 19–29.

74 Lillioja, S., Young, A. A., Culter, C. L., Ivy, J. L., Abbott, W. G. H., Zawadzki, J. K., Yki-Jarvinen, H., Christin, L., Secomb, T. W., and Bogardus, C. (1987) *J. Clin. Invest.* **80**, 415–424.

75 Saltin, B., Houston, M., Nygaard, E., Graham, T., and Wahren, J. (1979) *Diabetes* **28 (Suppl. 1)** 93–99.

76 Kang, S. S., Price, R. G., Bruckdorfer, K. R., Worchester, N. A., and Yudkin, J. (1977) *Biochem. Soc. Trans.* **5**, 235,236.

77 Papachristodoulou, D. and Heath, H. (1977) *Exp. Eye Res.* **25**, 371–384.

78 Papachristodoulou, D., Heath, H., and Kang, S. S. (1976) *Diabetologia* **12**, 367–374.

79 Singh, S. D. (1986) *Early Human Development* **14**, 89–98.

80 Bohlen, H. G. and Hankins, K. D. (1928) *Diabetologia* **22**, 344–348.

81 Unthank, J. L. and Bohlen, H. G. (1988) *Circ. Res.* **63**, 429–436.

82 Korthuis, R. J., Pitts, V. H., and Granger, D. N. (1987) *Am. J. Physiol.* **253**, G20–G25.

83 Korthuis, R. J., Benoit, J. N., Kvietys, P. R., Laughlin, M. H., Taylor, A. E., and Granger, D. N. (1987) *Am. J. Physiol.* **253**, G26–G32.

84 Yrle, L. F., Smith, J. K., Benoit, J. N., Grange, D. N., and Korthuis, R. J. (1988) *Am. J. Physiol.* **255**, G542–G546.

85 Carrier, G. O. and White, R. E (1985) *J. Pharmacol. Exp. Ther.* **232**, 682–687.

86 Krotkiewski, M. (1983) *Scand. J. Rehab. Med. Suppl.* **9**, 55–70.

87 Seals, D. R. and Hagberg, J. M. (1984) *Med. Sci. Sport. Exercise* **16(3)**, 207–215.

88 Liang, I. Y. S. and Stone, H. L. (1982) *J. Appl. Physiol.* **53(3)**, 631–636.

89 Laughlin, M. H. and Ripperger, J. (1987) *J. Appl. Physiol.* **62(2)**, 438–443.

90 Snell, P. G., Martin, W. H., Buckey, J. C., and Blomqvist, C. G. (1987) *J. Appl. Physiol.* **62(2)**, 606–610.

91 Poole, D. C., Mathieu-Costello, O., and West J. B. (1989) *Am. J. Physiol.* **256**, H1110–H1116.

92 Krotkiewski, M., Bylund-Fallenius, A.-C., Holm, J., Bjorntorp, P., Grimby, G., and Mansroukas, K. (1983) *Eur. J. Clin. Invest.* **13**, 5–12.

93 Mack, G. W., Shi, X., Nose, H., Tripathi, A., and Nadel E. R. (1987) *J. Appl. Physiol.* **63(1)**, 105–110.

94 Bedford, T. G. and Tipton, C. M. (1987) *J. Appl. Physiol.* **63(5)**, 1926–1932.

95 Ekblom, B., Kilbom, A., and Soltysiak, J. (1973) *Scand. J. Clin. Lab. Invest.* **32**, 251–256.

96 Overton, J. M., Tipton, C. M., Matthes, R. D., and Leininger, J. R. (1986) *J. Appl. Physiol.* **61(1)**, 318–324.

97 Pagani, M., Somers, V., Furlan, R., Dell'Orto, S., Conway, J., Baselli, G., Cerutti, S., Sleight, P., and Malliani, A. (1988) *Hypertension* **12**, 600–610.

98 Hagberg, J. M., Goldring, D., Ehsani, A. A., Heath, G. W., Hernandez, A., Schechtman, K., and Holloszy, J. O. (1983) *Am. J. Cardiol.* **52**, 763–768.

99 Ressl, J., Chrastek, J. J., and Andova, R. (1977) *Acta Cardiol.* **2**, 121–133.

100 Cox, R. H., Hubbard, J. W., Lawler, J. E., Sanders, B. J., and Mitchell, V. P. (1985) *Hypertension* **7**, 747–751.

101 Crisman, R. P., Rittman, B., and Tomanek, R. J. (1985) *Microvasc. Res.* **30**, 185–194.

102 Edwards, J. G., Tipton, C. M., and Matthes, R. D. (1985) *J. Appl. Physiol.* **58(5)**, 1683–1688.

103 Ganrot, P. O., Curman, B., and Kron, B. (1987) *Medical Hypotheses* **24**, 77–86.

104 Lithell, H., Krotkiewski, M., Kiens, B., Wroblewski, Z., Holm, G., Stromblad, G., Grimby, G., and Bjorntrop, P. (1985) *Diabetes Res. Clin. Pract.* **2**, 17–21.

105 Lash, J. M., Sherman, W. M., and Hamlin, R. L. (1989) *Diabetes* **38**, 854–860.

106 Ivy, J. L., Sherman, W. M., Culter, C. L., and Katz, A. L. (1986) *Am. J. Physiol.* **251**, E299–E305.

107 Lash, J. M., Sherman, W. M., Betts, J. J., and Hamlin, R. L. (1989) *Int. J. Obesity* **13**, 777–789.

108 Miller, B. G., Overhage, J. M., Bohlen, H. G., and Evan, A. P. (1985) *Microvasc. Res.* **29**, 56–69

Chapter 22

Acute-Microvascular-Injury Mechanisms

J. Jeffrey Marshall and Hermes A. Kontos

Introduction

A number of vascular-disease processes damage resistance microvessels. These include acute[1] and chronic hypertension,[2] diabetes mellitus,[3,4] some connective tissue diseases,[5] burn[6] and frostbite[7] injuries, and injury from ischemia followed by reperfusion.[8–11] Damage to microvessels poses a precarious insult to the organism because the microvasculature is ultimately responsible for oxygen delivery to the tissues. It has recently been shown that autocoid products of the endothelium, such as endothelium-derived relaxing factor(s) (EDRF), play an important regulatory role in microvascular tone.[12,13] This endothelial control of vascular smooth-muscle function is now known to be an active component of microvascular resistance and is therefore an important regulator of blood flow for critical metabolic processes. It is now postulated that, in a number of disease states, microvascular endothelial dysfunction, structural damage to the endothelium of microvessels, or abnormal endo-

From *The Resistance Vasculature*, J. A. Bevan et al., eds. ©1991 Humana Press

thelium–blood-component interactions (e.g., of endothelial cells of the microcirculation with platelets and neutrophils) are crucial pathophysiologic events that lead to microvascular damage and the death of cells dependent on the microvasculature blood flow. Oxygen radicals have been implicated as a mechanism of vascular injury in a number of diverse disease states. Recently the interaction between oxy radicals and microvascular endothelium-dependent function has gained considerable interest. Below, we review some of the acute mechanisms that damage resistance blood vessels.

Fundamental Oxygen-Radical Chemistry

Nomenclature

The term *free oxygen radical* carries with it some historical baggage; namely, the word "free." This dates to a physical chemistry debate over whether or not molecular radicals could exist in a free form. The term oxygen radical, or oxy radical, is probably less ambiguous than terminology including the expression "free."

Definition

Molecules having one or more unpaired electrons in their molecular orbitals are called radicals.[14] Molecular oxygen (O_2) is actually a biradical, because each of its outer orbitals has a single unpaired electron. The spin directions of the electrons in each of these outer orbitals are the same.[15] This orbital arrangement of electrons makes molecular oxygen a very unreactive molecule. However, oxygen can exist in a number of other oxidation states.[16,17] If oxygen loses a single electron (e^-) the oxidation state changes and the superoxide anion ($\cdot O_2^-$) is formed. Loss of an electron from superoxide again changes the oxidation state and hydrogen peroxide (H_2O_2) can be formed. Loss of a single electron from hydrogen peroxide results in another oxidation state and the hydroxyl radical ($\cdot OH$) is formed. The loss of another electron from the hydroxyl radical forms water (H_2O). This process is termed the univalent reduction of oxygen[16,17] and is represented as follows (formula 1).

$$O_2 \xrightarrow{e^-} \bullet O_2^- \xrightarrow{e^-} H_2O_2 \xrightarrow{e^-} \bullet OH \xrightarrow{e^-} H_2O \qquad (1)$$

These intermediate oxidation forms of oxygen—superoxide, hydrogen peroxide, and the hydroxyl radical—are much more unstable, and therefore more reactive, than oxygen and water.

Superoxide anion can act as either an oxidizing or a reducing agent. Its preferred reaction at physiological pH, in aqueous media, is the dismutation reaction producing hydrogen peroxide (formula 2).[16–18]

$$2 \bullet O_2^- + 2H^+ \rightarrow H_2O_2 + O_2 \qquad (2)$$

Superoxide is water-soluble and can cross cell membranes throuah the anion channel in cultured endothelial cells,[19,20] red blood cells,[21] and in cerebral microvessels in vivo.[22] The passage of superoxide through this channel can be blocked with anion-channel inhibitors.[22]

Hydrogen peroxide is not a radical, but is a strong oxidizing agent in biological systems, and can thus can be a harmful entity, much like a true radical. Since it is lipid-soluble, it can cross cell membranes easily.[15] It is usually formed in biological systems by the dismutation of superoxide anion (formula 2).

The hydroxyl radical is a strong oxidizing agent and is extremely reactive with biomolecules.[23] However, its high reactivity leads to its characteristic, short half-life and limits the physical distance that the molecule can travel before it reacts with other molecules.[24] Once formed, the hydroxyl radical can wander only a few molecular radii; thus, its half-life is only a few microseconds.

Sources of Oxygen-Radicals in the Microvasculature

The sources of oxygen radicals that can alter the normal function of the microvasculature are numerous and are shown in Table 1. The oxygen transport chain within the mitochondria of eukaryotic cells oxidizes oxygen to carbon dioxide and water through a number of oxidation–reduction pairings. Only a small percentage of the total oxygen consumption from mito-

Table 1
Sources of Oxygen Radicals
in the Microvasculature

Mitochondrial leak
Autooxidation of biomolecules
Xanthine oxidase
Unsaturated fatty acid metabolism
    via cyclooxygenase
Polymorphonuclear leukocytes

chondria produces superoxide.[16,25] Most of this small percentage comes from autoxidation of redox compounds within the electron transport chain. Specifically, autoxidation of ubisemiquinone produces the majority of superoxide from the mitochondria source.[26–28] However, NADH dehydrogenase,[26] and dihydroorotic dehydrogenase[28] also produce a small portion of the mitochondrial superoxide. In nonpathological states, this superoxide production proceeds at a very small rate and probably has no deleterious effects. However, in states of anoxia, with reperfusion, or during hyperbaric oxygen conditions, the rate of superoxide production from this source can be significant.[29]

Biomolecules containing iron and other transition metals can also produce superoxide from autoxidation. These include the hemoproteins hemoglobin[30,31] and myoglobin,[30] as well as catecholamines,[32] flavens,[33] and thiols.[34,35]

One of the most notable sources of superoxide production is the oxidative enzyme xanthine oxidase.[36] This enzyme, which is found in fairly high concentration in endothelial cells,[37] from a number of different organs, normally exists in the dehydrogenase form. The dehydrogenase form of the enzyme catayzes the transfer of one electron from xanthine to NAD, whereas the oxidases form of the enzyme oxidizes xanthine to hypoxanthine and uric acid.[37–39] This latter form of the enzyme is capable of transferring electrons directly to molecular oxygen, resulting in superoxide production. The conversion of xanthine dehydrogenase to xanthine oxidase (termed D-to-O conversion) occurs under conditions of ischemia or severe hypoxia that are followed by reperfusion. This conversion appears to be the result of a proteolytic action on thiol groups of the enzyme.[40–42] The xanthine oxidase enzyme can also catalyze the production of hydrogen peroxide via two electron reduction of molecular oxygen.[36]

Another intriguing mechanism of oxygen-radical production is the metabolism of unsaturated fatty acids such as arachidonate. Arachidonic acid can be metabolized via three metabolic pathways: cyclooxygenase,[43] the P450 monooxygenase system,[44,45] and lipoxygenases.[43] All three of these enzyme systems are capable of generating superoxide by a side chain reaction. The best-studied of these is the cyclooxygenase reaction. The cyclooxygenase enzyme complex has two enzymatic functions. The first is the endoperoxidase activity that converts arachidonate to the endoperoxide PGG_2. PGG_2 is then acted on by the peroxidase activity of the enzyme to produce PGH_2. It is the peroxidase activity of the enzyme that produces superoxide by a side chain reaction.[43] When PGG_2 reacts enzymatically with the peroxidase function of the enzyme, the enzyme molecule is converted to an active intermediate that is a radical form of the enzyme with the unpaired electron either on or near the iron molecule in the enzyme.[46] This radical intermediate is capable of reacting with NADH or NADPH, which converts them to their radicai forms (NAD• or NADP•).[47] These radicals in turn can then react directly with molecular oxygen to produce superoxide. This side chain reaction has a number of important biological characteristics:

1. The yield of superoxide from this reaction can approach 1 mol of superoxide per 10 mol of arachidonate metabolized.[43]
2. This production occurs in a burst-like manner and then rapidly subsides to a fraction of its initial high rate as a result of self-inactivation of the enzyme.
3. NADH and NADPH are rate-limiting constituents of this superoxide production. It is not known, however, what concentrations of these reduced nicotinamide adenine dinucleotides are in close proximity to the cyclooxygenase.
4. Nonarachidonate fatty acids (e.g., linolenic acid or PGG_2) can serve as substrates for this reaction.

Polymorphonuclear leukocytes are another source of oxygen radicals.[48] These white blood cells generate superoxide from the NADPH oxidase enzymes found on the surface of these cells. This enyme produces superoxide that is intended for the bactericidal action of neutrophils. Neutrophils also produce other long-lived oxygen-based radicals from myeloperoxidase reactions.

The production of hydrogen peroxide and hydroxyl radical are dependent on the production of superoxide anion. Hydrogen peroxide is formed by the spontaneous dismutation reaction between two superoxide anion molecules (formula 2). This reaction occurs spontaneously at physiological pH, but proceeds much more rapidly in the presence of the enzyme superoxide dismutase.[16–18] Since superoxide dismutase is a ubiquitous enzyme in the cells and fluid of the microcirculation, it is probably present during times of superoxide anion production. Hydroxyl radical production is dependent on both the presence of superoxide and hydrogen peroxide. These two reactive oxygen species can react spontaneously to generate hydroxyl radical by the uncatalyzed Haber-Weiss reaction (formula 3), but the rate constant for this reaction under physiological conditions predicts that the reaction occurs at insignificant rates.

$$H_2O_2 + \bullet O_2^- \rightarrow \bullet OH + OH^- + O_2 \qquad (3)$$

However, in the presence of a catalytic transition metal like ferrous iron, this reaction occurs at significant rates via the iron-catalyzed Haber-Weiss reaction (formula 6).[49–51] The reaction sequences shown below depict the chemistry of iron acting as a Fenton reagent to catalyze the Haber-Weiss reaction. Ferrous iron reacts with hydrogen peroxide to form the hydroxyl radical and ferric iron (formula 4).

$$H_2O_2 + Fe^{2+} \rightarrow \bullet OH + OH^- + Fe^{3+} \qquad (4)$$

Superoxide reduces ferric iron back to ferrous iron (formula 5)

$$Fe^{3+} + \bullet O_2^- \rightarrow Fe^{2+} + O_2 \qquad (5)$$

making it available for hydroxyl radical production by formula 4, completing the catalytic cycle. The summation of these two formulas (4 and 5) results in the iron catalyzed Haber-Weiss reaction (formula 6).

$$Fe^{2+} \quad Fe^{3+}$$

$$H_2O_2 + \bullet O_2^- \rightarrow \bullet OH + OH^- + O_2 \qquad (6)$$

This reaction can also be catalyzed by transferrin or ferritin. Interestingly, superoxide can release iron from ferritin in vitro;[51] however, the exact mechanism by which iron becomes available to act as a Fenton reagent for the Haber-Weiss reaction is not known with certainty. One important aspect of this reaction is that the reactivity of the hydroxyl radical is so high that this molecule probably reacts within one or two nuclear radii from the point at which it is generated.[23,52] This high reactivity dictates that the hydroxyl radical must be produced at or near the cellular elements it is to oxidize.[24]

Oxygen-Radical Defense Mechanisms

Since low levels of oxygen-radicals are being formed at baseline from autoxidation, mitochondrial leak, and activation of neutrophils, it is important that a cellular defense mechanism against these damaging species exist.

Superoxide anion is eliminated by superoxide dismutases, which catalyze the conversion of superoxide to hydrogen peroxide and oxygen (formula 2). In mammalian cells there are two types of this enzyme. The first is a cytosolic copper- and zinc-containing enzyme; the other is a manganese-containing enzyme found only in mitochondria. Additionally, extracellular fluid also contains a high mol wt superoxide dismutase (SOD) known as EC-SOD.[53]

Cellular protection from hydrogen peroxide-mediated damage is accomplished by two enzyme systems.[25] The first, glutathione peroxidase, is present at high concentration and is effective for very low levels of hydrogen peroxide.[25] The second, catalase, is present at lower concentration, but is more important when the concentration of hydrogen peroxide is higher.

The hydroxyl radical has no specific biological scavengers. However, a number of cellular antioxidants are capable of extinguishing the hydroxyl radical. These would include α-tocopherol, or vitamin E; beta carotenes; and ascorblc acid.[52,54]

Oxygen-Radical-Induced Injury of Cellular Components

Radical-mediated cellular injury depends on a number of interrelated factors, including (1) the rate and site of radical production and (2) the level of antioxidant protective mechanisms. If the amount of oxygen-radical production outstrips the cellular protective mechanisms, then cellular damage occurs. Oxygen radicals are known to have a number of damaging effects on cellular components (Table 2).

The lipid bilayers of mammalian cells contain large amounts of polyunsaturated fatty acids.[55] Oxygen radicals react with the double bonds in these unsaturated fatty acids to cause lipid peroxidation and fragmentation products, such as malondealdehyde. Once an oxygen radical reacts with one of these unsaturated fatty acids, it produces a chain reaction within the lipid bilayer. The oxy radical produces a lipid radical in the bilayer, which then reacts with an adjacent lipid to form a second lipid radical, leaving the first lipid molecule oxidized. This process propagates destruction in a chain-reaction fashion throughout the membrane. Oxidation of these molecules alters their inherent stereospecificity, which is necessary for appropriate alignment within the lipid bilayer.[58] This alteration increases cellular permeability, which results in abnormal cell-ular ion fluxes that can greatly reduce the functional integrity of the cell. In addition, membranes of other intracellular org-anelles can be affected by the same mechanism.[59]

Oxy radicals also react with essential proteins, usually by way of sulfydryl oxidation[60] or hydroxylation[61] of certain amino acids. This type of radical-mediated damage alters the stereospecificity of proteins and enzymes, rendering them inactive. There is evidence that a number of critical cellular enzymes are inactivated by oxygen radicals (e.g., the glycolytic enzymes, glyceraldehyde 3-phosphate dehydrogenase, and sodium potassium ATPase).[62] In addition, oxygen radicals are

Table 2
Effects of Oxygen Radicals
on Microvascular Cellular Components

Lipid peroxidation
Protein denaturation
Inactivation of enzymes
Release of calcium ions
    from intracellular stores
Damage to cytoskeleton
Damage to tissue structural
    components
Chemotaxis
Nucleic acid and DNA damage

capable of destroying the protein scavengers that protect the cell from these injurious agents (e.g., superoxide inactivates catalase and glutathione peroxidase;[63] hydrogen peroxide inactivates superoxide dismutase[64]). Oxy radicals can also damage proteins indirectly. Radicals are capable of causing calcium release from intracellular stores.[23] This is known to activate proteolytic enzymes, such as phospholipases, that can inactivate critical enzymes.

Oxygen radicals are capable of causing structural damage to cells. They can polymerize hyaluronic acid[65] and damage collagen.[66] Oxy radicals can also damage the cellular cytoskeleton either by interacting with thiol groups on actin[67] or indirectly through the calcium-activation process of phospholipases that can attack actin.[67] These processes are thought to contribute to the bleb formation seen by electron microscopy in microvessels following radical damage.

Finally, oxygen radicals stimulate further cellular damage through their chemotactic actions on leukocytes.[68] The interaction of superoxide and tissue components appears to produce a powerful chemotactic factor for neutrophils, which then become activated and further damage tissues with radicals and proteases. The rapidity with which leukocytes are attracted after radical injury is a function of the different microcirculatory beds.[22,68,69]

Oxygen–Radical–Induced Dysfunction and Injury of Microvessels

The univalent reduction of oxygen results in the production of the biologically active oxygen species superoxide anion radical, hydrogen peroxide, and hydroxyl radical. These products have been shown to have profound effects on the structure and function of the resistance vasculature (Table 3).

In the cerebral microcirculation of the cat, superoxide anion, hydrogen peroxide, and hydroxyl radical all dilate cerebral microvessels.[70,71] Oxygen radicals also dilate cerebral arterioles in the mouse[72] and rat,[71] as well as in the microcirculation of the rat cremaster muscle and the intestinal microcirculation of the cat.[73] However, in each of these tissues, it appears that different oxygen-based radicals are the ultimate mediators of the dilation. For instance, in the skeletal muscle of the rat, it appears that hydrogen peroxide is the important species,[71] whereas in cerebral arterioles it appears that hydroxyl radical is the primarily active agent,[70,72] and in the intestinal microcirculation of the cat superoxide radical is most important at physiological pH.[74] In the pial circulation of the cat, the effects of hydrogen peroxide cause some interesting results. In low concentrations, hydrogen peroxide does not affect vessel tone.[75] However, in higher concentrations (3 μM) vasodilation occurs and there is evidence of irreversible functional damage to the endothelium. The case for hydrogen peroxide in the cerebral arterioles of the cat alludes to the complexity of the regulation of microvascular tone and of the interactions oxygen radicals have in these mechanisms.

The interactions between oxy radicals and endothelium-dependent dilation are of particular interest, since loss of a potent vasodilator could cause vasoconstriction and potentiate tissue ischemia. Furchgott and Zawadzki[76] discovered endothelium-derived relaxing factor (EDRF) in 1980 and, since then, investigators have been trying to determine how EDRF might participate in vascular physiology and pathophysiology. The classical EDRF from acetylcholine has been shown to be a highly diffusible,[77] short-lived substance,[77–79] generated by the intact

Table 3
Effects of Oxygen Radicals on the Microvasculature

Alterations in vessel tone
Alterations in vessel reactivity
Alterations in endothelium-dependent functions
Increased platelet aggregability
Increased neutrophil adhesion
Increased endothelial permeability
Destructive lesions of the vessel wall
Reduced oxygen consumption of the vessel wall

endothelium, that rapidly diffuses into surrounding smooth-muscle, where it causes relaxation of vascular smooth muscle. EDRF activates the soluble guanylate cyclase enzyme[80–83] within the vascular-smooth-muscle cytoplasm leading to increased concentrations of cyclic guanosine monophosphate (cGMP). Through an as yet undefined process, increased intracellular cGMP levels result in relaxation and vasodilation of smooth muscle (Fig. 1).

The presence of EDRF in the microcirculation has now been clearly shown in the pial circulation of the cat,[13] the cerebral microcirculation of the mouse[17,84-88] and rat,[89] the microcirculation of the rat[90,91] and rabbit[92] cremaster muscle, the microcirculation of the heart,[31] and the microcirculation in the hindlimb of a number of animals.[93,94] In addition, EDRF may play an important role in the response of microvessels to increased flow by modulating vessel tone and reactivity. Thus, EDRF may play an important role in the regulation and distribution of blood flow through the microcirculation.[95,96] In the microcirculation of the rat cremaster muscle the endothelium has been shown to mediate the dilation as a result of increased flow.[12] However, in this preparation, the response to increased flow is mediated by a product of the cyclooxygenase enzyme and not the classic EDRF.[97] Therefore many investigators are now beginning to determine whether pathophysiological states alter endothelium-dependent mechanisms. This is especially true of EDRF–free-radical interactions, since it was shown, early on, that

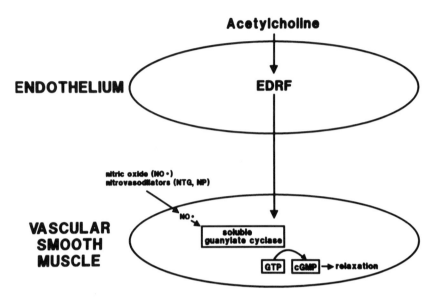

Fig. 1. Mechanism of vasodilation from the endothelium-dependent dilator acetylcholine and direct acting nitrodilators. GTP, guanosine triphosphate; cGMP, cyclic 3', 5' guanosine monophosphate.

oxygen radicals could destroy EDRF[13,98,99] and promote vasoconstriction.

EDRF–oxygen-radical interactions have been well-studied in the pial microcirculation of the cat. Oxygen radicals can interfere with endothelium-dependent function by two major mechanisms: First, oxy radicals can destroy the endothelium, resulting in inability to produce or secrete EDRF. Second, the oxygen radicals can inhibit vasodilation by reacting with EDRF and inactivating it in the extracellular space prior to relaxing the vascular smooth muscle. There is evidence for both of these mechanisms. Acetylcholine normally results in vasodilation when applied topically to the surface of pial microvessels in the cat.[13] However, after simultaneous application of or pretreatment of cerebral microvessels with agents known to generate oxygen radicals (e.g., methylene blue[13,100] or hemoalobin[30,31]) the relaxation from acetylcholine is abolished.[13] The vasodilation can be restored by application of superoxide dismutase (SOD) and catalase or by the iron-chelating agent deferoxamine.[13] In addition, removal of the oxygen-radical-generating agent, methylene blue or hemoglobin, restores the ability of these vessels to dilate in response to acetylcholine.[13] This oxy radical effect is specific for EDRF-mediated dilation, since it does not affect the vasodilation from the nitrodilators.[13] These studies show that the EDRF from acetylcholine is destroyed by the hydroxyl radical generated by the iron-catalyzed Haber-Weiss reaction from precursor superoxide and hydrogen peroxide.[49] The inhibition of acetylcholine-mediated dilation is immediate and reversible. Once the radicals are removed, either by scavenging the radicals produced or by removing the generating system, responses to acetylcholine normalize. Additionally, since SOD and catalase are large molecules and lack intracellular access, this scavenging of exogenously generated free radicals must be occurring in the extracellular spaces. Therefore, in this series of experiments the endothelium is capable of EDRF production, but the EDRF cannot initiate vasodilation because it is oxidized before it can enter the smooth-muscle cells to activate the soluble uanylate cyclase enzyme.

In other experiments, high-dose arachidonate appears to damage the endothelium permanently, rendering it unable to produce or secrete any EDRF. Topical application of high-dose arachidonate, which generates free radicals via cyclooxygenase, abolished the vasodilation from acetylcholine.[101] This, however, cannot be reversed by topical application of scavengers and, even when the arachidonate is removed, the vessels no longer dilated in response to acetylcholine. Electron microscopy reveals that high-dose arachidonate damages the endothelium of these microvessels.[78,102] Blebs and vacuoles are seen on the endothelial glycocalyx after the application of arachidonate.[70] This physical damage can be attenuated by pretreatment with free-radical scavengers; thus, it is a radical-mediated process. Hence, oxygen radicals can destroy endothelium-dependent dilation by irreversibly damaging the microvascular endothelium.

It is also noteworthy that the amount of radical generated can dictate the effects of the radicals on endothelium-dependent dilation. This is certainly the case with hydrogen peroxide. In low doses (1 μM), hydrogen peroxide inhibits acetylcholine-mediated dilation, and no EDRF can be demonstrated by in vivo bioassay experiments (a technique that can assay for the presence or absence of the EDRF).[77] Addition of SOD plus catalase restores the dilation from acetylcholine, even in the presence of hydrogen peroxide. On the other hand, high doses of hydrogen peroxide (3 μM) eliminate the vasodilation seen from acetylcholine, and SOD and catalase cannot reverse this.[75] In addition, EDRF is absent by bioassay, despite the presence of radical scavenging agents SOD and catalase.[75]

Oxygen radicals also impair the physiological responsiveness of vascular smooth muscle to the vasoconstrictor effects of hypocapnia[70,71,103] and the vasodilator effects of hypercapnia.[1] Furthermore, oxy radicals diminish the physiological vasodilation seen in cerebral arterioles in response to hypotension.[1] These important physiological rssponses are necessary to maintain autoregulation of the cerebral circulation.

Exposure of brain microvessels to oxygen-radical-generating agents promotes platelet aggregability. This effect is probably mediated through destruction of EDRF[104] and prostacycline,[105,106] which are potent inhibitors of platelet aggregation.[107]

Oxy-radical-mediated damage to endothelial cells can also alter important neutrophil–endothelial-cell interactions. This can promote further damage to endothelial cells, increase vascular permeability around damaged endothelial cells, and may lead to plugging of microvessels with inflammatory cells.

Oxygen radicals can also impair the metabolic energy-producing systems within cells.[108] Reduced oxygen consumption of cerebral microvessels following radical exposure is probably attributable to damage of mitochondria.

Cell–Cell Interactions

As early as 1968 Ames and his colleagues[109] showed that microvascular sludging and obstruction occurred following ischemia. This was termed the no-reflow phenomenon and has been observed by others studying different vascular beds.[110] This phenomenon brings to light an important mechanism of micro-vascular damage: cell–cell interactions within the microcirculation. The no-reflow phenomenon was shown to be injurious when tissue damage was reduced by making animals neutropenic prior to a pathophysiologic insult or by inhibiting neutrophil adhesion with antibodies.[111,112]

The interface of this cell–cell interaction is the microvascular endothelium. Therefore, studies into these interactions have investigated endothelial derangements caused by the cellular components of blood. The most frequently studied and probably the most important is the interaction between neutrophils and the endothelium. The initial step in this deleterious interaction is the adherence of neutrophil to the endothelial cell surface.[113] The factors that promote this include decreased blood velocity, hypoxic endothelial cells, and, probably, changes in surface receptors on both endothelial and white blood cells.[9] Once neutrophils have attached to damaged endothelial cells, a number of destructive neutrophil processes can be activated. A burst of oxygen-radical generation can occur if oxygen is present or if it is reintroduced.[114] This leads to radical production in close proximity to the endothelium. If the neutrophils become activated by chemotactic factors, then not only are oxy radicals released in close proximity to vulnerable endothelial lipid bilayers, but also a number of destructive enzymes are released upon neutrophil degranulation. Lysosomal enzymes, such as elastase, can destroy basement membrane and alter the structural integrity of the microvessel.[115] Hypochlorous acid formed by myeloperoxidase is a particularly toxic compound capable of causing further vascular injury. Other proteases can also denature proteins in the lipid bilayer of the cell membrane. Furthermore, radical-mediated lipid peroxidation can alter the foundation that proteins need to maintain their stereo-specificity within the membrane. Thus, damage to the bilayer matrix can cause shifting in protein confirmations and destroy the activity of the proteins without denaturing them.[58]

The endothelial cells have competing mechanisms that are aimed at preventing neutrophil adherence and resultant damage. These include production of vasodilator substances, such as prostacyclin and EDRF, as well as adenosine and cyclic AMP.[113,116,117] Furthermore, the intercalated membrane scavengers, such as beta carotene, α-tocopherol, and the glutathione reductase system, resist neutrophil-mediated oxy-radical damage.[52,54]

Thus, the overall extent of damage to the microvessel by neutrophil–endothelium interactions that could result in total occlusion of the microvessel and anoxia to distal tissues is the aggregate of the destructive neutrophil component and the protective endothelial components. The complex balance between these two opposing forces will determine the degree of microvascular damage. This complex interaction varies among different microvascular beds in its time of onset and severity. For instance, it appears to be an early phenomenon in the microvascular damage in cardiac beds,[111,112] but appears to be a much later occurrence in microvascular beds of the central nervous system .[22,69] The reasons for these differences are not known. It is, however, known that continued neutrophil–endothelial cell interactions result in further destruction of endothelial cells, thus, agents aimed at reducing or preventing these interactions may have therapeutic potential.[111,112] In addition to the destructive effects of neutrophil–endothelial-cell interactions on endothelial cells, the simple physical adherence of multiple neutrophils to microvessels is likewise detrimental, since it can obstruct flow.

Acute Microvascular-Injury Syndromes

Acute Hypertension

Acute, severe hypertension causes damage to microvessels in the kidney, retina, and brain. In the clinical setting, malignant hypertension often causes altered mental status, probably as a result of damage of cerebral microvessels, hematuria from renal microvascular damage, and retinal hemorrhages from injured retinal microvessels. The mechanisms of this type of microvascular damage have been extensively studied in the cerebral microcirculation of the cat. Acute severe hypertension caused by drug infusion[118] or fluid-percussion brain injury[119] has been shown to cause damage to cerebral microvascular endothelium, which consists of blebs and crypts on the surface of the cells. There is very little damage to the vascular smooth muscle.[1] The pial arterioles respond to acute severe hypertension with prolonged vasodilation that can be minimized by topical application of SOD and catalase. Similar vasomotor disturbances and morphologic changes seen with electron microscopy are produced when high-dose arachidonate[22,70] or bradykinin[1,70] is applied topically to these pial vessels. Both the morphologic and functional changes in these vessels can be inhibited with indomethacin. These studies demonstrate that acute severe hypertension causes the production of oxygen radicals, which are generated from the action of cyclooxygenase on arachidonic acid. The mechanism of this oxy-radical generation is from the peroxidase function of the cyclooxygenase enzyme complex on arachidonate.[43] Since bradykinin is known to generate oxygen radicals following topical application on pial microvessels and since kallikrein can activate the same mechanism by generating bradykinin, we have proposed that acute hypertension stimulates kallikrein release that activates bradykinin. Bradykinin-stimulated arachidonate release from cell membranes can then be acted on by the cyclooxygenase system to produce superoxide anion and ultimately the hydroxyl radical. This is shown diagrammatically in Fig. 2. This schema is further supported by the fact that detailed light microscopic and electron microscopic studies of pial vessels damaged by acute hypertension failed to show infiltration of white blood cells at the time when the cell damage had already occurred. This suggests that neutrophil–endothelial-cell interactions are not responsible for the oxygen-radical-mediated damage in acute hypertension in the cerebral microcirculation of the cat. Neutrophil infiltration into these damaged vessels does not occur until 4–6 h after the acute hypertension, and does not peak until almost 24 h after the injury.[22,69]

In order for oxy radicals to damage the luminal side of endothelial cell membranes, they must enter the extracellular space. The superoxide formed from accelerated arachidonate metabolism by cyclooxygenase exits the endothelial cell through

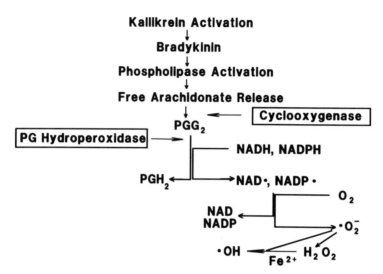

Fig. 2. Proposed mechanism of hydroxyl-radical generation from brady-kinin, arachidonic acid, and kallikrein activation. PG, prostaglandin; NADH, NADPH, reduced nicotinamide adenine dinucleotide and nicotinamide adenine dinucleotide phosphate, respectively; NAD•, NADP•, free-radical intermediate of nicotinamide adenine dinucleotide and nicotinamide adenine dinucleotide phosphate, respectively; NAD, NADPH, oxidized nicotinamide adenine dinucleotide and nicotinamide adenine dinucleotide phosphate, respectively; O_2, molecular oxygen; •O_2^-, superoxide anion; •OH, hydroxyl radical; H_2O_2, hydrogen peroxide; Fe^{2+}, ferrous iron.

the anion channel. This is known to occur in acute hypertension since extracellular superoxide production has been demonstrated by the SOD-inhibitable portion of nitro-blue tetrazolium (NBT) reduction.[120] Blockade of the anion channel with phenylglyoxal prevented the escape of superoxide into the extracellular space, resulting in no NBT reduction.[120] Additionally, superoxide production has been demonstrated histochemically in endothelial peroxide production by these cells after a brief episode of acute hypertension proceeds for up to 1 h after the original insult. The alterations in vasomotor tone and edema formation, and alterations in the blood–brain barrier, could contribute to the clinical entity of hypertensive encephalopathy.

Ischemia/Reperfusion Injury

Ischemia followed by reperfusion is a pathophysiological state that is responsible for a number of clinical syndromes associated with vasoocclusive atherosclerotic diseases. Stroke, myocardial infarction, and intestinal ischemia all demonstrate the pathophysiology of ischemia/reperfusion injury.

CEREBRAL MICROCIRCULATION. In the cerebral microcirculation of cats, superoxide production has been demonstrated in ischemia/reperfusion injury by the SOD-inhibitable portion of NBT reduction.[121] Additionally, superoxide production has been shown histochemically.[122,123] Using ESR and spin-trapping techniques,[124] radicals have been shown in the cerebral spinal fluid after cerebral ischemia. A number of lines of information support an oxygen-radical mechanism of cellular dysfunction in ischemia reperfusion of the central nervous system.

In the cerebral circulation, ischemia/reperfusion injury has been shown to cause prolonged vasodilation of pial microvessels,[125] alterations in neurologic functions, and breakdown of the blood–brain barrier,[126,127] along with extravasation of proteins, edema formation, and loss of endothelium-dependent relaxation.[10,128] After this early phase, hypoperfusion[129] may replace the hyperemia by a mechanism that is not entirely clear, but that may include the no-reflow phenomenon.[129]

A number of radical-scavenging interventions support the role of oxygen radicals in the damage from ischemia reperfusion. Application of SOD plus catalase or deferoxamine has been shown to minimize the morphologic damage seen from ischemia/reperfusion injury.[10] In addition, two investigators have shown that the endothelium-dependent relaxation from acetylcholine is abolished by ischemia/reperfusion injury[10,128] and can be protected by oxygen-radical-scavenging agents. Oxygen radicals, most probably hydroxyl radicals, destroy EDRF secreted in response to acetylcholine before it can cause vasodilation.[13] Although functional and morphologic changes can be reversed or minimized with scavenging agents, the neurologic sequelae of ischemia reperfusion are more difficult to quantitate. Some have shown that deferoxamine can improve survival following total cerebral ischemia in rats,[130] whereas in dogs deferoxamine did not influence the neurologic outcome.[131] Systemically administered SOD has been shown to improve mortality in gerbils injured by ischemia/reperfusion injury.[132] However, others have shown that SOD did not alter the secondary hypoperfusion and ischemia seen in rats following ischemia/reperfusion injury.[129,133]

The sources of the radical generation in cerebral ischemia/reperfusion injury seem to vary with the species studied. In cats, it appears that accelerated arachidonate metabolism via the cyclooxygenase system is the source.[43,134] Indomethacin preserved the functional characteristics of the microvasculature and prevented morphologic changes in these vessels. In gerbils, indomethacin improved recovery following ischemia/reperfusion injury. In rats, the conversion of xanthine dehydrogenase to xanthine oxidase with subsequent oxygen-radical production appears to play a major role. Animals fed tungsten, which replaces the molybdenum at the active site of the xanthine oxidase enzyme, rendering the enzyme inactive, reduced the ischemic damage from ischemia/reperfusion injury.[132] Others have shown that leukocytes may be important in ischemia/reperfusion injury in rats. Administration of an antineutrophil antibody reduced the cerebral hypoperfusion seen in the late stages of ischemia/reperfusion injury.[129]

Thus, in cerebral ischemia/reperfusion injuries there is evidence for oxy radical-mediated functional, morphologic, and neurologic damage. The sources of the oxygen radicals include the D-to-O conversion of xanthine dehydrogenase, accelerated arachidonate metabolism by cyclooxygenase, and neutrophil-mediated injury.

INTESTINAL MICROCIRCULATION. The microcirculation of the small bowel is a well-studied model of ischemia/reperfusion injury. An oxygen-radical-mediated mechanism of microvascular injury in the intestinal microcirculation has been demonstrated by the use of radical-scavenging agents. Pretreatment of cats with systemic SOD resulted in marked reduction in intestinal capillary permeability after ischemia/reperfusion injury.[135] Not only was SOD effective, it also demonstrated a dose-dependent protective effect on increases in radical-mediated vascular permeability. The protective effect of SOD was potentiated when the half-life of SOD was prolonged by functional nephrectomy (SOD is cleared by the kidneys), resulting in further reductions in intravascular fluid losses. In addition, dimethyl-sulfoxide (a hydroxyl-radical scavenger) and catalase both significantly inhibited the increased permeability mediated by ischemia/reperfusion injury.[73] These inhibitor data implicate the hydroxyl radical as the primary oxygen radical mediating this microvascular permeability alteration. The injury pattern seen with ischemia/reperfusion injury was reproduced when the lumen of the intestine was exposed to the xanthine/xanthine oxidase radical-generating system, and this could be inhibited with SOD.[136] Furthermore, intraarterial injection of xanthine/xanthine oxidase produced increased mucosal permeability and histologic changes similar to that seen with ischemia/reperfusion injury. This could be abolished by SOD.[137] Neither indomethacin, antihistamines, nor steroids affected the increased permeability from intestinal ischemia followed by reperfusion.[135] These data suggested that vasodilator prostaglandins, oxy radicals from accelerated arachidonic metabolism, histamine, and lysosomal enzymes had no role in this form of injury in the small intestine.[135] Histologic studies revealed that neutrophil infiltration was not an early event in intestinal ischemia/reperfusion injury.[138] However, it was known that the concentration of xanthine dehydrogenase in intestine was very high compared to other organs. It was also shown that the D-to-O conversion occurred very rapidly in the small intestine.[139] Experiments utilizing inhibitors of xanthine oxidase showed that the increased permeability from ischemia/reperfusion injury, as well as the histologic changes seen in the mucosa, could be inhibited with these agents.[140] Finally, the functional significance of these mucosal injuries was established when SOD and an inhibitor of D-to-O conversion attenuated the loss of disturbed intestinal water absorption.[73]

Oxygen radicals generated by ischemia reperfusion have been shown to alter the vasoreactivity of intestinal microvessels. Topical application of xanthine/xanthine oxidase oxy radical-generating system resulted in prolonged vasodilation that was inhibited by SOD and mannitol.[74] In rat mesenteric arterioles, intestinal ischemia damaged endothelial cells and inhibited the hyperemic response.[141] This suggests that ischemia inhibited or destroyed an endogenous vasodilator substance and that this could adversely effect vasoregulatory function in this vascular bed.

Thus, there are data that suggest that the ischemia/reperfusion injuries in the small intestine are mediated by the hydroxyl radical generated via the xanthine oxidase mechanism and that this injury damages microvessels, resulting in altered microvascular reactivity, cellular damage, and organ dysfunction. Unfortunately the oxy radical-mediated injury from ischemia reperfusion is not the sole mechanism of intestinal injury in this process. Total ischemia also causes mucosal injury, which is not affected by radical scavengers and of inhibitors xanthine oxidase; thus anoxia may also play an important role in the pathophysiology of intestinal ischemia and reperfusion.[142]

Acknowledgments

Supported by grants HL 21851, HL 07580, NS 19316, and NS 25630 form the National Institutes of Health, by grant VA-90F38 from the Virginia Affiliate of the American Heart Association and by a grant from the Jeanette and Eric Lipman Foundation.

References

[1] Kontos, H. A., Wei, E. P., Dietrich, W. D., Navari, R. M., Povlishock, J. T., Ghatak, N. R., Ellis, E. F., and Patterson, J. L, Jr. (1981) *Am. J. Physiol.* **240**, H511–H527.

[2] Hazama, F., Amano, S., and Ozaki, T. (1978) *Adv. Neurol.* **20**, 359–369.

[3] Hori, S., Nishida, T., Mukai, Y., Pomeroy, M., and Mukai, N. (1980) *Res. Comm. Chem. Pathol. Pharm.* **29**, 211–228.

[4] Bohlen, H. G. and Hankins, K. D. (1983) *Blood Vessels* **20**, 213–220.

[5] Jayson, M. I. V. (1983) *J. Royal Soc. Med.* **76**, 635–642.

[6] Boykin., J. V., Eriksson, E., and Pittman, R. N. (1980) *Plastic Reconstr. Surg.* **66**, 191–198.

[7] Daum, P. S., Bowers, W. D., Jr., Tejada, J., and Hamlet, M. P. (1987) *Cryobiology* **24**, 66–73.

[8] Sadrzadeh, S. M., Graf, E., Panter, S. S., Hallaway, P. E., and Eaton, J. W. (1984) *J. Biol. Chem.* **259**, 14354–14356.

[9] Forman, M. B., Puett, D. W., and Virmani, R. (1989) *J. Am. Coll. Cardiol.* **13**, 450–459.

[10] Kontos, H. A, (1989) *Cerebrovascular Diseases* (Ginsberg, M. D. and Dietric, W. D., eds.), Raven, New York, pp. 365–371.

[11] Granger, D. N., Hollwarth, M. E., and Parks, D. A. (1986) *Acta Physiol. Scand.* **548(Suppl.),** 47–63.

[12] Koller, A. and Kaley, G. (1990) *Am. J. Physiol.* **258,** H916–H920.

[13] Marshall, J. J., Wei, E. P., and Kontos, H. A. (1988) *Am. J. Physiol.* **255,** H847–H854.

[14] Slater, T. F. (1984) *Biochem. J.* **222,** 1–15.

[15] Taube, H. (1965) *J. Gen. Physiol.* **49,** 29–50.

[16] Fridovich, I. (1978) *Science* **201,** 875–880.

[17] Halliwell, B. and Gutteridge, J. M. C. (1984) *Biochem. J.* **219,** 1–14.

[18] Del Maestro, R. F. (1980) *Acta Physiol. Scand.* **492,** 153-168.

[19] Rosen, G. M. and Freeman, B. A. (1984) *Proc. Natl. Acad. Sci. USA* **81,** 7269–7273.

[20] Freeman, B. A. and Cunningham, K. (1986) *Anesthesiology* **3,** A90 (abstract).

[21] Hong, K. W., Rhim, B. Y., Lee, W. S., Jeong, B. R., Kim, C. D., and Shin, Y. W. (1989) *Am. J. Physiol.* **257,** H1340–H1346.

[22] Korltos, H. A., Wei, E. P., Ellis, E. F., Jenkins, L. W., Povlishock, J. T., Rowe, G. T., and Hess, M. L. (1985) *Circ. Res.* **57,** 142–151.

[23] Dorfman, L. M. and Adams, G. E. (1973) *Reactivity of the Hydroxyl Radical in Aqueous Solutions, National Bureau of Standards.* US Department of Commerce, Washington, DC, June.

[24] Halliwell, B. and Gutteridge, J. M. C. (1986) *Arch. Biochem. Biophys.* **246,** 501–514.

[25] Chance, B., Sies, H., and Boveris, A. (1979) *Physiol. Rev.* **59,** 527–605.

[26] Turrnes, J. F., Freeman, B. A., and Crapo, J. D. (1982) *Arch. Biochem. Biophys.* **217,** 411–421.

[27] Turrnes, J. F., Freeman, B. A., and Levitt, J. G. (1982) *Arch. Biochem. Biophys.* **217,** 401–410.

[28] Forman, H. J. and Kennedy, J. (1976) *Arch. Biochem. Biophys.* **173,** 219–224.

[29] Fridovich, I. (1979) *Adv. Neurol.* **26,** 255–259.

[30] Misra, H. P. and Fridovich, I. (1972) *J. Biol. Chem.* **247,** 6960–6962.

[31] Marshall, J. J. and Kontos, H. A. (1988) *FASEB J.* **2,** A710(abstract).

[32] Cohen, G. and Heikkila, R. (1974) *J. Biol. Chem.* **259,** 2447–2452.

[33] Massey, V. Palmer, G., and Ballou, D. (1971) *Flavins and Flavorproteins,* (Kamin, H., ed.), University Park, Baltimore, MD, pp. 349–361.

[34] Hamed, M. Y., Silver, J., and Wilson, M. T. (1983) *Inorg. Chim. Acta* **80,** 237–244.

[35] Misra, H. P. (1974) *J. Biol. Chem.* **249,** 2151–2155.

[36] Fridovich, I. (1970) *J. Biol. Chem.* **245,** 4053–4057.

[37] Betz, A. L. (1985) *J. Neurochem.* **44,** 574–579.

[38] Batelli, M. G., Corte, E. D., and Stirpe, F. (1972) *Biochem. J.* **126,** 747–749.

[39] Engerson, T. D., McKelvey, T. G., and Rhyne, D. B. (1987) *J. Clin. Invest.* **79,** 1564–1570.

[40] McCord, J. M. and Roy, R. S. (1982) *Can. J. Physiol.* **60,** 1346–1352.

[41] Bulkley, G. B. (1987) *Br. J. Cancer* **55,** 66–73.

[42] Beckman, J. S., Marshall, P.A., and Freeman, B. A. (1987) *Fed. Proc.* **46,** 417–421.

[43] Kukreja, R. J., Kontos, H. A., Hess, M. L, and Ellis, E. F. (1986) *Circ. Res.* **59,** 612–619.

[44] Karneda, K., Ono, T., and Imai, T. (1979) *Biochim. Biophys. Acta* **572,** 77–82.

[45] Kuthan, H. and Ullrich, V. (1982) *Eur. J. Biochem.* **126,** 583–588.

[46] Lambeir, A.–M., Markey, C. M., and Dunford, H. B. (1985) *J. Biol. Chem.* **260,** 14894–14896.

[47] Durford, H. B. (1979) *Biochemical and Clinical Aspects of Oxygen* (Caughy, W. S., ed.), Academic, New York, pp. 167–176.

[48] Badwey, J. A. and Karnovsky, M. L. (1980) *Ann. Rev. Biochem.* **49,** 695–726.

[49] Haber, F. and Weiss, J. (1934) *Proc. R. Soc. (Lond.),* **A147,** 332–351.

[50] McCord, J. M. and Day, E. D., Jr. (1978) *FEBS Lett.* **86,** 139–142.

[51] Aust, S. D., Morehouse, L. A., and Thomas, C. E. (1985) *J. Free Radicals Biol. Med.* **1,** 3–25.

[52] Pryor, W. A. (1978) *Free Radicals in Biology* (Pryor, W., ed.), Academic, New York, pp. 1–49.

[53] Marklund, S. (1982) *Proc. Natl. Acad. Sci. USA.* **79,** 7634–7638.

[54] Halliwell, B. (1978) *Cell Biol. Int. Rep.* **2,** 113–128.

[55] Rouser, G., Nelson, G. J., Fleicher, S., and Simon, G. (1968) *Biological Membranes. Physical Fact and Function* (Chapman, D., ed.), Academic, New York, pp. 5–69.

[56] Frankel, E. N. (1980) *Prof. Lipid Res.* **19,** 1–22.

[57] Vladimirov, Y. A. (1980) *Adv. Lipid Res.* **17,** 173–249.

[58] Demopoulos, H. B., Flamm, E. S., Pietronigro, D. D., and Seligman, M. L. (1980) *Acta Physiol. Scand.* **492,** 91–119.

[59] Hillered, L. and Ernster, L. (1983) *J. Cereb. Blood Flow Metab.* **3,** 207–214.

[60] Pryor, W. A. (1976) *Free Radicals in Biology* (Pryor, W. A., ed.), Academic, New York, pp. 1–49.

[61] Trelstad, R. L, Lawley, K. R., and Holmes, L. B. (1981) *Nature* **289,** 310–312.

[62] Maridonneau, I., Braquet, P., and Garay, R. P. (1983) *J. Biol. Chem.* **258,** 3107–3313.

[63] Blum, J. and Fridovich, I. (1985) *Arch. Biochem. Biophys.* **240,** 500–508.

[64] Hodgson, E. K. and Fridovich, I. (1975) *Biochemistry* 14, 5294–5299.
[65] Greenwald, R. A. and Moy, W. W. (1980) *Arthritis Rheum.* 23, 455–463.
[66] Greenwald, F. A., Moy, W. W., and Lazarus, D. (1976) *Arthritis Rheum.* 19, 799.
[67] Orrenius, S., McConkey, D. J., and Nicotera, P. (1988) *Oxy-Radicals in Molecular Biology and Pathology* (Cerutti, P. A., Fridovich, I., and McCord, J. M., eds.), Liss, New York, pp. 327–339.
[68] Del Maestro, R. F. (1982) *Can. J. Physiol. Pharmacol.* 60, 1406–1414.
[69] Christman, C. W., Wei, E. P., and Kontos, H. A. (1984) *Am. J. Physiol.* 247, H631–H637.
[70] Kontos, H. A., Wei, E. P., Povlishock, J. T., and Christman, C. W. (1984) *Circ. Res.* 55, 295–303.
[71] Wei, E. P., Christman, C. W., and Kontos, H. A. (1985) *Am. J. Physiol.* 248, H157–H162.
[72] Rosenblum, W. I. (1983) *Am. J. Physiol.* 245, H13–H142.
[73] Parks, D. A. and Granger, D. N. (1983) *Am. J. Physiol.* 245, G285–G289.
[74] Okabe, E., Todoki, K., Odajima, C., and Ito, H. (1983) *Jpn. J. Pharmacol.* 33, 1233–1239.
[75] Wei, E. P. and Kontos, H. A. (1990) *Hypertension* 16, 162–169.
[76] Furchgott, R. F. and Zawadzki, J. V. (1980) *Nature (Lond.)* 288, 373–376.
[77] Kontos, H. A., Wei, E. P., and Marshall, J. J. (1988) *Am. J. Physiol.* 255, H1259–H1262.
[78] Griffith, T. M., Edwards, D. H., Lewis, M. J., Newby, A. C., and Henderson, A. H. (1984) *Nature* 308, 645–647.
[79] Rubanyi, G. M., Lorenz, R. R., and Vanhoutte, P. M. (1985) *Am. J. Physiol.* 249, H95–H101.
[80] Furchgott, R. F., Cherry, P. D., Zawadzki, J. V., and Jothianandan, D. (1984) *J. Cardiovasc. Pharmacol.* 2(Suppl.), 5336–5343.
[81] Holzmann, S. (1982) *J. Cyclic, Nucleotide, Res.* 8, 409–419.
[82] Ignarro, W., Burke, T. M., Wood, K. S., Wolin, M. S., and Kadowitz, P. J. (1984) *J. Pharmacol. Exp. Ther.* 228, 682–690.
[83] Rapoport, R. M. and Murad, F. (1983) *Circ. Res.* 52, 352–357.
[84] Rosenblum, W. I., Nelson, G. H., and Povlishock, J. T. (1987) *Circ. Res.* 60, 169–176.
[85] Rosenblum, W. I. (1988) *Stroke* 17, 494–497.
[86] Watanabe, M., Rosenblum, W. I., and Nelson, G. H. (1988) *Circ. Res.* 62, 86–90.
[87] Rosenblum, W. I., McKonald, M., and Wormley, B. (1989) *Stroke* 20, 1391–1395.
[88] Rosenblum, W. I. and Nelson, G. H. (1988) *Stroke* 19, 1379–1382.
[89] Mayhan, W. G., Faraci, F. M., and Heistad, D. D. (1987) *Am. J. Physiol.* 253, H1435–H1440.

[90] Koller, A., Messina, E. J., Wolin, M. S., and Kaley, G. (1989) *Am. J. Physiol.* **257,** H1485–H1489.

[91] Koller, A., Messina, E. J., Wolin, M. S., and Kaley, G. (1989) *Am. J. Physiol.* **257,** H1966–H1970.

[92] Yamamoto, H., Bossaller, C., Cartwright, J., Jr., and Henry, P. D. (1988) *J. Clin. Invest.* **81,** 1752–1758.

[93] Holtz, J. Forstermann, U., Pohl, U., Giesler, M., and Bassenge, E. (1984) *J. Cardiovasc. Pharmacol.* **6,** 1161–1169.

[94] Pohl, U., Holtz, J., Busse, R., and Bassenge, E. (1986) *Hypertension* **8,** 37–44.

[95] Griffith, T. M., Edwards, D. H., Davies, R. U., Harrison, T. J., and Evans, K. T. (1987) *Nature* **329,** 442–445.

[96] Griffith, T. M., Edwards, D. H., Davies, R. U., and Henderson, A. H. (1989) *Microvasc. Res.* **37,** 162–177.

[97] Koller, A. and Kaley, G. (1990) *Circ. Res.* **67,** 529–534.

[98] Rubanyi, G. M. and Vanhoutte, P. M. (1986) *Am. J. Physiol.* **250,** H822–H827.

[99] Gryglewski, R. J., Palmer, R. M. J., and Moncada, S. (1986) *Nature* **320,** 454–456.

[100] Beauchamp, C. and Fridovich, I. (1971) *Anal. Biochem.* **44,** 276–287.

[101] Kontos, H. A., Wei, E. P., Povlishock, J. T., Kukreja, R. C., and Hess, M. L. (1989) *Am. J. Physiol.* **256,** H665–H671.

[102] Forstermann, U., Trogisch, G., and Busse, R. (1985) *Eur. J. Pharmacol.* **106,** 639–643.

[103] Kontos, H. A., Wei, E. P., and Povlishock, J. T. (1980) *Science* **209,** 1242–1245.

[104] Vanhoutte, P. M. (1988) *N. Eng. J. Med.* **319,** 512–513.

[105] Gryglewski, R. J., Bunting, S., and Moncada, S. (1976) *Prostaglandins* **12,** 685–713.

[106] Moncada, S., Gryglewski, S., and Bunting, S. (1976) *Prostaglandins* **12,** 713–715

[107] Furchgott, R. F. (1983) *Circ. Res.* **53,** 557–573.

[108] Levasseur, J. E., Kontos, H. A., and Ellis, E. F. (1985) *Am. J. Physiol.* **248,** H534–H539.

[109] Ames, A., III., Wright, R. L, Kowada, M., Thurston, J. M., and Majno, G. (1968) *Am. J. Pathol.* **52,** 437–453.

[110] Kloner, R. A., Ganote, C. E., and Jennings, R. B. (1974) *J. Clin. Invest.* **54,** 1496–1508.

[111] Engler, R. L., Dahlgren, M. D., Morris, D., Peterson, M. A., and Schmid-Schönbein, G. (1986) *Am. J. Physiol.* **251,** H314–H323.

[112] Engler, R. L., Dahlgren, M. D., Peterson, M. A., Dobbs, A., and Schmid-Schönbein, G. (1986) *Am. J. Physiol.* **251,** H93–H100.

[113] Harlan, J. M. (1985) *Blood* **65,** 513–525.

[114] Babior, B. M. (1978) *N. Eng. J. Med.* **298**, 659–668, 721–725.

[115] Smedly, L. A., Tonnesen, M. G., Sandhous, R. A., Haslett, C., Guthrie, L. A., and Johnson, R. B., Jr. (1986) *J. Clin. Invest.* **77**, 1233–1243.

[116] Boxer, L., Allen, J., Schmidt, M., Yoder, M., and Baehner, R. (1980) *J. Lab. Clin. Med.* **95**, 672–678.

[117] Cronstein, B. N., Levin, R. I., Belanoff, J., Weissmann, G., and Hirschorn, R. (1986) *J. Clin. Invest.* **78**, 760–770.

[118] Wei, E. P., Kontos, H. A., Christman, C. W., DeWitt, D. S., and Povlishock, J. T. (1985) *Circ. Res.* **57**, 781–787.

[119] Wei, E. P., Dietrich, W. D., Navari, R. M., and Kontos, H. A. (1980) *Circ. Res.* **46**, 37–47.

[120] Kontos, H. A. (1985) *Circ. Res.* **57**, 508–516.

[121] Kontos, H. A. and Wei, E. P. (1987) *Physiologist* **30**, 122(abstract).

[122] Povlishock, J. T., Wei, E. P., and Kontos, H. A. (1988) *FASEB J.* **2**, A–835(abstract).

[123] Kontos, H. A. and Wei, E. P. (1986) *J. Neurosurg.* **64**, 803–806.

[124] Kirsch, J. R., Phelan, A. M., Lange, D. G., and Traystman, R. J. (1987) *Fed. Proc.* **46**, 799 (abstract).

[125] Kukreja, R. C., Kontos, H. A., Hess, M. L., and Ellis, E. F. (1986) *Circ. Res.* **59**, 612–619.

[126] Kuroiwa, T., Ting, P., Martinez, H., and Klatzo, I. (1985) *Acta Neuropathol.* **68**, 122–125.

[127] MacKenzie, E. T., Strandgaard, S., Graham, D. I., Jones, J. V., Harper, A. M., and Farrar, J. K. (1976) *Circ. Res.* **39**, 33–37.

[128] Mayhan, W. G., Amundsen, S. M., Faraci, F. M., and Heistad, D. D. (1988) *Am. J. Physiol.* **255**, H879–H884.

[129] Grogaard, B., Schurer, L, Gerdin, B., and Arfors, K. E. (1986) *Superoxide and SuperoxideDismutase in Chemistry* (Rotilio, G., ed.), Elsevier, New York, p. 608.

[130] Babbs, C. F. (1985) *Ann. Emerg. Med.* **14**, 777–779.

[131] Fleischer, J. E., Lanier, W. L., Milde, J. H., and Michendelder, J. D. (1985) *Stroke* **18**, 124–126.

[132] Patt, A., Harken, A. H., Burton, L. K., Rodell, T. C., Piermattei, D., Schorr, W. J., Parker, N. B., Berger, E. M., Horesh, I. R., Terada, L. S., Linas, S. L., Cheronis, J. C., and Repine, J. E. (1988) *J. Clin. Invest.* **81**, 1556–1559.

[133] Kontos, H. A. (1989) *N. Eng. J. Med.* **314**, 711–718.

[134] Ross, R. (1986) *N. Eng. J. Med.* **314**, 711–718.

[135] Granger, D. N., Rutili, G., and McCord, J. M. (1981) *Gastroenterology* **81**, 22–29.

[136] Grogaard, B., Parks, D. A., Granger, D. N., McCord, J. M., and Forsberg, J. (1982) *Am. J. Physiol.* **5**, 448–454.

[137] Cook, B. H., Wilson, E. R., and Taylor, A. E. (1971) *Am. J. Physiol.* **22**, 1494–1498.

[138] Bounous, G. (1982) *Gastroenterology* **82,** 1457–1467.

[139] Roy, R. S. and McCord, J. M. (1983) *Proceedings of the Third International Conference on Superoxides and Superoxide Dismutase* (Greenwald, R. and Cohen, G., eds.), Elsevier/North Holland Biomedical, New York, pp. 141–153.

[140] Parks, D. A., Bulkley, G. B., Granger, D. N., Hamilton, S. R., and McCord, J. M. (1982) *Gastroenterology* **82,** 9–15.

[141] Altura, B. M., Gebrewold, A., and Burton, R.W. (1985) *Mircrocirc. Endothelium Lymphatics* **2,** 3–14.

[142] Parks, D. A., Grogaard, B., and Granger, D. N. (1982) *Surgery* **92,** 869–901.

Chapter 23

Chaos in the Fractal Arteriolar Network

James B. Bassingthwaighte

Why Should the Arteriolar Network Be Fractal?

There is no sure answer to such a philosophical question, but it is a way of phrasing the question of whether or not it might be beneficial to the overall system to have an arteriolar network that is fractal in structure and chaotic in behavior. Remembering that there is no such thing as proof that a model is a correct representation of the system, and that only disproof is possible, we can regard fractal models and chaotic dynamic behavior as potential models for the system, among other models. When are they good models? Why might it be advantageous to the system to be fractal or to have unpredictable behavior?

Fractals are for correlation. Whereas the word derives from Mandelbrot's definition from the Latin *fractere*, to fragment, the whole story of fractals and chaotic behavior is a story of correlation in space and time. It makes sense that a biological system is not random, but correlated in many features and functions, and this is the main tenet of this chapter.

From *The Resistance Vasculature*, J. A. Bevan et al., eds. ©1991 Humana Press

Parsimony in the instruction set for growth, the genes, is a prime reason for considering whether or not simple recursion rules not only give efficiency, but are adequate to the task. The recursion rule is a repetitive operation: Do something, take the result of that and do something again to it, repeat until the task is accomplished. The rule can be deterministic, as in taking successive powers of two, or probabilistic. A probabilistic rule might be instead of multiplying by exactly two on the next iteration to multiply by two plus or minus some variation. Thus, when one starts with one and generates a set of numbers by multiplying by a number that is something around two, one develops a family of numbers that are not exactly predictable, and never exactly alike but are similar to each other. Furthermore, they fulfill the fractal requirement of being "self-similar," even if not exactly. The self-similarity is that in going from any one generation of a recursive rule to the next generation, the magnitude approximately doubles. This idea of self-similarity is a hallmark of both fractal structures and chaotic behaviors.

We have only a few genes, on the order of 100,000. In the heart, on which I focus, there are 10^8 muscle cells and 3×10^9 endothelial cells, and the vascular patterns are complex, so it is obvious that only a few of the 100,000 genes can be used as the instruction set for vascular growth because otherwise there would be none available for guiding the growth of even more complex structures, such as the limbs or the brain. We shall show later that a simple single recursion rule can provide for the design of a vascular network that will fill two- or three-dimensional space. Further, one of the interesting features of fractals is that there is a kind of automatic self-regulation that reduces variation and tends toward uniformity.

Fractals in Vascular Anatomy

The arterial network is composed of a series of segments of cylindrical tubes joined at branch points. Downstream, daughter branches are smaller than the parent. Many vascular beds show anastomoses or arcades, but the importance of these varies from tissue to tissue and from animal to animal. In the hearts of pigs and humans, the coronary artery system is basically one of dichotomous branching, two daughters from a parent, repeated recursively down to the terminal arterioles. At the terminal arteriolar endings the story changes. Instead of dichotomous branching, there tends to be what I call "arteriolar bursts," by which I imply that the arteriole goes through two to four sets of branchings within a very short distance, 50–100 μ, and that the branchings are not necessarily dichotomous, but the parent may give rise to three or four branches. The "burst" or "flower" arrangement on the end of the arteriolar stalk then feeds a multitude of capillaries. Capillaries are *not* a part of the fractal network, but may be regarded as a swamp. In the heart they lie in long parallel arrays, one capillary per cell in cross-section, and a single capillary may be traced for many millimeters, even centimeters. This hugely dense network composed of 2000–4000 capillaries/mm^2 in cross-section is fed by many arteriolar bursts scattered in three-dimensional space and drained by roughly twice as many venular confluences. Venules, larger and commonly oval in cross-section, travel with the arterioles, two venules per arteriole, through much of the network. This is not so on the epicardium, where there are many venules unaccompanied by arterioles. But what is the evidence that the arteriolar network is fractal?

In the heart the evidence is scanty, simply because more measurements need to be made. Van Bavel[1] observed approximately logarithmic relationships between the diameters of parent and daughter vessels in dog myocardium. Strikingly detailed and persuasive studies were accomplished by Suwa et al.[2] and Suwa and Takahashi[3] in the mesenteric and renal arterial beds. They observed log–log relationships that were apparently linear over 2–3 orders of magnitude in dimensions for vessel radii, branch lengths, and wall thick-nesses. The logarithmic relationships illustrate the approximate constancy of the ratio between parent and daughter dimensions. Such data have not been obtained on other vascular beds. The relationship (which is typical of many of those illustrated in their papers) is shown in Fig. 1. The fact that there is a single log–log slope without an obvious break would suggest that either there is a single recursion rule, or if there is more than one recursion rule then all of the rules are repeated at each generation. In the figure shown, the rule might be: At each branch point, make two daughter branches of 70% of the length of the parent branch. This of course provokes the question "How in the world would the daughter

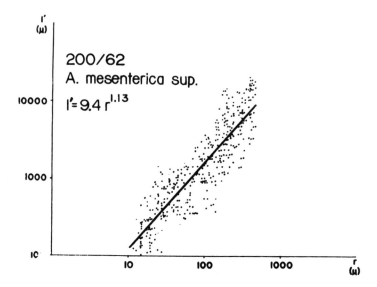

Fig. 1. Relationship between the length of the branches plotted against radius of the branch on a log–log scale. The approximate constancy of the ratio over the large range of lengths is compatible with a fractal recursion rule. (Reprinted from ref. 2, with permission.)

branch know what was the length of the parent branch?" So obedience to such a rule is no trivial matter and the mechanisms for adhering to such simple descriptive rules require much research.

The vascular beds of organs are not very tree-like. As a generality, the vascular beds of plants tend toward filling a sunlight-gathering surface rather than a three-dimensional space. Even so, there are similarities and in some ways the circulation of the mesentery can be equated to that in the leaf: There are dichotomous or trichotomous branching systems with anastomosing arcades. In organs as in leaves, the vascular system grows as the organ grows. The primary branches enlarge, but their topology is maintained, even as they undergo remodeling with respect to the details of their form, length, wall thickness, and so on. It is likely that the major arteries to an organ develop in some nonfractal fashion just as the aorta and pulmonary artery develop from the primitive branchial arches. At the other extreme, the end of the network instead of the beginning, the form of the capillary bed is determined by the nature of the tissue, the arrangements of the cells and the requirement for close proximity to the functioning cells, so the geometry is secondary to the tissue cells. So there is really only the intermediate network of arteriole feeding networks and venular draining networks that can follow simple recursion rules.

Statistical Measure of Flow Heterogeneity Are Fractal

The observed variations in flow per unit mass of tissue in the heart and lung are quite well described by fractal relationships, not at the causative level, but at the level of statistical observation. This is evident from the logarithmic relationships between measurements of the heterogeneity of regional blood flows and the size of the tissue samples, and has been observed for both the heart[4] and the lung.[5] The example is taken from the study of Bassingthwaighte et al.[4] Regional myocardial flows have been measured using the microsphere technique,[6] in which the deposition of the spheres locally is taken to be in proportion to the local blood flow. The adequacy of this assumption has been affirmed by comparisons of the microsphere deposition techniques with that of a molecular microsphere the deposition of which is not influenced by rheological biases affecting particulate spheres, and the extraction of which is complete during a single passage.[7] Note that the degrees of heterogeneity observed in normally functioning hearts in the sheep (illustrated in Fig. 2) are similar to those in awake baboons described by King et al.[6] and there is roughly a sixfold range of flows in pieces of myocardium about 150 mg in size, i.e., flows range from about one third of the mean flow to about twice the mean flow. The right panel of Fig. 2 shows plots of the relative dispersion, RD (which is SD/mean), vs the mass, m, of the pieces in each of which the average flow was measured. The log–log relationships are linear, and the slope is 1-D where D is the fractal dimension:

$$RD(m) = RD(m_0) \bullet (m/m_0)^{1-D} \qquad (1)$$

where m_0 is a chosen reference mass. A D of 1.0 means that the pieces are internally uniform so that dividing them further reveals

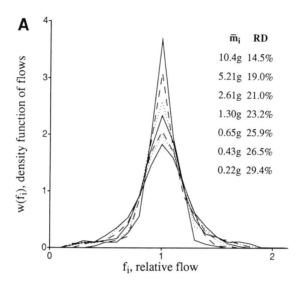

Fig. 2. Panel **A**: Distributions of regional myocardial blood flows in a sheep heart at six levels of observation. Higher resolution, smaller pieces give broader measures of dispersion. Panel **B**: Fractal regression lines for the degree spatial dispersion (RD, relative dispersion) for regional myocardial blood flows in eleven sheep. The relative dispersion is the standard deviation of regional blood flows per unit mass of tissue divided by the mean flow for the whole heart. In each case, the whole heart is being examined but six or seven levels of resolution of observation (different masses of

(continued)

no further heterogeneity. A D = 1.5 indicates complete randomness without correlation. Observed Ds are about 1.2 for myocardial blood flows and 1.15 for pulmonary blood flows. Why the range of blood flows is so broad in a normally functioning heart is not known. In resting skeletal muscle the range of flows is far greater.[8] Presumably, local metabolic requirements are the final determining factor for regional blood flow.

However, without knowing about the metabolic drive governing local flows, it appears that geometry alone can provide an adequate explanation. Van Beek et al.[9] found that a simple dichotomous branching tree with a mild degree of asymmetry at each branch point can give rise to the observed heterogeneities of flows (Fig. 3). In this case, the only assumption was that the branching

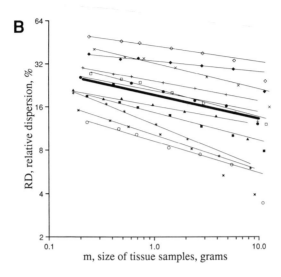

B

tissue volume elements) are shown. At the highest level of resolution (the smallest pieces), the greatest degree of heterogeneity can be observed. When the observations are made in larger chunks of tissue, the apparent heterogeneity is less. Each symbol type represents an individual animal. The straight line (log–log) relationships from about 200 mg to 4 g show that a good description is provided by a fractal power law relationship. The equation $RD_s = 19.4m^{-0.16}$ $r = 0.997$ is for an average relationship and the correlation coefficient, r, gives the average goodness of fit. The average fractal $D = 1.16$. At the largest pieces the data deviate from the straight-line relationship.

arteriolar network obeyed a fractal branching rule, and was matched by a venular branching rule. Fractal dichotomous networks with a deterministic asymmetry in which the flow in a daughter branch deviated by only three or four percent away from the mean 50% at each branch point gave good descriptions to the data, as shown in Fig. 3. A probabilistic model, where the deviation was not by a fixed percentage from the mean, but by a variable Gaussian distribution around the mean, served equally well. In this case, even though the fractal branching network follows a precisely fractal rule (either deterministic or probabilistic), the resultant plot of the relative dispersion (to the measure of heterogeneity) vs the generation number of the volume of tissue served is curvilinear.

A   Flows

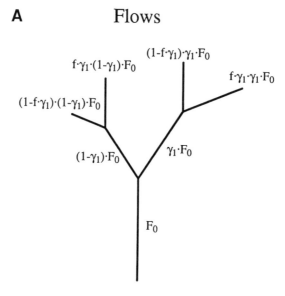

Fig. 3. Dichotomous asymmetric branching can explain the observed variation in regional myocardial flows. Panel **A**: A simple branching algorithm
for flows. Panel **B**: Heterogeneity of regional blood flows fitted with fractal
network models (hearts of 10 baboons, microsphere data). Fitting of the data
(continued)

This means that a fixed rule fractal dichotomous branching gives a
distinctly different result than a straight line power law relationship. It is also a very slightly better descriptor, in that it gives cur-
vature that tends to fit the observations at large piece sizes better
than did the straight line relationship of Fig. 2. In any case, results
for the heart of the sort shown in Fig. 2 and 3, and which are similar to the observations in the lung,[5] suggest that either type of fractal
relationship is suitable.

The degree of asymmetry at the bifurcations that was found
by van Beek et al.[9] was less than that estimated by Horsfield and
Woldenberg,[10] who appeared to find much more asymmetry. This
may be explicable in terms of the differences in the anatomy
between heart and lung. The lung appears to have an arterial tree
that is more or less like a fir tree for the vessels of a few millimeters
down to about a millimeter in diameter and thereafter becomes

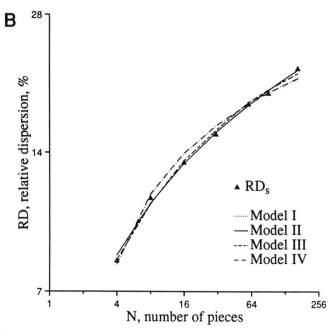

on flow heterogeneity in one heart with four different variants of the branching model, using fixed or stochastic estimates of γ. All models give mild asymmetry, equivalent to γs around 0.46.

dichotomously branching, whereas in the heart the dichotomous branching may start at a higher level.

Temporal Fluctuations Contribute to the Observed Spatial Flow Variance

In vivo microscopic observations of diameters and flow velocities in arterioles practically always reveal fluctuations. In fact, when the fluctuations are absent it is usually considered that the tissue is not in a normal physiological state and is not likely to show autoregulation, so fluctuations are normal. However, they are never exactly periodic. Frequency power spectral analysis tends to show a wide range of frequencies or to show two or three fairly broad peaks. How do these observations at the microscopic level compare with those at the macroscopic level? The question is relevant since practically all observations made in microvessels of 20–200-μ diameter are giving information on tissue regions that are very small compared with those in which the broad scale heterogeneity was illustrated for Fig. 2 and 3. King and Bassingthwaighte[11] found that the temporal fluctuations in 200 mg pieces were small compared with the range of flows of the stable spatial patterns. The temporal fluctuations caused only small deviations from the dominant spatial patterns. What this means in the heart is that those low-flow regions of the left ventricular myocardium having flow less than half the mean flow showed fluctuations that never brought their flow up to equal the mean flow. Likewise, those regions with flows greater than 150% of the mean flow fluctuated, less percentagewise than did low-flow regions, but these fluctuations never reduced their flow to the mean flow for the left ventricle. Whereas it is understandable that the left atrial flows should be much lower than the mean flow for the heart (about 40%) and right ventricular flows also less (about 70% of the mean), it is not so obvious why regional left ventricular flows are so strikingly varied.

Because temporal fluctuations are very modest, then it is evident that those regions of the left ventricle that have low blood flows must have lower metabolic requirements. This must be so, because the oxygen extraction is about 70% across the whole heart. If a low-flow region had metabolic requirements equaling those of the average for the heart, it would have to extract more oxygen than was delivered to the tissue, and would have to do it continuously since the fluctuations are not large enough to make up for transient deficiencies.

"Twinkling"[12] is what we call these temporal fluctuations. Like the twinkling of stars, the fluctuations are irregular. Analysis of continuous signals of arteriolar diameters in rabbit tenuissimus muscle obtained by Slaaf et al.[13] illustrate chaotic dynamics. What this means is that the signals show nonrandom fluctuations with correlation over time. Just as the fluctuations in the flows in rivers are tempered by "memory" resulting from slow fluctuations in the water table as well as nonrandom levels of annual rainfall, so do microvessels show chaotic, partially coordinated, yet unpredictable but nevertheless nonrandom, behavior.

Yamashiro et al.[14] has shown that the tenuissimus microvascular diameters and flow velocities show characteristics of a low-order nonlinear dynamical system. The functional order of the coordinated "chaotic" system appears to be approx 3 or 4. This is not surprising since a few processes may dominate: smooth muscle contraction, membrane potential, calcium levels, levels of extracellular regulatory substances, and so on. Although there are many processes superimposed on these dominating ones, it makes sense that there should be only a few control points in a complex system, and that the system may therefore be observed to be of relatively low-order, even though of high complexity. What is most relevant is that the system is not random.

Taking the same observations and randomizing their order gives rise to a more steeply sloped relationship between the estimated functional order of the system, the so-called Grassberger-Procaccia Dimension[15] vs embedding dimension, the dimension of the phase base in which the signal is being traced. (A phase space simply uses a succession of delayed copies of the signal. In a planar phase space, one plots the signal vs the signal delayed by a time interval, τ. In three-dimensional phase space, the signal is plotted against itself delayed one τ and against itself delayed by two τ. The traces in n-dimensional space are unraveled so that there are fewer intersections or lines crossing one another. When n is sufficiently large to unravel the signal completely, then the dimensionality of the system is revealed. This is the so-called "minimal order" of the system.) Vasomotion signals, by this approach, appear to have a minimal order dimension of between 3 and 4, implying that the system is dominated by 3 or 4 control points. By this kind of analysis the numbers do not turn out to be specific integers; this is what you would expect in a system in which there are some influences of secondary controllers, even though they may not be dominating.

Fractal Measures of Heterogeneity

The technology of fractals provides several ways to obtain measures of heterogeneity of any property over space or time. The relative dispersion illustrated in Fig. 2 is just one of these. The interesting feature about estimating fractal dimensions of signals is that the same dimension is estimated from any of the moments of the observed signal. This means that measuring means or standard deviations gives the same estimate of the fractal dimension, as described by Voss.[16] Therefore, if a system is truly fractal, the estimate of the dimension will not be dependent on the particular property that is examined. Practical approaches are listed in Fig. 4. There is considerable overlap between the techniques.

There is no space constant! When a system is fractal, the degree of correlation between neighbors falls off in a particular fashion. The application of an exponential space constant or of a Gaussian expectation for correlation is inappropriate. Figure 5 shows that the falloff in correlation is dependent on the number of units into which the domain is divided and examined, rather than on the actual distance. This means that there is no characteristic space constant. The point is that there is fluctuation at all macroscopic levels of observation, but there is an overriding correlation that dominates. Van Beek et al.[9] worked out the relationship between the correlation coefficient of nearest neighbors, r_1, and the fractal dimension, D:

$$r_1 = 2^{3-2D} - 1 \text{ or } r_1 = 2^{2H-1} - 1 \qquad (2)$$

where H is the Hurst coefficient and H = 2–D. r_1 is found experimentally by measuring the two-point autocorrelation between nearest neighbors. Thus, by determining the correlation coefficient between neighbors of a given size, one can make an estimate of the fractal dimension. The same thing can be done for nonadjacent neighbors to find r_n, where n denotes the number of integer units separation. This is useful because it then gives an idea of the spread of the correlation throughout the organ. The general expression for n^{th} neighbors is:

$$r_n = 1/2 \left[|n + 1|^{2H} - |2n|^{2h} + |n - 1|^{2H} \right] \qquad (3)$$

The falloff in correlation is initially fast and then relatively slow, much slower than an exponential or a Gaussian falloff. This gives rise to a test to determine whether or not a system is fractal. For example, when using an autoradiographic technique provid-

Measures of Spatial Heterogeneity	Measures of Temporal Fluctuations
1. Range/SD, the Hurst measure	1. Range/SD, the Hurst measure
2. D_b, box dimension	2. D_b, box dimension
3. D_{RD}, relative dispersion	3. D_{RD}, relative dispersion
4. Mass-radius method, $L_T(\rho) \cong \rho^D$	4. D_{gp}, correlation dimension
5. Autocorrelation, D_r	

Fig. 4. List of approaches to estimating the fractal dimension of spatial and temporal signals. Left panel: Measures usually applied to spatial signals. Right panel: Measures commonly applied to temporal signals.

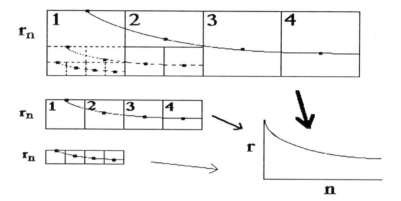

Fig. 5. Extended correlation, r_n, where r_n is independent of voxel size. Correlation falls off fractally. When correlation between units is determined for large units then the correlation coefficient falls off slowly, in this case falling to about one-quarter in 3 unit distances. If the correlation is examined between neighbors of one-half that size (middle left), then the falloff in correlation is faster with respect to distance, but still is about 3 units to fall to one-quarter. When units of 1/16 the original size are used, the correlation falloff is still 3 units to drop down to 1/4. When scaled to the number of units, the three curves of correlation are found to be superimposed.

ing estimates of the distribution of flows in the microvasculature, one can look at the falloff in correlation as a function of the number of pixels of a given size in the image using the two-point auto-correlation technique. Obtaining the estimates of the falloff in correlation by using two or three different pixel sizes will demonstrate whether or not the system is fractal vs an exponential or other relationship. With the fractal relationship the falloff will be the same in terms of the numbers of units no matter what the unit size. This appears to be the case for the heart. If there were a stable space constant independent of unit size, then the apparent falloff would be dependent on distance explicitly and not on the number of units of differing pixel sizes used to make the measure.

Fractal relationships are not forever. Since the fractals can extend only over a finite size range in the heart, from a maximum of some fraction, such as 10% of the heart, down to a minimum of something encompassing at least a few microvascular units or a few hundred micrograms of tissue, the range of applicability of the fractal model is limited. Even so, the idea that there is correlation between neighbors at all levels or sizes then generates the next question, "How is coordination between units effected?"

All the statistical considerations provided above now have to be translated into the biology. First, it is evident that there are anatomical and functional linkages between neighboring units in the heart, as in other organs. Neighboring units can be expected to be alike with respect to their flows because neighboring regions are supplied, for the most part, by an artery that feeds the arterioles supplying both. Moreover, because neighboring contractile bundles must have similar mechanical stresses and must shorten to similar degrees, they might be expected to have similar metabolic requirements. This idea is in concert with the observations of van Beek et al.[9] shown in Fig. 2, where a dichotomous branching vascular tree gives rise to the correct degree of heterogeneity. However, the model has not been tested to see if it gives the right degree of spatial correlation. Before this can be tested, the anatomic measures of branch length and radii at each generation need to be made.

Many additional factors contribute to local coordination. Autoregulation is one, simply because diameter reduction at a core level controls the total flow to the supplied region. Another is the seeming coordination of arteriolar diameter that is propagated along the vessel, both upstream and downstream, from a stimulated point.[17]

Dichotomous Growth

Fractal algorithms for dichotomous growth patterns are still at a very primitive stage. The work of Nelson[18] illustrates that realistic looking patterns for dichotomous arterial or bronchial trees can be made using fractal algorithms. Wang and Bassingthwaighte[19] improved on these by using an algorithm that was proven to be space-filling in two dimensions, but the three-dimensional algorithm that is truly space filling has not yet been defined.

The Wang algorithm is exceedingly primitive: In an enclosed planar space, run a line from a point on the circumference toward the opposite side so that the area is split evenly. Then project this line for a specified fraction of the total distance to the next obstacle, e.g., one-third. Repeat ad infinitum. Of course, none of these things really repeat ad infinitum and the fractal relationship dissolves in the swamp of the capillary bed, or rather, it ceases at the level at which a single terminal arteriole can feed the capillary bed adequately. In the heart, this is about 20–50 terminal arteriolar units per mm^3. In the heart and lung, organs richly endowed with capillaries, there are huge numbers of interconnections among capillaries and so it is probably legitimate to consider them as a continuous swamp, fed here and there by arterioles and drained at intermediate points. This relationship does not hold for other organs with a sparser vascularity. But even in skeletal muscle it is observed that one venule will drain the capillary bed supplied from more than one arteriole, and vice versa. So growth patterns in a fractal sense should be sought in looking at arteriolar and venular trees but not in the capillary bed. Algorithms for three-dimensional space filling fractals can be developed in the same fashion and are only somewhat more complex geometrically.

Growth algorithms that simply fill a previously defined space are inappropriate, for real growth occurs in the vascular system in parallel with the growth of the cellular structures of the tissue. There is both growth and remodeling,[20] so vascular growth needs to be considered on three levels:

1. Large vessels of preordained form of phylogenetic origin.

2. Fractal branching levels at the middle level of vessel size.
3. The capillary network independently formed and is structured around the cells of the tissue served.

Can We Explain the Observed Heterogeneities?

The answer at the moment is no. Whereas the observed spatial heterogeneities are describable in terms of these fractal statistical considerations, and the "twinkling" or temporal fluctuations super-imposed on the spatial pattern can be described in terms of a low order of nonlinear dynamical process, the whys and wherefores of these are undiscovered. The easiest question to solve is why the spatial heterogeneity is so great, or stated otherwise, why tissues are not more uniform in their flow requirements. Clearly, if the metabolic needs of the tissues were uniform everywhere then one would expect flow to be uniform, and so the question evolves to the question of why the metabolic requirements are different in different portions of the seemingly homogeneously structured organ, such as the heart or skeletal muscle. The answer must be that some parts are metabolically different than others.

Speculations on System Design

If portions of an organ are working harder than another or metabolizing more than another for one one reason or another, then in an efficiently designed system all the mechanisms for providing delivery of energy to those working parts should have capacities proportional to the need. This sounds like a socialist manifesto, "from each according to his ability, to each according to his need." There is at this point not much direct evidence that metabolic requirements and flows are locally matched. In the myocardium, Caldwell et al.[21] have found that the transport capacity of the membranes is proportional to the local flows and therefore express the idea that there is an optimality in substrate transport whereby the processes of flow, transmembrane transport, and metabolic turnover are all matched regionally. This is in tune with the general perception that there should be optimality in the design of a vascular system in terms of the energy required for delivery by pressure driven flow, and the energy required for building the structure of the vascular walls, as outlined in a very nice way by Lefèvre.[22] Whereas the concept of universal impedence matching for each part of the process makes sense from the point of view of system design, there is little direct evidence for it as yet. However, one knows that when metabolic requirements are increased, as in cardiac hypertrophy,[23] capillarity is increased, and large vessel growth occurs as well. Whereas the large vessel growth that occurs in collaterals in the heart[24] is different from that shown by Langille et al.,[25] the observations of large vessel growth in response to peripheral need is striking. The response appears to be mediated via endothelial cell shear stress, and presumably, an endothelially derived growth factor. Langille observed when a kidney was removed while leaving the renal artery supplying only the adrenal gland that the renal artery shrank down to almost the size of the adrenal artery (reported in ref. 26). EDRF plays an important role in acute responses to a flow increase,[27] but there is little specific evidence on the roles of endothelin and EDRF on long-term responses. These observations on the striking growth capacity of a large vessel[28] should be kept in mind when examining relationships between local metabolic requirements and regional flow. It is apparent that local metabolic needs must dominate in an important way, and how the effects are mediated is an area of intense study.

Conclusion

Fractal growth patterns and chaotic vascular dynamics are probably real. However, whereas fractals can allow for a parsimonious instruction set in providing for growth over many generations, it can be regarded as a simple genetic algorithm rather than a dominating factor. The dominating factors will be the overall size of the tissue and its metabolic requirements. Likewise, chaotic dynamics may provide descriptors of the apparent order of a complex nonlinear system of dynamical events, that are really minor fluctuations in the control system rather than dominating events. Again, the overall regulation will be tissue metabolic requirements and the chaotic dynamics of the microvasculature are simply a functional detail. Thus fractals and chaos can be regarded as useful thinking tools for examining the physiological state and asking old questions in a new way.

Just as the fractal branching helps us to understand correlation among regions, the chaotic behavior also is evidence of correlated events. "Chaos" is distinctly nonrandom. Just as a fractal instruction set for branching makes for efficiency in genetic control, the chaotic dynamics are evidence for the existence of a relatively relaxed control system. Extremely tight control of a chemical system is expensive, requiring high gain controllers and sensitive detectors of imbalance. Such features are quite unnecessary in a system that has no set point and works via interactions among various components that tend to balance things out.

References

[1] Van Bavel, E. (1989) *Metabolic and Myogenic Control of Blood Flow Studied on Isolated Small Arteries*, Ph.D. Thesis, Physiology, University of Amsterdam, The Netherlands.

[2] Suwa N., Niwa, T., Fukasawa, H., and Sasaki, Y. (1963) *Tohoku J. Exp. Med.* **79**, 168–198.

[3] Suwa, N. and Takahashi, T. (1971) *Morphological and Morphometrical Analysis of Circulation in Hypertension and Ischemic Kidney*, Urban and Schwanenberg, Munich-Berlin, Germany.

[4] Bassingthwaighte, J. B., King, R. B., and Roger, S. A. (1989) *Circ. Res.* **65**, 578–590.

[5] Glenny, R. W. and Robertson, H. T. (1990) *J. Appl. Physiol.* **69**, 532–545.

[6] King, R. B., Bassingthwaighte, J. B., Hales, J. R. S., and Rowell, L. B. (1985) *Circ. Res.* **57**, 285–295.

[7] Bassingthwaighte, J. B., Malone, M. A. Moffett, T. C., King, R. B., Chan, I. S., Link, J. M., and Krohn, K. A. (1990) *Circ. Res.* **66**, 1328–1344.

[8] Tripp, M. R., Meyer, M. W., Einzig, S., Leonard, J. J., Swayze, C. R., and Fox, I. J. (1977) *Am. J. Physiol. (Heart. Circ. Physiol. 1)* **232**, H173–H190.

[9] Van Beek, J. H. G. M., Roger, S. A., and Bassingthwaighte, J. B. (1989) *Am. J. Physiol. (Heart Circ.Physiol. 26)* **257**, H1670–H1680.

[10] Horsfield, K. and Woldenberg, M. J. (1989) *Anat. Rec.* **223**, 245–251.

[11] King, R. B. and Bassingthwaighte, J. B. (1989) *Pflugers Arch. (Eur. J. Physiol.)* **413/4**, 336–342.

[12] Yipintsoi, T., Dobbs, W. A., Jr., Scanlon, P. D., Knopp, T. J., and Bassingthwaighte, J. B. (1973) *Circ. Res.* **33**, 573–587.

[13] Slaaf, D. W., Tangelder, G. J., Teirlinck, H. C., and Reneman, R. S. (1987) *Microvasc. Res.* **33**, 71–80.

[14] Yamashiro, S. M., Slaaf, D. W., Reneman, R. S., Tangelder, G. J., and Bassingthwaighte, J. B. (1990) *Mathematical Approaches to Cardiac Arrhythmias, Ann. NY. Acad Sci.* vol. S91, (Jalife, J., ed.), pp. 410–416.

[15] Grassberger, P. and Procaccia, I. (1983) *Physica* **9D**, 189–208.

[16] Voss, R. F. (1988) *The Science of Fractal Images*, (Peitgen, H. O. and Saupe, D., eds.), Springer-Verlag, New York, pp. 21–70.

[17] Segal, S. S., Damon, D. N., and Duling, B. R. (1989) *Am. J. Physiol. (Heart. Circ. Physiol.* **25**) **256**, H832–H837.

[18] Nelson, T. R. (1988) *SPIE* **914**, 326–333.

[19] Wang, C. Y. and Bassingthwaighte, J. B. (1990) *J. Math. Comput. Modeling.* **13**, 27–33.

[20] Hudlická, O. (1984) *Handbook of Physiology. Section 2: The Cardiovascular System*, vol. IV, (Renkin, E. M. and Michel, C. C., eds.), American Physiological Society, Bethesda, MD, pp. 165–216.

[21] Caldwell, J. H., Martin, G. V., and Bassingthwaighte, J. B. (1989) *FASEB J.* **3**, A404 (abstract).

[22] Lefèvre, J. (1983) *J. Theor. Biol.* **102**, 225–248.

[23] Hudlická, O. and Tyler, K. R. (1986) *Angiogenesis: The Growh of the Vascular System*. Academic, London.

[24] Schaper, W., and Wüsten, B. *The Pathophysiology of Myocardial Perfusion*, (Schaper, W., ed.), Elsevier, Amsterdam, pp. 415–470.

[25] Langille, B. L., Bendeck, M. P., and Keeley, F. W. (1989) *Am. J. Physiol. (Heart Circ. Physiol.* **25**) **256**, H931–H939.

[26] Duling, B. R., Hogan, R. D., Langille, B. L., Lelkes, P., Segal, S. S., Vatner, S. F., Weigelt, H., and Young, M. A. (1987) *Federation Proc.* **46**, 251–263.

[27] Griffith, T. M. and Edwards, D. H. (1990) *Am. J. Physiol. (Heart Circ. Physiol.* **27**) **258**, H1171–H1180.

[28] Langille, B. L. and O'Donnell, F. (1986) *Science* **231**, 405–407.

Chapter 24

Scaling the Resistance Vessels

Architecture of the Mammalian Arterial Tree

F. Eugene Yates

What Is Invariant in the Mammalian Body Plan?

Approaches to Classification

The General Problem

Classifying members of the biosphere has been a major problem for biologists and remains so.[1] It is still with a sense of awe that we look at the classification proposed by Linnaeus in his *Systema Naturae* in 1735, which was adopted by botanists until the ultimate establishment of the system founded by Jussieu in 1789. All plants were divided by Linnaeus into 24 classes: The first 20 were based on the number or arrangement of the stamens; the next three classes included plants having monoecieus, diecieus, and polygammodiecieus flowers, respectively; and the last class included all known cryptogams. These classes were then divided into orders, based on characters of the gynoecium or the fruit. The system became known as the "artificial system," because it gave no clue to the relation-

From *The Resistance Vasculature*, J. A. Bevan et al., eds. ©1991 Humana Press

ship of species or genera, but merely afforded a key to their rapid determination. Other artificial systems had been proposed before Linnaeus, but none was so extensively adopted. Even today with sophisticated category and hierarchy theories we are unable to state clearly the bounds of a biological species.

Gross Definition of Mammals

Class Mammalia has been defined as the class of vertebrates, including humans and all animals that nourish their young with milk. All are warm-blooded and, except for the monotremes, they are viviparous; their embryos develop an amnion and an allantois. Characters peculiar to the class are the following: Skin more or less covered with hairs of peculiar structure, although hairs are almost lacking in cetaceans (whales); mammary glands; a mandible articulating directly with the squamosal; a crurotarsal ankle joint (when the joint is present); a chain of small, separate ear bones; a brain with four optic lobes; a muscular diaphragm separating the heart and lungs from the abdominal cavity; a left aortic arch only; and red blood corpuscles without nuclei (except in the fetus). The body size range covers seven orders of magnitude, from the small, 3 g shrew to the largest animal that has ever lived, the now-living blue whale (120,000 kg in air). It is remarkable that certain anatomical or physiological features are even approximately invariant across that great range.

DNA Fingerprinting

Representative members of different taxa can be compared crudely using techniques of DNA hybridization. It is too early to state with certainty the similarities of the genome across all members of class Mammalia. A crude estimate is that all members have genomes of the same size (50,000–100,000 genes) and that about 85% of these may be in common. (Humans and the chimpanzees share the same class and order — Mammalia and Primata — and our genomes overlap by about 98%.) More detailed DNA "fingerprinting" will no doubt be able to determine classifications more sharply than has been done so far by gross features or crude DNA hybridization.

Similarities

The empirical, classical equation proposed by Snell and Huxley expressed the changes in anatomic proportions that occurred nonlinearly during growth of an individual animal:

$$y = ax^b \quad x = x(t) \quad 0 < t < T \tag{1}$$

where y is the feature of interest (e.g., weight of the liver); x is body size; b is the "allometric" exponent derived from the slope of the plot of log y vs log x; and a is a constant for the species, obtained from the intercept of the log–log plot when $x = 1$ (in the units chosen). T is a terminal time, such as life-span, and t is time. This same equation form, without time, has been used for cross-species comparisons of body plans (*see* Yates and Kugler[2] for discussion). The empirical allometric equation fits many data sets, but provides no explanation of underlying processes. It is not fundamental. In the classical, linear allometric plots, the independent variable is the log of body mass (usually as weight [W] in kg). The dependent variable is the log of some form or physiological function within a "typical" individual in that size range. In the discussion below, the representative individual is taken to be synonymous with a species of class Mammalia. In this kind of analysis, similarity appears in three different guises:

1. Absolute constancy of a value for form or function across species ($y = k, b = 0$);
2. Relative constancy (i.e., per unit weight, meaning $y = aW$ and $b = 1.0$); and
3. Linear relation in the allometric log–log plot for a form or function across all orders of magnitude of body size considered (i.e., constant allometric exponent b, $b \neq 0$ or 1.0, in Eq. 1. The coefficient "a" is usually assumed to be constant, not always correctly.)

If we confine our attention to the terrestrial mammals that must have a bony skeletal structure sufficient to bear their weight because they are not significantly buoyant in air, we still have an effective size range of six orders of magnitude from the shrew to the elephant. Since Galileo's brilliant analysis, it has been recognized that the relative fraction of the body weight serving as supporting skeletal structure must be a nonlinear function of overall size according to mechanical principles. Arguments from elastic, mechanical, or static stress similitude, or from the nonlinear relationship between surface area and volume (of a sphere), all can be linearized on a log y vs log W plot.[3-5] The three kinds of similarity mentioned above can be tested by seeing whether a unit of form or function is relatively or absolutely constant across a large size range. If it is not constant either way, then the log–log linearization can be tested for allometric similarity (constant exponent b, $b \neq 0$ or 1.0) instead of constancy similarity. The data, however, are apt to be noisy; thus, when testing is done for variables relevant to oxygen consumption and cardiovascular function for class Mammalia, it is found that (for example) heart size in units of g/100 g scales erratically in small samples. Specifically, in increasing rank order of body size from shrew (*Blarina*), mouse, Arctic weasel, dog, human, zebra, cow, to elephant, the hearts scale as (in these units), respectively, 1.05, 0.68, 1.71, 0.85, 0.42–0.81, 1.42, 0.37, and 0.39 (Table 133 in ref. 6). In larger samples, however, the heart scales symmorphically ($b = 1.0$). In contrast to the seemingly erratic scaling of heart with size, the allometric assumption that surface area is proportional to $W^{2/3}$ holds very well even in small samples, as can be seen by calculating the proportionality constant based on the assumption expressed in the equation below:

$$A = K(W)^{2/3} \qquad (2)$$

When A is surface area in cm^2 and W is body weight in g, the K values for mouse, dog, cow, and whale are, respectively, 10.5, 11, 11, and 11 (Table 146 in ref. 6). Thus, surface area obeys the log–log linear relationship with only a small residual effect from nonspherical shapes. Many details of allometric scalings can be found in ref. 4.

The scaling of oxygen consumption in vitro (37.5°C in Krebs-Ringer phosphate buffer usually), expressed as $mm^3 \, O_2$/mg dry wt/h measured on each tissue with glucose substrate in the media in nonlimiting amounts, indicate size-independent constancy. For example, data from tissues of mouse, rat, rabbit, cat, monkey, dog, pig, and humans give values of 2–4 for both skeletal muscle and cardiac muscle, and values of 8–10 for cerebral cortex in all cases. Thus, the specific energetics of metabolically active tissue (in vitro) expressed as rate of oxygen consumption per unit of tissue weight is independent of the size of the animal donating the tissue. In contrast, cardiac output and total oxygen consumption scale allometrically (b constant, but $\neq 1.0$). Iberall[7] reexamined some of the metabolic and cardiovascular flow data from mammals and estimated that the allometric exponent for cardiac output (Q_o) is approx 0.83 and that for oxygen consumption (metabolic rate) is approx 0.79, rather than 2/3 or 3/4 as previously thought to be the case for both variables. The discrepancy is thought to be beyond observational error. Dimensional analysis based on geometric, static stress, or elastic similarity arguments predicts only a rational exponent. The range of uncertainty on these allometric exponents is approx 0.04. Therefore it appears that total oxygen consumption and cardiac output scale allometrically with exponents that may be identical, at about 0.81, which is significantly different from 3/4, and that therefore some nonmechanical similarity influence is operating, perhaps thermodynamic.[2]

Intensive thermodynamic vaiables, such as core temperature, mean arterial pressure (MAP), or cardiovascular "grounded" pressure (right atrial pressure, RAP), tend to be absolutely constant over the whole mammalian size range ($b = 0$). The pressure drop across the systemic cardiovascular circuit (MAP minus mean RAP) is about 100 mmHg in mammals, regardless of heart rate or body size. The ranges for heart rates (beat/min) include: deer mouse (not clearly at rest), 324–858; shrew, 588–1320; dog (at rest), 50–70; human (rest), 40–84; elephant, 22–53; and whale (*Beluga*), 12–23. Physiological times, such as life spans or heart-beat intervals, scale allometrically with an exponent b of 1/4 or 1/5 over a wide range.[4]

Scaling Total Peripheral Resistance

Simple Force–Flux Model

Thermodynamic Phenomenological Form

The simple, linear thermodynamic force–flux model is given.

$$J = L \ (\partial\mu/\partial x) \ _{T,\ldots} \tag{3}$$

where J = flow (flux); L = conductance; μ = a thermodynamic potential; x = distance; and T = temperature.

In this form it is usually assumed that the conductance term L is constant and independent of time or changes in thermodynamic potentials. In the nonlinear case in which $L = f(\mu_i, t)$, which may be more realistic, the conductance values must be experimentally estimated repeatedly under steady-state conditions at various operating points of the system to gain insight into the nonlinearity. The subscript i designates a particular chemical species.

Simplest Cardiovascular Model: A Sketch of the Problem

The classical, simple, linear, global hemodynamic version of Eq. 3 is:

$$Q_o = (\text{MAP}-\text{RAP})/\text{TPR} = \text{MAP}/\text{TPR} \tag{4a}$$

where RAP $\cong 0$ and therefore is not shown at right.

Scaling Steady-State TPR (Rest State)

In the simplest possible, linear, lumped model of steady-state cardiovascular pressures and flows (shown in Eq. 4a), we can substitute a constant 100 mmHg for MAP and a_i $(W)^{0.83}$ for cardiac output, as noted above. Therefore,

$$\text{TPR} = 100/[a_1(W)^{0.83}] = a_2/(W)^{0.83} \tag{4b}$$

Equation 4b shows that as size increases, total peripheral resistance decreases. To explain that decrease we might first simply look to the increased number of parallel circuits open in large mammals compared to the smaller, all of which have one aorta carrying systemic flows away fom the heart and one right atrium gathering the inflows. The number of normally open, terminal arterioles, the chief resistance vessels (diameter $\cong 5$–7μm) that are open at rest, has been estimated for various mammalian species: shrew, 2.68×10^5; rat, 1.13×10^7; dog, 4.4×10^8; human, 1.08×10^9; and whale, 3.1×10^{11}.

The capillary vascularity (capillaries/mm^2 cross-section) in the gastrocnemius muscle of the cat is 2341; in the guinea pig it is 1136; in the mouse, 3060; and in the rabbit, 1344 (Table 274 in ref. 6). (The conditions were supposedly maximal vasodilation.) For cardiac muscle, representative values are: human, 3343; guinea pig, 1970;

and rabbit, 3420. In summary, in common experimental animals the number of capillaries/mm^2 of skeletal muscle is about 1000 at rest and 4000 under conditions of maximal dilation or activity. For cardiac muscle the range is 2000–3500. Assume, then, that 2000 capillaries/mm^2 is an overall representative number for open, parallel circuits, and that this number is a local, microscopic characteristic of tissues, independent of the global size of the mammal. (This independence may not be strictly true; smaller mammals may have a somewhat higher local density of capillaries.) In that case we would expect TPR to decrease with size as a function of the parallel resistance law shown in the crude oversimplification below, on the assumption that the hydraulic resistance is the same, on average, along each of the parallel paths. The lumped model is:

$$R_t = R/n \tag{5}$$

where R_t is the total resistance of the parallel network, R is the identical resistance in each of the parallel pathways and is species-independent; and n is the number of parallel pathways. Setting R_t = TPR, we conclude that n scales roughly according to size as shown below:

$$TPR = R/n = a_2/(W)^{0.83} \tag{6a}$$

$$\therefore \ n \propto W^{0.83} \tag{6b}$$

where R is a constant (at rest) Poiseuille resistance $(8\lambda\eta/\pi r^4)$ in each (identical) parallel path; λ is the length of resistance vessels in each path; r is the radius of the resistance vessels [~2.5–5 μm in diameter]; η is the viscosity of blood in small tubes; and n is the number of parallel paths. Of course, this is an absurd model that assumes that the aorta branches only once, into n identical channels; however, it can be used as an equivalent-circuit model of the many-tiered, branching arterial tree in order to explore the influence of gross compositional change with size.

Assuming that the n parallel open channels exist only in non-bone tissues, and using the scaling for the skeleton based on the data of Prange et al.[8]

$$W_s = 0.061(W)^{1.09} \tag{7a}$$

$$W = W_s + W_p \tag{7b}$$

where: W = weight of the animal; W_s = weight of supporting, skeletal tissues (poorly perfused); and W_p = weight of well-perfused tissues (especially skeletal muscle, spleen, and heart), we determine that the nonskeletal (perfused) mass scales as shown below:

$$W_p = W - 0.061(W)^{1.09} \tag{8}$$

If the integument fraction, which decreases with increasing size, and the skeleton fraction, which increases slightly with increasing size, are considered together as supporting structures, then much of the rest of the tissues (for example, skeletal muscle + heart + lungs + spleen) remains a nearly constant 60% of total body weight, regardless of size.[6] This fact suggests that perhaps $n \propto 0.6W$. However, with increasing mass, gut, liver, kidney, and brain decrease substantially as a percent of body mass.

There are three obvious expressions for n. They are summarized below as Models A, B, and C.

Model A

This model assumes parallel, identical circuits; constant local perfusion per unit of cross-sectional area; Poiseuille's law applying to each parallel channel; and the empirical allometric exponent for cardiac output scaled against body weight, 0.83. This model, as described above, gives $n = a_3(W)^{0.83}$.

Model B

This model assumes constant local perfusion per gram, as does Model A. However it assumes that the perfused tissue weight (W_p) is a constant fraction of total weight: $W_p = a_4W$ and $a_4 = 0.60$. Finally, it assumes that n is proportional to W_p. This model gives $n = a_5W$.

Model C

This model recognizes that total weight (W) is made up of two parts, W_p (that is perfused) and W_s that is supporting, or skeletal, and not rich in blood vessels), as given by Eq. 8.

The model assumes that $n \propto W_p$. Therefore, this model gives $n = a_6[W-0.061\ (W)^{1.09}]$.

Model B can be rejected immediately, because it forces TPR to be proportional to $W^{-1.0}$ via Eq. 6, whereas the allometric data on cardiac output indicate that it scales as $W^{-0.83}$. The question is then whether Model C, simply showing a compositional change in proportions with increasing size, is sufficient to account for the allometric scaling of cardiac output through simple changes of TPR at constant MAP. Although the constants of proportionality are not known, this question can be answered by allowing body weight to change 10×, examining a ratio of TPR values before and after the change, and making the tenfold tests over a wide size range ($W = 0.01, 0.1, 1, 10, 100, 10,000, ...$). When this is done, the data reveal that by the allometric scaling for actual cardiac output (Model A), $TPR_{10} = 0.15\,TPR_1$ for each interval of tenfold increase in total body weight. From the hypothesized effect of compositional change (Model C), $TPR_{10} \cong 0.10\,TPR_1$. Thus TPR actually falls more slowly with increasing size than is suggested by assuming an increasing contribution of poorly perfused skeletal and supporting structures with increasing size ("Galileo effect") and constant microscopic perfusion of nonsupporting tissues, independent of size. A different approach is needed.

Branching of the Arterial Tree

Optimality Approach

The branching structure of the vascular system was treated as a duct system of minimum work by Murray in 1926.[9,10] He minimized the cost function, which was the sum of the frictional power losses and an energy term proportional to charging up blood volume under a pressure. Murray derived the optimal conditions for a single branching (dichotomization):

$$r_0^3 = r_1^3 + r_2^3 \tag{9a}$$

where r_0 is the radius of the mother vessel, and $r_{1,2}$ are the radii of the daughter branches. (This expression is equivalent to assuming that $m = 3.0$ in Eq. 10a, as discussed below.) The branching angles were given by

$$\cos \theta_1 = (r_0^4 + r_1^4 - r_2^4)/(2\,r_0^2 r_1^2) \tag{9b}$$
$$\cos \theta_2 = (r_0^4 + r_2^4 - r_1^4)/2r_0^2 r_2^2)$$

His analysis was limited to a local, single level of branching, not necessarily symmetrical. It has been extended by Kamiya and his associates.[11,12] They construed the optimality problem for the vascular system as being the determination of the branching structure of a tree that minimizes its volume under the restriction of given location, pressure, and flow at the origin and at the terminals. They began with an analysis of biterminal branching of cylindrical and straight ducts lying on a plane and with laminar flow. The solution was aimed at volume minimization, and they showed that the minimal-work model of Murray was a special case of their minimum-volume model.

Fractal Architectures

Recently it has been fashionable to note that tree-like structures in nature, from plants to bronchial trees in lungs or vascular trees, have fractal characteristics, self-similarity, and $1/f$ distributions.[13–15] These are discussed elsewhere in this volume by Bassingthwaighte. Here it is sufficient to point out that the fractal-branching vascular architectures have been anticipated by earlier work discussed below.

Groat-Suwa Branching Rule

A branching-tree description of the mammalian arterial network was provided in 1950 by Green,[16] based on data from F. Mall in 1888 and 1905. Improved, more recent data from Suwa et al.,[17] based on casting techniques (which slightly overestimate diameter and require correcting, as Suwa did), suggest the following branching rule for the mammalian arterial tree:

$$d_i = (1/N_i)^{1/m_i} d_k \qquad (10a)$$

$$A_i = (1/N_i)^{2/m_i} A_k \qquad (10b)$$

where d = diameter of a cylindrical vessel, i = branching level index (aorta = 1, capillary = 11), K = any branching level upstream from i (i.e., larger vessels), m = Groat-Suwa branching exponent; N = total number of parallel branches at level i; A_i = cross-sectional area of a single vessel at level i. The branching exponent m_i is an empirical exponent that is 2.0 for the first three levels of branching and 2.4 (after some correction for artifacts) thereafter, up to the last branching, to capillaries, which have a much larger exponent. With these values of m, Eqs. 10a and b describe a geometry in which the total area of all vessels at level i is increasing *more* than it would if the bifurcation conserved the cross-sectional area. That is, for conserved area at a bifurcation ($N = 2$), $A_{i+1}/A_i = 0.50$; $m = 2.0$. When m = 2.4 at a bifurcation, $A_{i+1}/A_i = 0.56$; therefore, total area is increased ($2 \times 0.56 = 1.12$) by 12% at each bifurcation to level i. (Side branching is ignored in this demonstration calculation.)

Iberall's Extension

Iberall[18] computed the diameters and areas for each level of the arterial system for data from a 23-kg dog. The dog is at about the logarithmic mean for the size scale of all mammals, from the shrew to the largest whales. From the data of Suwa et al.[17] it appears that λ/d (length/diameter) is constant (i.e., independent of i) and approx 25 ± 5. Then, using the Hagen-Poiseuille law for laminar flow, with the viscosity term corrected for suspension of red cells, the pressure drop at each level can be calculated from the geometry at that level and the total flow, which is equal to the cardiac output. It is possible to determine the total area for each branching level and the number of tubes at each level. I have discussed these calculations of Iberall and presented some of the data elsewhere.[19] In the dog there are 11–12 levels of branching, starting with a single, tapered aorta with an average cross-sectional area of 3.2 cm², and ending in ~2–4×10^8 open capillaries (at rest), with a total cross-sectional area of ~80 cm².

Iberall provided cross-checks on his calculations, and his several different approaches to the architecture of the arterial tree and to the distribution of the velocities and pressures in it provide perhaps the best estimate we have. The discussion that follows summarizes and paraphrases his work, which can be found in several important refs.[7,18,20–22]

Some Important Arterial-System Parameters

In Table 1 are estimates of some standard arterial cardiovascular parameters and variables referenced to a conscious, resting

Table 1
Estimates of Some Standard Arterial Cardiovascular Parameters
and Variables for a Conscious, Resting Dog[6,7,17,18,20–22]

Item	Approximate Value
W	23 kg
$Q_b = Q_o$ (Total tissue blood flow = cardiac output)	50 mL/s; 3 L/min; $Q_b \propto W^{0.83}$
Branching levels in arterial tree (to capillaries), n	11–12
Terminal arterioles, n	4.4×10^8
Diameter of those terminal arterioles normally open[a]	5–7 μm
Blood viscosity (η) in small tubes[a]	0.035–0.040 P
Suwa branching exponent, first levels (aorta = level 1)[a]	2.0
Suwa branching exponent, higher levels (to capillaries)[a]	2.4
Side branches per equal bifurcation branching level, n[a]	5–6
MAP[a]	100 mmHg
Δp across high resistance, arteriolar level[a,b]	60–75 mmHg
λ/d (all arterial system levels)[a]	22
Rest mean velocity in aorta	15 cm/s
Peak mean velocity in aorta	50 cm/s
Velocity in terminal arteriole[a]	6 mm/s
Velocity (v) in any bifurcated branch[a]	~80% v in parent vessel
V_{O_2}	0.21 L/min
Tissue q_{O_2}	mm³/mg dry wt/h
Skeletal muscle, cardiac muscle[a]	2–4
Cerebral cortex[a]	7

[a]These items may be independent of size (i.e., of species of mammal).
[b]p = pressure; Δp = pressure drop.

dog. Because physiologically these parameters and variables interrelate pressures, flows, oxygen delivery, and geometry, there is a critical necessity for internal consistency of the values. Difficulties arise at many points, e.g., determining to what extent the viscosity of whole blood is anomalous in small tubes (diameters < 10 μm), as suggested by the Fahraeus-Lindquist effect (which will not be discussed here). Questions concern the physiological state of animals during investigations of pressures and flows in microcirculations and the representativeness of any particular vascular bed chosen for observation. Effects of stress, anesthetics, or observational methods are often unknown. Casting techniques to identify branching levels in the arterial tree require pressure injections that may dilate vessels, distorting the estimate of diameters. I have attempted to take these various factors into account in choosing the values to be shown in Table 1.

The superscripts *a* in Table 1 identify entries that may be characteristic of mammals in general and not peculiar to the dog. The intensive pressure variable, MAP, obtains its remarkable constancy across the seven orders of magnitude of size for animals belonging to class Mammalia, possibly through the special characteristics of renal resistances.[19]

Regulation of Arterial Resistances

By combining a statement of continuity of flow for any branching level, i, in the arterial tree ($Q_i = N_i (\pi/4) d_i^2 v_i$, where d is diameter, v is average velocity, and N is number of identical parallel channels), with the Poiseuille laminar-flow law and the Suwa branching rule relating diameters at two different levels, and letting n = the number of side branches between bifurcations, the following relationship can be derived:

$$\Delta p = 32 \, \eta(\lambda/d) \, [(v/d)_{ta}]/[1 - n^{(m-3)/m}] \tag{11}$$

Where p = pressure; the index "ta" means "terminal arteriole"; λ/d is constant at all levels, at about 22; and n is level-independent and is approx 5.6. It is possible to use this equation for a parameter-sensitivity analysis under various assumptions about the degree of vasoconstriction or vasodilation under specified conditions of total flow and an associated pressure drop. Under the resting, reference conditions ($\Delta p = 70$ mmHg; $\lambda/d = 22$; $n = 5.6$; $\eta = 0.04$ P; and $m = 2.4$, as in Table 1), $d_{ta} = 4.7\mu$m and $v_{ta} = 6.5$ mm/s.

By modulating the resistance vasculature through the index m, we find that if pressures and flows are (artificially) held constant, decreasing m to 2.1 (vasoconstriction) yields $d_{ta} = 4.1$ μm, and

v_{ta} = 8.5 mm/s. Increasing m to 2.9 (vasodilation) yields d_{ta} = 8.6 μm and v_{ta} = 2.0 mm/s. This limited-range sensitivity analysis $(2.0 < m < 3.0)$ suggests a physiologically reasonable domain over which neurohumoral influences can presumably easily adjust the arterial microcirculation: Changing the index m means that one is changing the ratio of daughter-vessel diameter to parent-vessel diameter (Eq. 10) at the level of the resistance vessels. The resistance adjustments noted could be obtained with very small changes in tension in the walls of the vessels (as calculated by use of the law of Laplace), so the discontinuity of "critical closing" would not seem to arise. (I recognize that the critical-closure concept, as described by Alan Burton and his colleagues, is no longer in favor, although Alexander has shown that closure of small arteries might occur because of the helical orientation of the smooth-muscle cells, if the vessels have high tone.[23])

The picture of the organization of the arterial tree presented here has been hierarchical. *Heterarchically*, across any given level there are many regional variations in flow for the same overall total flow, as has long been recognized. Very local factors affect perfusion within different organs and tissues, leading to a distribution of flow to active tissues without necessitating a change in MAP. For a useful example, see the discussion of the pressure–flow relations in the local coronary circulation by Hoffman and Spaan.[24] My intention here has been merely to highlight the essential and primitive global characteristics of the arterial circulation, with special emphasis on the resistance vessels, by bringing together scaling factors common to mammals.

Although the scaling approach I have used starts with classical similitude analysis, which is no longer considered applicable to biological branching morphologies by some[25,26] I have not invoked the exponetial branching rule that usually results. The rule is:

$$d\,(z,a) = d_0 e^{-a} \tag{12}$$

where $d(z,a)$ is average diameter of tubes in the zth generation, d_0 is the diameter of the parent vessel, and a is the scale factor between successive generations. On a log-linear plot of ln$d$ vs z, a straight line of slope–*a* should result. When such a plot is made for the mammalian bronchial tree, after $z = 10$ the data depart from this model. The departure has supported a fractal model of branching that assumes heterogeneity, self-similarity, and absence of a characteristic scale. According to the fractal model the data should be linearized on a log-log plot because of an inverse power law for the branching:

$$d(z) \propto (1/z^{\mu}) \tag{13}$$

where μ is the (constant) power law index.[25,26]

The scaling I suggest for arterial branching (Eqs. 10a,b) indeed describes a fractal geometry by following the form of Eq. 13, where $\mu = (1/m)$ and N_i is a function of z. Thus the Suwa/Iberall calculations and data anticipated the fractal character of mammalian vascular tree branching now attracting renewed attention. The fractal geometry of the resistance vasculature implies that the relative distribution of flow between asymmetric branchings does not depend simply on the absolute value of the resistance or the branch level.[14]

Conclusion

Over its full, normal operating range, the mammalian cardiovascular system achieves a wide range of total flows and regional distributions of flows. It does all this while conserving a grounded, essentially zero RAP and a steady MAP, driving blood through a distributional tree with a specific fractal geometry and branching rule. It has been my aim to survey the chief invariances in the system across seven orders of magnitude of body size among the many species of class Mammalia. The dog was chosen as a "representative" mammal because it is at the logarithmic mean of the size range of the class (and also happens to be a common subject for cardiovascular research). The values presented for the representative parameters and variables have been checked for internal consistency and therefore offer a suitable set for further discussions of scaling of the adult arterial system. (Not all of the data and calculations could be shown here but references are provided.)

A major assumption behind this presentation is that the resting, unstressed performance of the mammalian cardiovascular system must be homeodynamically effective because of its geometry. Although the analysis does not assume that the system is passive at rest, it does suppose that the physics of the design is such that very little neurohumoral or metabolic signaling is required to sustain normal operations in the unstressed condition. In the discussion, the numbers of vessels operating in the microcirculation are considered from a functional viewpoint, not an anatomical viewpoint. That is, physical constraints are used to calculate the number of channels open at rest, rather than the maximal number of anatomically possible channels under maximal vasodilation of every parallel pathway—a state that presumably would cause lethal hypotension (shock).

Much of the excitement in microvascular research now concerns processes that can adjust local flows, since these may involve chemical influences arising from nearby endothelial cells or brought to the region by the inflow of blood or by neural signals. Many of these processes are discussed elsewhere in this volume. However, whatever the flexibility of local circulations and their instabilites, overall, the system operates under severe constraints, which I have tried to address in the scaling analysis presented here. This is the framework on which any regulatory phenomena adjusting regional flows must work. Furthermore, this framework surely must express the morphogenetic rules that operate genetically and epigenetically during the growth and development to maturity of any mammal.

Acknowledgment

This work was supported in part by a gift to UCLA in honor of Mary Ruth Belt for research in medical engineering.

I am indebted to Arthur Iberall for providing me with the information on which the above discussion is based, and for the privilege of seeing prepublication copies of some of his work.

References

[1] Ruse, M. (1988) *Philosophy of Biology Today* (State University of New York Press, Albany) pp. 51–62.

[2] Yates, F. E. and Kugler, P. N (1986) *J. Pharm. Sci.* **75**, 1019–1027.

[3] McMahon, T. A. and Bonner, J. T. (1984) *On Size and Life*, Scientific American Library, New York, pp. 139–161.

[4] Calder, W. A., III (1984) *Size, Function, and Life History*, Harvard University Press, Cambridge, MA.

[5] Boxenbaum, H. and D'Souza, W. (1990) *Adv. Drug Res.* **19**, 139-196.

[6] Spetor, W. S., ed. (1956) *Handbook of Biological Data*, W. B. Saunders, Philadelphia, PA.

[7] Iberall, A. S. (1972) *Ann. Biomed. Eng.* **1**, 1–8.

[8] Prange, H. T., Anderson, J. F., and Rahn, H (1979) *Am. Naturalist* **113**, 103–122.

[9] Murray, C. D. (1926) *Proc. Nat. Acad. Sci. USA* **12**, 204–214.

[10] Murray, C. D. (1926) *Proc. Nat. Acad. Sci. USA* **12**, 299–304.

[11] Kamiya, A. and Togawa, T (1972) *Bull. Math. Biophys.* **34**, 431–438.

[12] Kamiya, A., Togawa, T., and Yamamoto, A (1974) *Bull. Math. Biophys.* **36**, 311–323.

[13] Lefèvre, J. (1983) *J. Theor. Biol.* **102**, 225-248.

[14] Bassingthwaighte, J. B. and van Beek, J. H. G. M. (1988) *Proc. IEEE* **76**, 693–699.

[15] Rigaut, J. P. (1984) *J. Microsc.* **133**, 41–44.

[16] Green, H. D. (1950) *Medical Physics*, vol II (Glasser, O., ed.), Year Book Publishers, Chicago, IL, pp. 228-251.

[17] Suwa, N., Niwa, T., Fukasawa, H., and Sasaki, Y. (1963) *Tohoku J. Exp. Med.* **79**, 168–198.

[18] Iberall, A. (1967) *Math. Biosci.* **1**, 375–395.

[19] Yates, F. E. (1983) *Fed. Proc.* **42**, 3143–3149.

[20] Iberall, A. (1973) *Trans. ASME Series G, J. Dyn. Syst. Mes. Control* **95**, 291–295.

[21] Iberall, A., Cardon, S., and Young, E. (1973) *On Pulsatile and Steady Arterial Flow*. Upper Dauby, General Technical Services (Library of Congress Catalog Card No 72-96894).

[22] Iberall, A. (1974) *Flow—Its Measurement and Control in Science and Industry*, vol. 1, part 3, (Dowdell R., ed.), Instrument Soc. America, Pittsburgh, PA, pp. 50–63.

[23] Alexander, R. S. (1977) *Circ. Res.* **40**, 531–533.

[24] Hoffman, J. I. E. and Spaan, J. A. E. (1990) *Physiol. Rev.* **70**, 331–390.

[25] West, B. J. (1987) *Chaos in Biological Systems*, (Degn, H., Holden, A. V., and Olsen, L. E., eds.), Plenum, New York, pp. 305–314.

[26] West, B. J., Bhargava, V., and Goldberger, A. L. (1986) *J. Appl. Physiol.* **60**, 1089–1097.

Index